T0202686

Springer Series in Reliability Engineering

Series Editor

Hoang Pham, Department of Industrial and Systems Engineering
Rutgers University, Piscataway, NJ, USA

More information about this series at http://www.springer.com/series/6917

Jan-Erik Vinnem · Willy Røed

Offshore Risk Assessment Vol. 1

Principles, Modelling and Applications of QRA Studies

Fourth Edition

 Springer

Jan-Erik Vinnem
Faculty of Engineering
Norwegian University of Science
and Technology
Trondheim, Norway

Willy Røed
Faculty of Science and Technology
University of Stavanger
Stavanger, Norway

ISSN 1614-7839   ISSN 2196-999X   (electronic)
Springer Series in Reliability Engineering
ISBN 978-1-4471-7446-2   ISBN 978-1-4471-7444-8   (eBook)
https://doi.org/10.1007/978-1-4471-7444-8

Preface to Fourth Edition

This is the fourth edition of the book; the first edition was published in 1999, the second edition in 2007 and the third edition in 2014. This version represents a new development where we have been two authors co-operating on the updating of the manuscript. My earlier M.Sc. and Ph.D. student and good friend, Willy Røed, Adjunct Professor at the University of Stavanger, Department of Safety, Economics and Planning, and consultant in Proactima in Stavanger, has shared the workload with me regarding the updating of the chapters and the new text.

The original author has since 2014 been professor at NTNU, Department of Marine Technology, Trondheim, in marine operational risk. The authors have been involved in several significant research projects during the last few years. Some few major accidents worldwide have had a considerable effect on the HES performance and awareness, in addition to the substantial fall of oil price in 2013. This prompted a need for a further update of the book.

Norwegian offshore installations have for more than 40 years been large and complex and manned installations with extensive maintenance needs. Manning levels have been reduced from a few hundred to around one hundred and considerably less for recent installations. There is now an interest in unmanned, simple wellhead installations, as they are less expensive to install than subsea production equipment. The first normally unmanned wellhead installation in the Norwegian sector will start to operate in stand-alone mode from the first half of 2019. It is envisaged that within a few years, also unmanned production installations, fixed and floating, will be developed. Unmanned installations have been used in other offshore sectors (UK, the Netherlands, USA, etc.), but these have been small, simple, shallow water installations. In future, we may find quite complex, unmanned production installations also in deep water if technology optimists are to be believed. An extensive new chapter (22) is devoted to discussion of some of the challenges that unmanned installations are faced with.

Gratitude is expressed to the Springer Nature Publishers (London), in particular Executive Editor Anthony Doyle, Project Coordinator Arun Kumar Anbalagan and their staff, for agreeing to publish the fourth edition of this book and for providing inspiring and valuable advice and assistance throughout the process.

Appendix A has been updated with the latest overview of some of the important software tools that are commonly used in offshore risk assessment. Thanks to all the consultancies and software suppliers who have provided the information required for the update of this appendix.

There are also several people who have kindly contributed to relevant information on various aspects or prepared small text sections for us to use; Dr. Haibo Chen, consultant to Lloyd's Register Consulting, China; Silje Frost Budde, Safetec Nordic; Sandra Hogenboom, DNV GL (Ph.D. student at NTNU); Petter Johnsen, Presight Solutions; Kenneth Titlestad, Sopra Steria; Dr. Xingwei Zhen, Dalian University of Technology, China; Odd W. Brude, DNV GL; Kaia Stødle, Ph.D. student at UiS; and Eldbjørg Holmaas, Proactima. Many thanks to all of you for valuable assistance.

Karoline Lilleås Skretting, M.Sc. student at the Department of Marine Technology, NTNU, has been our assistant during the final stages of the revision work. She has mainly been assisting with the updating of Appendix A, in addition to some other editorial tasks, and this has been very helpful to finishing the revision work in a timely manner. We are grateful for the excellent work done by Karoline. The Department of Marine Technology, NTNU, has supported the publishing of the book by paying the salary for Karoline.

Safetec Nordic has allowed us to use their QRAToolkit software, whereas ConocoPhillips, Equinor and Vår Energi have allowed us to use the results from a Joint Industry Project on collision risk modelling. We are also grateful to the RISP project and its contributors for providing access to relevant information and results before the RISP project is finalised.

April 2019

Jan-Erik Vinnem
Professor, Norwegian University
of Science and Technology
Trondheim, Norway

Willy Røed
Adjunct Professor, University of Stavanger
Stavanger, Norway

Preface to Third Edition

This is the third edition of the book; the first edition was published in 1999 and the second edition in 2007. The author has since then returned to an adjunct professorship at University of Stavanger, Norway, teaching a course in applied offshore risk assessment. Starting from January 2013, the author is also adjunct professor at NTNU, Trondheim in marine operational risk. The author has been involved in several significant research projects during the last few years. Several major accidents worldwide have had considerable effect on the HES performance and awareness, not the least the Macondo accident in 2010. This prompted a need for a further update of the book.

Norwegian offshore regulations were profoundly revised around the beginning of the new century, with new regulation becoming law from 2002. A limited revision was implemented from 2011, mainly limited to the integration of onshore petroleum facilities. This edition of the book captures some of the experience and challenges from the application of the new regulations. The important aspects of the new regulations are also briefly discussed, see Chapter 1.

About 30 major accidents and incidents are discussed at some length in Chapters 4 and 5 (Macondo accident), in order to demonstrate what problems have been experienced in the past. I have increased the emphasis on this subject in both the second and third editions, because it is essential that also new generations may learn from what occurred in the past. Where available, observations about barrier performance are discussed in addition to the sequence of events and lessons learned.

It is often claimed "what is measured will be focused upon". This implies that even if QRA studies have several weaknesses and limitations, quantification is the best way to focus the attention in major hazard risk management. This is also one of the lessons from the Macondo accident, in the author's view. It has therefore been surprising to realise how strong the opposition to QRA studies still is at the end of 2012 from many professionals in major international oil companies. This has to some extent given further inspiration to update this book, about a topic I consider crucial for improvement of major hazard risk management in the offshore petroleum industry.

Thanks are expressed to Springer London publishers, in particular Senior Editor Anthony Doyle and his staff, for agreeing to publish the third edition of this book, and for providing inspiring and valuable advice and assistance throughout the process.

Appendix A presents an overview of some of the important software tools that are commonly used in offshore risk assessment. Thanks to all the consultancies and software suppliers who have provided the information required for the update of this appendix.

There are also several people who have kindly contributed with relevant information on various aspects; Torleif Husebø, PSA; Prof. Stein Haugen, NTNU; Celma Regina Hellebust, Hellebust International Consultants, Prof. Bernt Aadnøy, UiS and Dr. Haibo Chen, Scandpower Inc. China. Many thanks for valuable assistance to all of you.

Meihua Fang has been my assistant during the final stages of the revision work, during her stay in Norway as the wife of a student at UiS in an international MSc. program in offshore risk management. Meihua has a M.Sc. degree in safety technology and engineering from China University of Geosciences in Beijing and HES management experience from SINOPEC in China and Latin America, and has been an ideal assistant. The main task has been the updating of Appendix A, in addition to several other editorial tasks, which has been very helpful to finish the revision work in a timely manner. I am very grateful for the excellent assistance provided by Meihua.

Last, but not least, I am very grateful to those companies that responded positively when asked for a modest support in order to cover the expenses involved in the production of this third revision. My warmest and most sincere thanks go to these companies:

- Faroe Petroleum Norway
- Norwegian oil and gas association
- Total E&P Norway
- VNG Norway

Bryne, May 2013   Jan-Erik Vinnem
                                                                   Adjunct Professor
                                                      University of Stavanger & NTNU

Preface to Second Edition

This is the second edition of the book; the first edition was published in 1999. The author has since then taken up a full professorship at University of Stavanger, Norway, teaching courses in offshore risk analysis and management. This prompted a need for an update of the book. The fact that several important developments have occurred since 1999 also implied that a major revision was required.

The oil price has reached its peak in 2006, at the highest level ever (nominally). But the economic climate is at the same time such that every effort is made to scrutinise how costs may be curtailed and profit maximised. This will in many circumstances call for careful consideration of risks, not just an 'off the shelf risk analysis', but a carefully planned and broad-ranging assessment of options and possibilities to reduce risk.

Norwegian offshore regulations were profoundly revised around the beginning of the new century, with new regulation becoming law from 2002. This second edition of the book captures some of the experience and challenges from the first 4–5 years of application of the new regulations. The important aspects of the new regulations are also briefly discussed, see Chapter 1.

The first Norwegian White Paper on HES management in the offshore industry was published in 2001, and the second in 2006. One of the needs identified in this paper was the need to perform more extensive R&D work in this field, and a significant programme has been running in the period 2002–06. Some of the new results included in the second edition of the book result from that R&D initiative.

About 20 major accidents, mainly from the North Sea, are discussed at some length in Chapter 4, in order to demonstrate what problems have been experienced in the past. I have put more emphasis on this subject in the second edition, because it is essential that also new generations may learn from what occurred in the past. Where available, observations about barrier performance are discussed in addition to the sequence of events and lessons learned.

When it comes to management of risk and decision-making based upon results from risk analyses, this is discussed separately in a book published in parallel with my colleague at University of Stavanger, Professor Terje Aven, also published by

Springer in 2007. Interested readers are referred to this work, 'Risk Management, with Applications from the Offshore Petroleum Industry'.

Thanks are also expressed to Springer London publishers, in particular Professor Pham and Senior Editor Anthony Doyle, for agreeing to publish the second edition of this book, and for providing inspiring and valuable advice throughout the process. Simon Rees has given valuable assistance and support during production of the camera-ready manuscript.

Appendix A presents an overview of some of the important software tools that are commonly used. Thanks to all the consultancies and software suppliers who have provided the information required for this appendix.

In preparing the second edition of the book, I have been fortunate to have kind assistance from many colleagues and friends, who have provided invaluable support and assistance. First of all I want to express sincere thanks and gratitude to my friend David R. Bayly, Crandon Consultants, who has also this time assisted with improvement of the English language, as well as providing technical comments and suggestions. I don't know how I could have reached the same result without David's kind assistance.

My colleague at UiS, Professor Terje Aven has contributed significantly to the discussion of statistical treatment of risk and uncertainty. I am very pleased that this important improvement has been made. Dr. Haibo Chen, Scandpower Risk Management Beijing Inc has contributed valuable text regarding the analysis of DP systems on mobile installations.

Safetec Nordic AS has allowed use of several of their tools as input to the descriptions and cases. I want to express my gratitude for allowing this, and in particular express thanks to the following; Thomas Eriksen, Stein Haugen and Arnstein Skogset.

There are also several people who have contributed with relevant information on various technical details; Finn Wickstrøm, Aker Kvaerner and Graham Dalzell, TBS[3]. My daughter, Margrete, has assisted in the editing of the manuscript. Many thanks to all of you.

Bryne, January 2007   Professor, Jan-Erik Vinnem
                                                     University of Stavanger

Preface to First Edition

From a modest start in Norway as a research tool in the late 1970s, Quantified Risk Assessment (QRA) for offshore installations has become a key issue in the management of Safety, Health and Environment in the oil and gas industries throughout the entire North Sea. While the initiatives in the early stages often came from the authorities, the use is now mainly driven by the industry itself. The QRA is seen as a vehicle to gain extended flexibility with respect to achievement of an acceptable safety standard in offshore operations. The models may be weak in some areas and the knowledge is sometimes limited, but studies are nevertheless used effectively in the search for concept improvement and optimisation of design and operation.

This book results from working with offshore QRAs for more than 20 years. The author has, during this period, had the opportunity to practice and evaluate the use of such studies from different perspectives; the consultancy's, the operating company's, the researcher's and the educator's point of view.

The author has for several years taught a course in risk analysis of marine structures at the Faculty of Marine Technology, NTNU, Trondheim, Norway. The starting point for the manuscript was the need to update the lecture notes.

It is hoped that this book in the future also may be a useful reference source for a wider audience. There has been for some years a rapid expansion of the use of risk assessments for the offshore oil and gas activities. It is expected that the expansion is going to continue for some time, as the offshore petroleum industry expands into new regions and meets new challenges in old regions.

The oil price reached its lowest level for many years, during the first quarter of 1999. One might be tempted to think that the economic climate may prohibit further attention to risk assessment and safety improvement. The opposite is probably more correct. As a friend in Statoil expressed not so long ago: 'Whenever the margins are getting tighter, the need for risk assessments increases, as new and more optimised solutions are sought, each needing an assessment of risk'.

In Norway, the beginning of 1999 is also the time when the Norwegian Petroleum Directorate is preparing a major revision of the regulations for offshore installations and operations, anticipated to come into effect in 2001. It has obviously not been possible to capture the final requirements of the new regulations, but an attempt has been made to capture the new trends in the regulations, to the extent they are known.

There have over the last 10–15 years been published a few textbooks on risk assessment, most of them are devoted to relatively generic topics. Some are also focused on the risk management aspects, in general and with offshore applicability. None are known to address the needs and topics of the use of QRA studies by the offshore industry in particular. The present work is trying to bridge this gap.

The use of QRA studies is somewhat special in Northern Europe, and particularly in Norway. The use of these techniques is dominated by offshore applications, with the main emphasis on quantification of risk to personnel. Furthermore, the risk to personnel is virtually never concerned with exposure of the public to hazards. Thus, the studies are rarely challenged from a methodology point of view. Most people will probably see this as an advantage, but it also has some drawbacks. Such challenges may namely also lead to improvements in the methodology. It may not be quite coincidental that the interest in modelling improvement and development sometimes has been rather low between the risk analysts working with North Sea applications.

This book attempts to describe the state-of-the-art with respect to modelling in QRA studies for offshore installations and operations. It also identifies some of the weaknesses and areas where further development should be made. I hope that further improvement may be inspired through these descriptions.

About the Contents

A Quantified Risk Assessment of an offshore installation has the following main steps:

1. Hazard identification
2. Cause and probability analysis
3. Accidental scenarios analysis
4. Consequence, damage and impairment analysis
5. Escape, evacuation and rescue analysis
6. Fatality risk assessment
7. Analysis of risk reducing measures

This book is structured in much the same way. There is at least one chapter (sometimes more) devoted to each of the different steps, in mainly the same order as mentioned above. Quite a few additional chapters are included in the text, on risk analysis methodology, analytical approaches for escalation, escape, evacuation and rescue analysis of safety and emergency systems, as well as risk control.

It is important to learn from past experience, particularly from previous accidents. A dozen major accidents, mainly from the North Sea, are discussed at the end of Chapter 4, in order to demonstrate what problems that have been experienced in the past.

The main hazards to offshore structures are fire, explosion, collision and falling objects. These hazards and the analysis of them are discussed in separate chapters. Risk mitigation and control are discussed in two chapters, followed by an outline of an alternative approach to risk modelling, specially focused on risk relating to short duration activities. Applications to shipping are finally discussed, mainly relating to production and storage tankers, but also with a view to applications to shipping in general.

Acknowledgements

Parts of the material used in developing these chapters were initially prepared for a course conducted for PETRAD (Program for Petroleum Management and Administration), Stavanger, Norway. Many thanks to PETRAD for allowing the material to be used in other contexts.

Some of the studies that have formed the main input to the statistical overview sections were financed by Statoil, Norsk Hydro, Saga Petroleum, Elf Petroleum Norge and the Norwegian Petroleum Directorate. The author is grateful that these companies have allowed these studies to be made publically available.

Direct financial support was received from Faculty of Marine Technology, NTNU, this is gratefully acknowledged. My part time position as Professor at Faculty of Marine Technology, NTNU, has also given the opportunity to devote time to prepare lecture notes and illustrations over several years. The consultancy work in Preventor AS has nevertheless financed the majority of the work, including the external services.

Thanks are also expressed to Kluwer Academic Publishers, Dordrecht, The Netherlands, for agreeing to publish this book, and for providing inspiring and valuable advice throughout the process.

Appendix A presents an overview of some of the important software tools that are commonly used. Thanks to all the consultancies and software suppliers who have provided the information required for this appendix. Appendix B is a direct copy of the normative text in the NORSOK Guideline for Risk and Emergency Preparedness analysis, reproduced with kind permission from the NORSOK secretariat.

Some of the consultancies have kindly given permission to use some of their material, their kind assistance is hereby being gratefully acknowledged. DNV shall be thanked for allowing their database Worldwide Offshore Accident Databank (WOAD) to be used free of charge, as input to the statistics in the book. The Fire Research Laboratory at SINTEF has given kind permission to use illustrations from their fire on sea research, and Scandpower has granted permission to use an illustration of the risk assessment methodology. Safetec Nordic has given kind permission to use results and illustrations from their software Collide.

I am particularly indebted to several persons who have offered very valuable help in turning this into a final manuscript. My colleague Dr. David Bayly, Crandon

Consultants, has reviewed the raw manuscript and contributed with many valuable comments of both a technical and linguistic nature. The importance of providing clear and concise text can never be overestimated, the efforts made in this regard are therefore of utmost importance. This unique contribution has combined extensive linguistic improvements with pointed comments and additional thoughts on the technical subjects. I am very grateful to you, David, for your extensive efforts directed at improvement of the raw manuscript.

My oldest son, John Erling, has helped me with several of the case studies that are used in the text, plus quite a few of the illustrations. My part time secretary, Mrs. Annbjørg Krogedal, has had to devote a lot effort to decipher a challenging handwriting, thank you for enthusiasm and patience. Assistance with the proof-reading has been provided by Ms. Kjersti G. Petersen, thanks also to Kjersti for enthusiastic and valuable assistance. Finally, M.Sc. Haibo Chen has also helped with the proof reading and checking of consistency in the text, your kind assistance is gratefully acknowledged.

Bryne, May 1999   Jan-Erik Vinnem

Contents

Abbreviations

AIBN	Accident Investigation Board Norway
AIR	Average Individual Risk
AIS	Automatic Identification System
ALARP	As Low As Reasonably Practicable
ALK	Alexander L. Kielland [accident]
ALS	Accidental limit state
ANP	National Petroleum Agency [Brazil]
AR	Accident Rate
ARCS	Admiralty Raster Chart Services
ARPA	Automated Radar Plotting Aid
ASCV	Annulus safety check valve
ASEA	Agency of Safety, Energy and Environmental enforcement [Mexico]
ASV	Annular safety valve
ATM	Air Traffic Management
bara	Bar absolute
barg	Bar gauge (overpressure)
BAST	Best Available and Safety Technology
BBD	Barrier Block Diagram
bbls	Barrels
BBN	Bayesian belief network
BD	Blowdown
BDV	Blowdown Valves
BF	Barrier Function
BFETS	Blast and Fire Engineering for Topside Systems
BHP	Broken Hill Proprietary Company Limited
BLEVE	Boiling Liquid Expanding Vapour Explosion
BOE	Barrels of Oil Equivalent
BOEMRE	Bureau of Offshore Energy Management, Regulation and Enforcement [USA]

BOP	Blowout Preventer
BORA	Barrier and Operational Risk Analysis
BP	British Petroleum (formerly)
BSEE	Bureau of Safety and Environmental Enforcement [USA]
CAA	Civil Aviation Authority
CAD	Computer Aided Design
CAPEX	Capital expenditure
CBA	Cost Benefit Analysis
CCA	Cause-Consequence Analysis
CCPS	Center for Chemical Process Safety
CCR	Central Control Room
CDSM	Cidade de São Mateus
CFD	Computational Fluid Dynamics
CNLOPB	Canada-Newfoundland and Labrador Offshore Petroleum Board
CNOOC	China National Offshore Oil Corporation
CNSOPB	Canada-Nova Scotia Offshore Petroleum Board
CO	Carbon monoxide
CO_2	Carbon dioxide
CPA	Closest Point of Approach
CPP	Controllable Pitch Propeller
CREAM	Cognitive reliability and error analysis method
CRIOP	Crisis intervention and operability analysis
CSE	Concept Safety Evaluation
DAE	Design Accidental Events
DAL	Design Accidental Load
DEA	Danish Energy Agency
DeAL	Design Accidental Load
DFU	Defined situations of hazard and accident (Definert fare- og ulykkessituasjon)
DGPS	Differential Global Positioning Systems
DHJIT	Deepwater Horizon Joint Investigation Team
DHSG	Deepwater Horizon Study Group
DHSV	DownHole Safety Valve
DiAL	Dimensioning Accidental Load
DNV GL	Det Norske Veritas—Germanischer Lloyd
DOL	Department of Labor [New Zealand]
DP	Dynamic Positioning
DSB	Directorate for Civil Protection and Emergency Planning (Direktoratet for samfunnssikkerhet og beredskap)
DSHA	Defined situations of hazard and accident (same as DFU)
DWT	Dead Weight Tonnes
E&P	Exploration and Production
E&P Forum	Previous name of organisation now called IOGP
EASA	European Aviation Safety Authority

EER	Escape, Evacuation and Rescue
EERS	Evacuation, Escape and Rescue Strategy
EESLR	Risk due to explosion escalation by small leaks
EFSLR	Risk due to fire escalation by small leaks
EGPWS	Enhanced Group Proximity Warning System
EIA	Environmental Impact Assessment
EIF	Environmental Impact Factor
EPIM	E&P Information Management [association]
EQD	Emergency Quick Disconnection [system]
EQDC	Emergency Quick Disconnect
ERA	Environmental risk analysis
ESD	Emergency Shutdown
ESREL	European Safety and Reliability
ESV	Emergency Shutdown Valve
ETA	Event Tree Analysis
Ex	Explosion [protected]
FAHTS	Fire And Heat Transfer Simulations
FAR	Fatal Accident Rate
FCC	Frigg Central Complex
FEM	Finite Element Method
FES	Fire and Explosion Strategy
Fi-Fi	Fire Fighting
FLACS	Flame Accelerator Software
FLAR	Flight Accident Rate
FMEA	Failure Mode and Effect Analysis
FMECA	Failure Mode, Effect and Criticality Analysis
f-N	Cumulative distribution of number of fatalities
FPPY	Fatalities per platform year
FPS	Floating Production System
FPSO	Floating Production, Storage and Off-Loading Unit
FPU	Floating Production Unit
FRC	Fast Rescue Craft
FSU	Floating Storage Unit
FTA	Fault Tree Analysis
GBS	Gravity Base Structure
GIR	Group Individual Risk
GIS	Geographical Information System
GoM	Gulf of Mexico
GPS	Global Positioning System
GR	Group Risk
GRP	Glass fibre Reinforced Plastic
GRT	Gross Register Tons
HAZAN	Hazard Analysis
HAZID	Hazard Identification
HAZOP	Hazard and Operability Study

HC	Hydrocarbon
HCL	Hybrid Causal Logic; Hydrocarbon Leak
HCLIP	Hydrocarbon Leak and Inventory Project
HCR	Hydrocarbon Release
HEP	Human error probability
HES	Health, Environment and Safety
HF	Human factors
HIPPS	High Integrity Pressure Protection System
HMI	Human–machine interface
HOF	Human and Organisational Factors
HOFO	Helicopter offshore operations
HP	High pressure
HR	Human Reliability
HRA	Human Reliability Analysis
HRO	High reliability organisation
HSE	Health and Safety Executive [UK]
HSS	Helicopter safety studies
IAEA	International Atomic Energy Agency
ICT	Information and communications technology
IEC	International Electrotechnical Commission
IMEMS	International Marine Environmental Modeling Seminar
IMO	International Maritime Organization
IO	Integrated operations
IOGP	International Oil and Gas Producers Association
IR	Individual Risk
IRF	International Regulators' Forum
IRIS	International Research Institute of Stavanger
IRPA	Individual Risk per Annum
ISO	International Organisation for Standardisation
JIP	Joint Industry Project
JU	Jack-up
KFX	Kameleon Fire Ex
kN	Kilonewton (10^3 N)
KNM	Royal Norwegian Navy
KPI	Key Performance Indicator
kW	Kilowatt (10^3 W)
LCC	Life Cycle Cost
LEL	Lower Explosion Level
LFL	Lower Flammability Level
LNG	Liquefied Natural Gas
LOPA	Layers of protection analysis
LP	Low pressure
M&O	Management and Operation
MGB	Main Gearbox
MIRA	Environmental risk analysis (Miljørettet risikoanalyse)

MIRMAP	Modelling of Instantaneous Risk for Major Accident Prevention
MISOF	Modelling of ignition sources on offshore oil and gas facilities
MJ	Megajoule (10^6 J)
MMI	Man–Machine Interface (see also HMI)
MMS	Minerals Management Service (now BEMRE)
MNOK	Million Norwegian kroner
MO	Human and organisational (or HO)
MOB	Man overboard
MOC	Management of Change
MODU	Mobile Offshore Drilling Unit
MOEX	Mitsui Oil Exploration
MP	Main (propulsion) Power
MSF	Module Support Frame; Main Safety Function
MTBF	Mean time between failures
MTO	Man, Technology and Organisation
MUSD	Million US Dollar
NCS	Norwegian Continental Shelf
NEA	National Environment Agency
NGOs	Non-governmental organizations
nm	Nautical Mile
NMD	Norwegian Maritime Authority
NOPSEMA	National Offshore Petroleum Safety and Environmental Management Authority [Australia]
NOROG	Norwegian Oil and Gas [association]
NORSOK	Norwegian Offshore standardisation organisation (Norsk Sokkels Konkurranseposisjon)
NOU	Norwegian Offshore Report (Norges Offentlige Utredninger)
NPD	Norwegian Petroleum Directorate
NPPs	Nuclear Power Plants
NPV	Net Present Value
NTNU	Norwegian University of Science and Technology
NUI	Normally unmanned installation
OCS	Outer Continental Shelf
OIM	Offshore Installation Manager
OMT	Organisational, Human and Technology
ONGC	Oil and Natural Gas Corporation
OPEX	Operational expenditure
OR	Overall risk
OREDA	Offshore and onshore reliability data
OSD	Offshore Division
OSPAR	Oslo and Paris Convention

OTS	Operational Condition Safety ('Operasjonell Tilstand Sikkerhet')
P&ID	Piping and Instrumentation Drawing
PCCC	Pressure-containing anti-corrosion cap
PDO	Plan for Development and Operation
PDQ	Production, Drilling and Quarters
PDS	Reliability of computer-based safety systems
PETRAD	Program for Petroleum Management and Administration
Petro-HRA	Human reliability analysis for petroleum industry
PFD	Process flow diagram; Probability of failure on demand
PFEER	Prevention of Fire and explosion and Emergency Response
PFP	Passive Fire Protection
PGS	Petroleum Geo-Services ASA
PHA	Preliminary Hazard Analysis
PLATO	Software for dynamic event tree analysis
PLL	Potential Loss of Life
PLOFAM	Process leak for offshore installations frequency assessment model
PLS	Progressive Limit State
PM	Position mooring
POB	Personnel On Board
PR	Performance Requirements
PRA	Probabilistic risk analysis
PRS	Position Reference System
PS	Performance Standards
PSA	Petroleum Safety Authority Norway; Probabilistic Safety Assessment
PSAM	Probabilistic Safety Assessment and Management
PSD	Process Shut Down
PSF	Performance shaping factor
PSV	Pressure Safety Valves
PWV	Production wing valve
QA	Quality Assurance
QC	Quality Control
QM	Quality Management
QP	[Frigg] Quarters Platform
QRA	Quantified Risk Assessment; Quantified Risk Analysis
R&D	Research and Development
RABL	Risk Assessment of Buoyancy Loss
RAC	Risk Acceptance Criteria
RACON	Radar signal amplification
RAE	Residual Accidental Events
RAMS	Reliability, Availability, Maintainability, Safety

RIDDOR	Reporting of Injuries, Diseases and Dangerous Occurrences Regulations [UK]
RIF	Risk Influencing Factor
RISP	Risk informed decision support in development projects
RNNP	Risk level in the Norwegian petroleum activity (Risikonivå i norsk petroleumsvirksomhet)
ROS	Risk and vulnerability study (Risiko- og sårbarhetsstudie)
ROV	Remote Operated Vehicle
RPM	Rotations per minute
RRM	Risk Reducing Measure
RTC	Risk tolerance criteria
SA	Shelter area
SAFOP	Safety and Operability Study
SAR	Search and Rescue
SBM	Single buoy mooring
SBV	Standby Vessel
SCR	Safety Case Regulations
SDS	Safe disconnection system
SERA	Safetec Explosion Risk Assessor
SIL	Safety Integrity Level
SIS	Safety instrumented systems
SJA	Safe Job Analysis
SLR	Risk due to small leaks
SNA	Snorre Alpha [installation]
SOLAS	Safety of Life at Sea
SOV	Service operations vessel
SPAR-H	Standardized plant analysis risk—human reliability analysis
SRA	Society of risk analysis
SSIV	Subsea Isolation Valve
SSM	State Supervision of Mines (Netherland)
ST	Shuttle tanker
STAMP	Systems-Theoretic Accident Model and Processes
STEP	Sequential time event plotting
SUPER-TEMPCALC	Software for 2D temperature analysis
TASEF-2	Software for 2D temperature analysis of structures exposed to fire
TCAS	Traffic Collision Avoidance System
TCP2	[Frigg] Treatment Platform 2
TCPA	Time to closest point of approach
TH	Thruster
THERP	Technique for human error-rate prediction
TLP	Tension Leg Platform
TP1	[Frigg] Treatment Platform 1
TR	Temporary Refuge

TRA	Total Risk Analysis
TST	Technical Safety Condition ('Teknisk Sikkerhets Tilstand')
TTS	Technical Condition Safety ('Teknisk Tilstand Sikkerhet')
UEL	Upper Explosive Limit
UFL	Upper Flammability Limit
UKCS	UK Continental Shelf
ULS	Ultimate limit state
UPP	Unmanned production platform
UPS	Underwater Production System; Uninterruptible Power Supply
US CSB	United States Chemical Safety Board
US GoM	United States Gulf of Mexico
US OCS	United States Outer Continental Shelf
USD	US Dollar
USFOS	Software for nonlinear and dynamic analysis of structures
VEC	Valued Ecological Component
VHF	Very High Frequency
VOC	Volatile Organic Compounds
VTS	Vessel Traffic System
W2W	Walk-to-walk
WCPF	Worst credible process fire
WHP	Wellhead platform
WOAD	Worldwide Offshore Accident Database (ref. DNV GL)
WP	Work Permit
X-MAS TREE	Christmas Tree (safety valve assembly)

Part I
Background and Risk Assessment Process

Chapter 1
Introduction

1.1 About 'QRA'

'QRA' is used as the abbreviation for 'Quantified Risk Assessment' or 'Quantitative Risk Analysis'. The context usually has to be considered in order to determine which of these two terms is applicable. Risk assessment involves (see Abbreviations, Page ???) risk analysis as well as an evaluation of the results. 'QRA' is one of the terms used for a type of risk assessment frequently applied to offshore operations. This technique is also referred to as:

- Quantitative Risk Assessment (QRA)
- Probabilistic Risk Assessment (PRA)
- Probabilistic Safety Assessment (PSA)
- Concept Safety Evaluation (CSE)
- Total Risk Analysis (TRA), etc.

In spite of many decades of use and development, no convergence towards a universally accepted term has been seen. QRA and TRA are the most commonly used abbreviations. The nuclear industry, with its origins in the USA, particularly favours the terms Probabilistic Risk Assessment or Probabilistic Safety Assessment.

Concept Safety Evaluation (CSE) has been used since 1981 in Norway and appears to have arisen as a result of risk assessment of new concepts. Total Risk Analysis (TRA), also originated in Norway as a term implying essentially a detailed fatality risk analysis.

It may be argued that all of these terms have virtually the same meaning. This book will concentrate on the term 'QRA' as an abbreviation for 'Quantitative Risk Analysis'. An alternative would be to use 'QRA' as an abbreviation for 'Quantitative Risk Assessment', the difference between these two expressions being that the latter includes evaluation of risk, in addition to the analysis of risk.

© Springer-Verlag London Ltd., part of Springer Nature 2020
J.-E. Vinnem and W. Røed, *Offshore Risk Assessment Vol. 1*,
Springer Series in Reliability Engineering,
https://doi.org/10.1007/978-1-4471-7444-8_1

Use of QRA studies in the offshore industry dates back to the second half of the 1970s. A few pioneer projects were conducted at that time, mainly for research and development purposes, in order to investigate whether analysis methodologies and data of sufficient sophistication and robustness were available.

The methodologies and data were mainly adaptations of what had been used for some few years within the nuclear power generation industry, most notably WASH 1400 [1], which had been developed 3–4 years earlier.

The next step in the development of QRA came in 1981 when the Norwegian Petroleum Directorate issued guidelines for safety evaluation of platform conceptual design [2]. These regulations required QRA be carried out for all new offshore installations in the conceptual design phase. The regulations contained a cut-off criterion of 10^{-4} per platform year as the frequency limit for accidents that needed to be considered in order to define design basis accidents, the so-called Design Accidental Events.

When the design basis accidents had been selected and protective measures implemented, the residual risk had to be assessed. These residual levels were to be compared to the cut-off limit as stated above. Figure 1.1 shows a typical set of results for a floating production concept where the annual frequency for events that impair the different safety functions is given.

For many years, Norway was the only country using QRAs systematically. The offshore industry and authorities in the UK persistently declared that such studies were not the right way to improve safety.

The next significant step in this development was the official inquiry, led by Lord Cullen in the UK, following the severe accident on the Piper Alpha platform

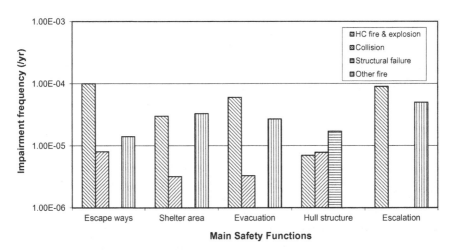

Fig. 1.1 Annual frequencies for residual accidental events

in 1988. Lord Cullen in his report [3] recommended that QRAs should be introduced into UK legislation in much the same way as in Norway nearly 10 years previously.

In 1991 the Norwegian Petroleum Directorate replaced the 1981 guidelines for risk assessment by Regulations for Risk Analysis, which considerably extended the scope of these studies [4].

In 1992 the Safety Case Regulations came into force in the UK, and since then the offshore industry in the UK has been required to perform risk assessments as part of the safety cases for both existing and new installations. The use of QRA studies was rapidly expanded under the new regulations. It is worth noting that the scepticism regarding the use of QRA studies which existed before the Piper Alpha disaster is still strong in some fora.

The next step in this brief historical review is the blast and fire research carried out as part of the (BFETS) programme [5], which was undertaken in the period 1996–1998, Blast and Fire Engineering for Topside Systems. This has focused attention on the high blast loads caused by possible gas explosion scenarios on the platforms. As a result of this work considerable attention is now being given to evaluating how explosion scenarios may be included probabilistically in QRA models.

NPD published a new set of regulations in 2001, which replaced the risk analysis and technical regulations from 1 January 2002. The requirement for risk analysis and other analyses are stipulated in the Health, Environment and Safety (HES) Management regulations. These regulations have requirements for analysis of risk as well as requirements for the definition of risk tolerance criteria. NPD was divided into two organisations from 1 January 2004, the safety division of NPD was separated as a new organisation, and given the name Petroleum Safety Authority (PSA) [Norway]. At the same time, PSA took over the responsibility for 6 onshore facilities in the petroleum sector, terminals and refineries. The Management regulations are controlled by PSA.

The Safety Case regulations were revised in 2005 and 2015. One of the motivations behind the last revision was to fully implement the EU Offshore Directive 2013/30/EU on the safety of offshore oil and gas operations and the amending Directive 2004/35/EC [6]. The EU directive has not been implemented in Norway. However, during recent years, there has been discussion of whether the EU directive should be implemented in Norway, since Norway is part of the European Economic Area and the EEA agreement [7].

The structure of the Norwegian regulations changes in 2007, due to the need to integrate more fully the regulations for offshore and onshore facilities. These changes were temporary, and were superseded by permanent changes, when the structure of the Norwegian regulations was changed again from 1 January 2011. There is little or no material change for offshore installations in these two last revisions, as the main purpose is to issue regulations for onshore petroleum installations.

1.2   QRA in Relation to Other Analysis Methods

'Risk analysis' has been the term used by Norwegian authorities for all systematic approaches to risk assessment, including qualitative as well as quantitative analysis [4]. This covers:

- Hazard and Operability Study (HAZOP)
- Safety and Operability Study (SAFOP)
- Safe Job Analysis (SJA)
- Preliminary Hazard Analysis (PHA)
- Failure Mode and Effect Analysis (FMEA)
- Quantitative Risk Analysis (QRA).

The first five items on this list are essentially qualitative approaches, although some of these techniques may be used in a semi-quantitative fashion. The last item is a quantitative approach. It has been a disadvantage that no specific term has been used in Norwegian legislation to differentiate QRA from the qualitative techniques. Discussion of requirements has often been rather unclear, as no distinction has been made between the different types of risk analysis. Only the quantitative approach, QRA, is discussed in this book.

1.3   Objectives

This book is essentially focused on applications of QRA in the offshore oil and gas industry. The objectives of this book are as follows:

1. To provide guidance about the performance of QRA studies for offshore installations and marine structures.
2. To show how tools, approaches and data may be used effectively to ensure that QRA studies provide useful input to risk based decision-making.
3. To demonstrate how the best practice is being carried out.
4. To demonstrate what is new knowledge from recent research activities during the last five years.
5. To provide some perspective on issues that have not yet been sufficiently resolved.

The discussion of modelling is also application oriented. Modelling of hazards is therefore related to the most prominent hazards for offshore installations:

- Fire
- Explosion
- Collision
- Marine hazards

Other hazards such as falling objects are also addressed, but somewhat more briefly.

Risk to personnel is addressed most thoroughly, but also risk to the environment and material damage risk are covered. The methodology for environmental risk assessment is briefly discussed. But it is still an area where several approaches are still being attempted.

Most QRA work has been devoted to risk assessment during the design phase. The use of risk assessment during the operations phase is also important and thus a significant part of this book is devoted to this phase. Recent research has focused on the operations phase, which will be discussed in some depth.

All illustrations and cases that are presented are mostly related to offshore installations and marine structures involved in offshore oil and gas exploration, production and transportation. Consideration is also given to aspects related to the transportation of personnel and supplies to the installations.

1.4 Relevant Regulations and Standards

There are several countries that have legislation that call for the use of QRA studies in the design and operation of offshore installations:

- United Kingdom
- Canada
- Australia
- Norway

The following is a brief summary of the requirements of the legislation in these countries except for UK and Norway, which are discussed in some depth throughout the remainder of this chapter:

- Canada (Newfoundland and Labrador offshore areas)

  - In association with new development proposals, a Concept Safety Analysis is required. The field development proposal needs to define how this will be met, and state the 'Target Levels of Safety' that have been set as acceptance criteria for risk.
  - The development proposal shall also define a 'Risk Assessment Plan' which should contain a listing of the various specific risk and safety analyses that may be required as detailed design proceeds. It should also provide a plan for the completion of these studies and analyses and an explanation of how this process is integrated into the design process. Finally, it should provide an explanation of the methodologies to be utilised and a discussion of their validity and relevance in the overall process.

- Australia
  - Petroleum (Submerged Lands) (Management of Safety on Offshore Facilities) Regulations 1996 [8].
  - These regulations call for Safety Cases to be prepared for all installations and to demonstrate that risks have been reduced to a level that is as low as reasonably practicable (ALARP).
  - The National Offshore Petroleum Safety and Environmental Management Authority (NOPSEMA) has also issued Safety Case Guidelines [9]. NOPSEMA was established from 1 January 2012 as a follow-up of the Montara accident in 2009, it was previously known as National Offshore Petroleum Safety Authority (NOPSA).

There are also several other countries where voluntary schemes are dominating, for instance due to company policies, such as by the Shell Group worldwide.

A thorough overview and discussion of offshore regulations is provided by Baram et al. [10]. The main emphasis in the following is on legislation in UK and Norway, where the relevant requirements with respect to risk assessment and risk management are briefly introduced.

1.5 Norwegian Regulations

PSA has four regulations which control the safety of design and operation of offshore installations. To a large extent, these have remained unchanged since 2011. However, minor changes are carried out occasionally. This section reflects the regulation text as it was at the end of 2017. Updated regulations are published on the PSA webpage. The five regulations mentioned are:

- Regulations relating to health, environment and safety in the petroleum activities and certain onshore facilities (the Framework regulations) [11].
- Regulations relating to management and the duty to provide information in the petroleum activities and at certain onshore facilities (the Management regulations) [12].
- Regulations relating to design and outfitting of facilities etc., in the petroleum activities (the Facilities regulations) [13].
- Regulations relating to conducting petroleum activities (the activities regulations) [14].

1.5.1 Framework Regulations

This is a high level regulation which has the overall principles that are spelled out in more detail in the other regulations. One of the requirements is not found in any other regulation, this is the Norwegian equivalent of the so-called ALARP evaluation (ALARP—As Low As Reasonably Practicable, see Sect. 1.6.1), see the copy of Section 11 below.

It is in particular the first and second paragraphs of Section 11 that define requirements for risk reduction that follow closely the interpretation of ALARP in UK regulations.

Section 11

Risk reduction principles

Harm or danger of harm to people, the environment or material assets shall be prevented or limited in accordance with the health, safety and environment legislation, including internal requirements and acceptance criteria that are of significance for complying with requirements in this legislation. In addition, the risk shall be further reduced to the extent possible.

In reducing the risk, the responsible party shall choose the technical, operational or organisational solutions that, according to an individual and overall evaluation of the potential harm and present and future use, offer the best results, provided the costs are not significantly disproportionate to the risk reduction achieved.

If there is insufficient knowledge concerning the effects that the use of technical, operational or organisational solutions can have on health, safety or the environment, solutions that will reduce this uncertainty, shall be chosen.

Factors that could cause harm or disadvantage to people, the environment or material assets in the petroleum activities, shall be replaced by factors that, in an overall assessment, have less potential for harm or disadvantage.

Assessments as mentioned in this section, shall be carried out during all phases of the petroleum activities.

This provision does not apply to the onshore facilities' management of the external environment.

Section 11 of the Framework regulations was not focused on significantly in the first few years after the regulations were stipulated. This was gradually changed, starting from 2006.

Another subject which is focussed in the Framework regulations is emergency preparedness. The overall requirements to emergency planning and dimensioning of systems are not spelled out in more detail in other regulations.

1.5.2   Management Regulations

There are several sections in the Management regulations that are important, with respect to analysis of risk, analysis of barriers, and risk tolerance (acceptance is used by PSA) criteria.

Two sections in the Management regulations are particularly important with respect to analysis of major accident risk and quantitative risk analysis, these two sections are given in full below:

Section 16
General requirements for analyses

The responsible party shall ensure that analyses are carried out that provide the necessary basis for making decisions to safeguard health, safety and the environment. Recognised and suitable models, methods and data shall be used when conducting and updating the analyses.

The purpose of each risk analysis shall be clear, as well as the conditions, premises and limitations that form its basis.

The individual analysis shall be presented such that the target groups receive a balanced and comprehensive presentation of the analysis and the results.

The responsible party shall set criteria for carrying out new analyses and/or updating existing analyses as regards changes in conditions, assumptions, knowledge and definitions that, individually or collectively, influence the risk associated with the activities.

The operator or the party responsible for operating an offshore or onshore facility shall maintain a comprehensive overview of the analyses that have been carried out and are underway. Necessary consistency shall be ensured between analyses that complement or expand upon each other.

Section 17
Risk analyses and emergency preparedness assessments

The responsible party shall carry out risk analyses that provide a balanced and most comprehensive possible picture of the risk associated with the activities. The analyses shall be appropriate as regards providing support for decisions related to the upcoming processes, operations or phases. Risk analyses shall be carried out to identify and assess what can contribute to, i.e., major accident risk and environmental risk associated with acute pollution, as well as ascertain the effects various processes, operations and modifications will have on major accident and environmental risk.

Necessary assessments shall be carried out of sensitivity and uncertainty.

The risk analyses shall

(a) identify hazard and accident situations,
(b) identify initiating incidents and ascertain the causes of such incidents,
(c) analyse accident sequences and potential consequences, and
(d) identify and analyse risk-reducing measures, cf. Section 11 of the Framework Regulations and Sections 4 and 5 of these regulations.

Risk analyses shall be carried out and form part of the basis for making decisions when e.g.:

(a) identifying the need for and function of necessary barriers, cf. Sections 4 and 5,
(b) identifying specific performance requirements of barrier functions and barrier elements, including which accident loads are to be used as a basis for designing and operating the installation/facility, systems and/or equipment, cf. Section 5,
(c) designing and positioning areas,
(d) classifying systems and equipment, cf. Section 46 of the Activities Regulations,
(e) demonstrating that the main safety functions are safeguarded,
(f) stipulating operational conditions and restrictions,
(g) selecting defined hazard and accident situations.

For larger discharges of oil or condensate, simulations of drift and dispersion shall be carried out.

Emergency preparedness analyses shall be carried out and be part of the basis for making decisions when e.g.

(a) defining hazard and accident situations,
(b) stipulating performance requirements for the emergency preparedness,
(c) selecting and dimensioning emergency preparedness measures.

The environmental risk and emergency preparedness analyses shall be updated in case of significant changes affecting the environmental risk or the emergency preparedness situation. In any case, updating needs shall be assessed every five years. The assessment shall be documented and made available to the Norwegian Environment Agency on request.

Section 5 deals with barriers or defences as they may be called. This section concerns design as well as operation of installations.

Section 5

Barriers

Barriers shall be established that at all times can

(a) identify conditions that can lead to failures, hazard and accident situations,
(b) reduce the possibility of failures, hazard and accident situations occurring and developing,
(c) limit possible harm and inconveniences.

Where more than one barrier is necessary, there shall be sufficient independence between barriers.

The operator or the party responsible for operation of an offshore or onshore facility, shall stipulate the strategies and principles that form the basis for design, use and maintenance of barriers, so that the barriers' function is safeguarded throughout the offshore or onshore facility's life.

Personnel shall be aware of what barriers have been established and which function they are intended to fulfil, as well as what performance requirements have been defined in respect of the concrete technical, operational or organisational barrier elements necessary for the individual barrier to be effective.

Personnel shall be aware of which barriers and barrier elements are not functioning or have been impaired.

Necessary measures shall be implemented to remedy or compensate for missing or impaired barriers.

Risk tolerance criteria are specified in Section 9 (called risk acceptance criteria), including personnel, main safety functions (see Sect. 1.5.3), pollution and damage to third party groups and facilities. The last aspect is not applicable for offshore installations, but is applicable to onshore facilities that also fall under the jurisdiction of the PSA.

Section 9
Acceptance criteria for major accident risk and environmental risk

The operator and the party responsible for operating a mobile facility, shall set acceptance criteria for major accident risk and for environmental risk associated with acute pollution.

Acceptance criteria shall be set for:

(a) the personnel on the offshore or onshore facility as a whole, and for personnel groups exposed to particular risk,
(b) loss of main safety functions for offshore petroleum activities,
(c) acute pollution from the offshore or onshore facility,
(d) damage to third party.

The acceptance criteria shall be used when assessing results from risk analyses, cf. Section 17. Cf. also Section 11 of the Framework Regulations.

1.5.3 Facilities Regulations

With respect to risk assessment, the main contributions from the Facilities regulations are the principles for maximum frequency of events that impair the main safety functions. The text of Sections 7 and 11 are shown below.

Section 7
Main safety functions

The main safety functions shall be defined in a clear manner for each individual facility so that personnel safety is ensured and pollution is limited.

For permanently manned facilities, the following main safety functions shall be maintained in the event of an accident situation:

(a) preventing escalation of accident situations so that personnel outside the immediate accident area are not injured,
(b) the capacity of main load-bearing structures until the facility has been evacuated,

(c) protecting rooms of significance to combatting accidents so that they remain operative until the facility has been evacuated,

(d) protecting the facility's safe areas so that they remain intact until the facility has been evacuated,

(e) at least one escape route from every area where personnel are found until evacuation to the facility's safe areas and rescue of personnel have been completed.

Section 11

Loads/actions, load/action effects and resistance

The design loads/actions that will form the basis for design and operation of installations, systems and equipment, shall be determined. When determining design loads/actions, the requirement to robust solutions, cf. Section 5, and the requirement to risk reduction, cf. the Framework Regulations Section 11, shall form the basis. The design loads/actions shall ensure that installations, systems or equipment will be designed such that relevant accidental loads/actions that can occur, do not result in unacceptable consequences, and shall, as a minimum, always withstand the dimensioning accidental load/action.

When determining design loads/actions, the effects of fire water shall not be considered. This applies to both fire loads/actions and explosive loads/actions.

Installations, systems and equipment that are included as elements in the realisation of main safety functions, cf. Section 7, shall as a minimum de designed such that dimensioning accidental loads/actions or dimensioning environmental loads/actions with an annual likelihood greater than or equal to 1×10^{-4}, shall not result in loss of a main safety function.

When determining loads/actions, the effects of seabed subsidence over, or in connection with the reservoir, shall be considered.

Functional and environmental loads/actions shall be combined in the most unfavourable manner.

Facilities or parts of facilities shall be able to withstand the design loads/actions and probable combinations of these loads/actions at all times.

The main safety functions are more closely associated with design characteristics, compared to for instance fatalities. But there are several aspects associated with how these requirements have been worded that are not as clear as would have been preferred. This is in particular associated with how to define areas and how to sum over different event categories and areas.

1.5.4 Activities Regulations

There are no relevant requirements in the Activities regulations with respect to risk assessment and management. From a broader HES management point of view, the most relevant aspects are emergency preparedness, working environment, external environment as well as drilling and well control and barriers.

1.5.5   NMD Risk Analysis Regulations

The Norwegian Maritime Directorate has issued 'Regulations for risk analysis of mobile units', which applies to all mobile units that shall be registered in the Norwegian register of ships.

The regulations apply to the owner of the unit, and have sections for execution and updating of risk analysis. It covers risk analysis of the concept, construction risk analysis, 'as built' risk analysis, in addition to reliability and vulnerability analysis as well as emergency preparedness analysis. The regulations also contain general risk tolerance criteria and design criteria for main safety functions.

1.6   UK and EU Regulations

The offshore regulatory regime was completely rewritten as a consequence of the Piper Alpha (see Sect. 4.7) in 1988, based on the recommendations from the inquiry chaired by Lord Cullen [3]. The following regulations have been issued:

- Safety Case Regulations (SCR) [15].
- PFEER (Prevention of Fire and Explosion, and Emergency Response) Regulations [16].
- Management and Administration Regulations [17].
- Design and Construction Regulations [18].
- The EU Offshore Directive 2013/30/EU on safety of offshore oil and gas operations [19].

1.6.1   Safety Case Regulations

The duty holder is required to identify hazards, evaluate risks and demonstrate that measures have been or will be taken to control the risks such that the residual risk level is as low as reasonably practicable (ALARP). The Safety Case should also demonstrate that the operator has a HES management system which is adequate in order to ensure compliance with all health and safety regulatory requirements.

There is no reference to QRA in the regulations themselves. QRA is mentioned in some of the schedules, listing the documentation to be submitted. Further discussion on the use of QRA is however, found in 'Content of Safety Cases—General Guidance'. The use of QRA under this legislation is mainly to analyse:

- The risk of impairment of the Temporary Refuge.
- The risk to personnel directly, expressed in terms of PLL and AIR, or some other fatality measures.

The main basis for the use of the QRA approach is actually implicit, as the duty holder is required to demonstrate through the safety case that the risk level for personnel on the installation is 'as low as reasonably practicable', abbreviated as ALARP. This can only be effectively done through the use of QRA.

The approach to QRA under the SCR is virtually the same as under the Norwegian regulations, with the exception that SCR applies to risk to personnel only, whereas the Norwegian regulations apply to a set of risk dimensions include personnel, environment and assets, as has been discussed in Chap. 2.

The Safety Case regulations were revised in 2005 and in 2015. The main background for revising the regulations in 2015 was to fully implement the EU Offshore Directive (see Sect. 1.6.5), as well as to include lessons learned from the Montara and Macondo accidents (see Chap. 4). The changes from the 2005 to the 2015 version are summarized in paragraphs 48–60 in a guidance document published by HSE [19] and will not be elaborated on in this book.

1.6.2   PFEER Regulations

The so-called PFEER (Prevention of Fire and Explosion, and Emergency Response) Regulations (HSE, 1995a) imply important requirements for active and passive safety systems, as well as emergency preparedness systems and functions. The purpose of these regulations is to ensure that measures to protect against fire and explosion result in a risk level which is as low as reasonably practicable, and that sufficient arrangements are in place in order to provide a good prospect of rescue and recovery for personnel in all reasonably foreseeable situations. Operators are according to these regulations required to:

- Take measures to prevent fires and explosions and provide protection from any which do occur;
- Provide effective emergency response arrangements.

The need for risks to be as low as reasonably practicable is the basis for using a risk based design in relation to fire and explosion.

The need to provide facilities which give a good prospect of rescue and recovery for personnel in all reasonably foreseeable situations may appear as a probabilistic framework, but this is questionable. The way this requirement appears to be implemented, is that any accidental situation which a lay person would consider as reasonably foreseeable, is a reasonably foreseeable event. The implication of this is that there is very little room for a probabilistic consideration, if the situation can occur, then the operator has to use the situation in a deterministic way as the basis for the provision of 'good prospects of rescue and recovery'. If this is not possible, then the activity has to be halted until such prospects may be restored. This is mainly associated with the possibility to provide such 'good prospects' during periods of severe environmental conditions.

1.6.3   Management and Administration Regulations

The Offshore Installations and Pipeline Works (Management and Administration) Regulations [17] set out requirements for the safe management and administration of an offshore installation, such as the use of permit to work systems. The requirements are essential provisions in order to comply with the legislation, but there are no requirements as such to risk assessment and management.

1.6.4   Design and Construction Regulations

The Offshore Installations and Wells (Design and Construction, etc.) Regulations (1996) (DCR) [18] are aimed at ensuring the integrity of installations, the safety of offshore and onshore wells, and the safety of the workplace environment offshore.

1.6.5   EU Regulations

As a follow-up of the Macondo Deepwater Horizon accident in 2010, a new EU directive [15] has been developed, relevant for all existing and future installations in the EU. This directive, which came into force in June 2013, requires operators to take all necessary measures to prevent major accidents as well as to have sufficient physical, human and financial resources to limit the consequences should an accident occur. The directive requires that the operator or owner of an installation must provide the authorities with a copy of the company's major accident prevention policy, the safety and environmental management system, as well as a report on major hazards. The directive requires the operators to prepare emergency preparedness plans to respond to major accidents, including an analysis of how to handle oil spills. National authorities are also required to develop such plans, covering oil and gas installations, including infrastructure.

1.7 National and International Standards

There is a small core group of international standards, by the International Organization for Standardization (ISO), reflecting a risk based approach to decision-making in the offshore industry. The following standards have been issued:

- ISO 10418; Analysis, design, installation and testing of basic surface safety systems for offshore production platforms [20].
- ISO 13702; Control and mitigation of fires and explosions on offshore production installations—requirements and Guidelines [21].
- ISO 15544; Requirements and guidelines for emergency response [22].
- ISO 17776; Guidelines on tools and techniques for identification and assessment of hazards [23].

The ISO organisation has the responsibility to revise and reissue vital API standards with respect to safety. For example, ISO 10418 replaces API RP 14C. No other international standard organisations have issued standards for risk assessment or risk based design. IOGP (formerly E&P Forum) has, however, issued guidelines on HES management (IOGP, 1994). Other ISO standards that are essential:

- Safety aspects—Guidelines for their inclusion in standards, ISO/IEC Guide 51:2014 [24].
- Risk management vocabulary, guidelines for use in standards, ISO/IEC Guide 73:2009 [25].
- ISO31000—Risk management—Guidelines [26].

The terminology used in this book is accordance with the terminology of ISO/IEC Guide 73 [25]. The definitions given at the back (see Page ???) are extracted from this standard, where relevant.

There are several national guidelines or standards for HES management, but these are not covered here. The only national standard for risk assessment is the Norwegian offshore standardisation organisation (NORSOK) document:

- Guidelines for Risk and Emergency Preparedness Analysis, NORSOK Z–013 [27].

The presentation in this book is based on the NORSOK standard, for instance in relation to terminology. There is some distinctions between the definitions adopted in the NORSOK standard and the current Norwegian legislation, however the NORSOK versions have been chosen.

1.8 New Topics

This section briefly outlines the topics that are new in the 4th Edition of this book. There are many places where new models have been included in existing topics, this is not mentioned here, only topics that are new.

1.8.1 Cyber Risks in the Petroleum Industry

In the early days of offshore oil and gas activities, all interventions on the installation had to be carried out locally on the offshore facility. If, for example, a valve was going to be closed or some properties of the production needed to be adjusted, this was carried out manually on the offshore facility. More recently, the trend is that more and more activities are carried out remotely from on shore. This is not only limited to real-time analysis of data from the offshore installation. There are also examples of control room activities being moved away from the offshore facility. This provides opportunities, for example in terms of costs saved and an increased ability to include expert opinions in decision making. However, it also introduces new threats in terms of a possibility that people located remotely from the offshore installation can intervene in the process systems either deliberately or accidentally. This is not necessarily limited to personnel in the organization. If firewalls and other IT equipment are not sufficiently set up and maintained, access may in theory be achieved from external locations as well, in principle from any location in the world. Sect. 15.11.1 gives some examples of incidents and near-misses, followed by an overview of standards and assessment approaches.

Since cyber threats have emerged during recent decades, they have not been included in traditional QRAs even though some of the cyber threats imply a major accident risk potential. The situation at present is that there is a need to improve the focus/attention on cyber threats as well as the methods being used to analyse such threats.

The regulations in the Norwegian petroleum industry do not address cyber threats specifically. However, since the regulations are functional, the operating companies are expected to manage cyber risk in the same way as other threats. This includes ensuring a sufficient level of redundancy in terms of independence between safety barriers [12, 13]. To some extent, cyber threats are covered by standards. For example, there are requirements for safety critical equipment: see for example IEC 51508 [28] and IEC 61511 [29]. Also Norwegian oil and gas have issued a recommended guideline on information security baseline requirements for process control, safety and support ICT systems [30]. In addition, there is a large number of standards covering IT and cyber issues that have not been elaborated on in this book.

1.8.2   Normally Unmanned Installations

Normally unmanned installations have existed for a long time now. New in Norway in the last couple of years are very stripped down normally unmanned wellhead installations, which are intended to be visited only a few times per year. The next generation will be normally unmanned production installations, with significantly more equipment, but still with relatively few visits per year. Chapter 22 is devoted to normally unmanned installations.

1.9   Activity Levels

The International Regulators Forum (IRF) is briefly described in Sect. 6.5. The data which is available from IRF may also be used in order to present what levels of offshore activity that the different member countries have. This has been compared against the total data reported by IOGP (see Sect. 17.1.2) for the period 2007–2016. The sum of manhours from the IRF member countries in 2010 (Denmark & New Zealand missing) is 442 million manhours. The total manhours reported from IOGP member companies for the same period is 886 million manhours, the IRF member countries is about half of the manhours reported by IOGP members (Fig. 1.2).

Almost half of the manhours reported by IRF members is from US. Brazil, Mexico, Norway and UK are all around 50 million manhours, whereas the rest have much lower activity levels.

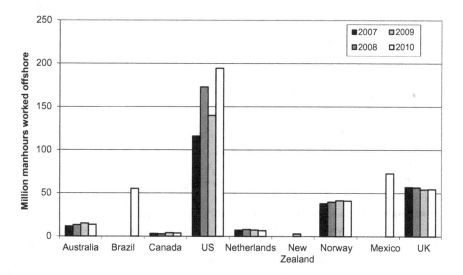

Fig. 1.2 Manhours worked per annum, IRF member countries, 2007–2010

1.10 Limitations

This book is focused on offshore risk assessment i.e., the analysis of offshore risks, and the presentation and evaluation of results. The emphasis is first of all on risk to personnel, secondly on risk to the environment and risk to assets is the least emphasised subject.

As a consequence of these priorities, there are some areas that are not focused on or may be not considered at all. This section provides brief overviews of some of these limitations.

1.10.1 Risk Management

Risk management is discussed in depth in the book Risk Management, with Applications from the Offshore Petroleum Industry [31]. The role of risk assessment in risk management is discussed in Chap. 3, within the context of the ISO31000 approach. Apart from the discussion in Chap. 3, this subject is not addressed in this book.

1.10.2 Emergency Response

There is a special Norwegian requirement for so-called emergency preparedness analysis (see Sect. 1.5.2), which is a tool for emergency response planning. The input to this process is partly from QRA studies, as discussed in Chap. 20. Apart from that discussion the topic of emergency preparedness analysis is not addressed in general in this book.

1.10.3 Subsea Production

Deep water production implies subsea production systems tied into floating pro-duction facilities or pipelines directly to onshore facilities. This concept has become increasingly popular on the Norwegian continental shelf during recent years. This book covers extensively the floating production facilities and the associated hazards.

The subsea production systems are usually at a significant distance from the surface facilities, in which case any failure of the subsea facilities is not a source of risk for the personnel on the surface installations. Such failures may cause hydrocarbon leaks, which may cause spill and subsequent oil pollution. This last aspect is within the scope of this book. One crucial aspect associated with such leaks is the detection of leaks, which may not be easy if they are not extensive.

The main challenge with subsea production facilities is the reliability of the production function, due to the extensive costs and sometime delays involved in maintaining subsea production facilities. This aspect is not considered at all in this book.

In the future, process equipment such as separators and compression units may be installed subsea. Such concepts have not been considered in this book.

1.10.4 Production Regularity

A subject which is closely associated with risk analysis is regularity analysis, either as production and/or transport regularity. This aspect is coupled with risk to assets, and has little direct connection to risk to personnel and risk to the environment. Some of the hazardous events that may lead to fatalities or spills may also cause production disruption, and thus have an impact on the regularity. Traditionally, however, production regularity studies disregard such rare events in any case. Production regularity is outside the scope of this book.

Moreover, the term 'RAMS' (Reliability, Availability, Maintainability, Safety) is not covered in full, in accordance with what is discussed above.

1.10.5 Resilience

Several scientists have in recent years stressed the need for a different approach within safety engineering that includes studying normal performance rather than failure. This has become known as 'resilience engineering'. This is said to represent a new way of thinking about safety. Woods [32] describes resilience engineering is a paradigm for safety management that focuses on how to help people cope with complexity under pressure to achieve success. Rather than view past success as a reason to reduce investments, resilient organisations continue to invest in anticipating the changing potential for failure because they appreciate that their knowledge of the gaps is imperfect and that their environments constantly change. One measure of resilience is therefore the ability to create foresight, namely to anticipate the changing shape of risk before failure and damage occurs, as discussed by Hollnagel et al. [33].

Resilience is an important topic, however, it has no obvious close connection to QRA studies, and is therefore not addressed in this book.

1.10.6 High Reliability Organisations

A High Reliability Organization (HRO) is an organisation that has succeeded in avoiding major accidents in an environment where normal accidents can be expected due to risk factors and complexity. There are several characteristics related to HRO. One is that they aggressively seek to know what they don't know, as discussed by Roberts & Bea [34]. They also emphasize that HRO organizations also use failure simulations to train everyone to be heedful of the possibility of accidents. Techniques similar to event trees may be used to simulate decision gates and different scenarios related to precursor incidents. Accident investigation of precursor incidents can be used to communicate organisational concern with accidents to reinforce the cultural values of safety, and identify parts of the system that should have additional barriers. All the elements are characteristics of an HRO organization. Organisations that have fewer accidents have developed systems and processes for communicating the big picture to everyone in the organization. This is a major challenge that begins with top management encouraging the culture to be supportive of open communication. The reward and incentive system has to reinforce an open flow of communication as well as support the open discussion of organisational purpose [34].

HRO is an important topic for risk management, however, it has no connection to QRA studies, and is therefore not addressed in this book.

1.10.7 STAMP

STAMP is the acronym for Systems-Theoretic Accident Model and Processes, introduced by Leveson [35]. The accident analysis, hazard analysis, and system engineering techniques built on STAMP can be used to improve the design, operation, and management of potentially dangerous systems or products.

STAMP has so far no known applications in the offshore petroleum industry, but may in the future be a possibility to refine analytical approaches. At the present time, however, this topic is not addressed in this book.

1.10.8 Inherently Safe

The person who is most closely associated with the term 'inherently safe' is Trevor Kletz, Loughborough University [36]. The best way of dealing with a hazard is to remove it completely; this is significantly better option than attempting to control a hazard. Principles to remove hazards completely are the main scope of inherently safe. Some main principles of inherently safe design are discussed by Lees [37]:

1. Intensification
2. Substitution
3. Attenuation
4. Simplicity
5. Operability
6. Fail-safe design
7. Second chance design.

Inherently safe design is very important in order to reduce risk to a level which is ALARP, which is important in a risk management context. However, inherently safe design has no connection to QRA studies, and is therefore not addressed in this book.

References

1. NRC (1975) Reactor safety study, WASH 1400. Nuclear Regulatory Commission, Washington DC
2. NPD (1980) Guidelines for conceptual evaluation of platform design. Norwegian Petroleum Directorate, Stavanger
3. Lord Cullen (The Hon) (1990) The public inquiry into the piper alpha disaster. HMSO, London
4. NPD (1990) Regulations relating to implementation and use of risk analysis in the petroleum activities. Norwegian Petroleum Directorate, Stavanger
5. SCI (1998) Blast and fire engineering for topside systems, phase 2. Ascot, SCI. Report no. 253
6. European Union (2013) The safety of offshore oil and gas operations directive 2013/30/EU and amending directive 2004/35/EC. http://eur-lex.europa.eu/homepage.html
7. European Free Trade Association (1994) EEA agreement: agreement on the European economic area. Issued January 1994, last amended in August 2016
8. NOPSEMA (1996) Petroleum (submerged lands) (management of safety on offshore facilities) regulations 1996. National offshore petroleum safety and environmental management authority, Canberra, Statutory Rules 1996 No. 298
9. NOPSEMA (2004) Safety case guidelines. National Offshore Petroleum Safety and Environmental Management Authority, Canberra
10. Baram M, Lindøe PH, Renn O (2013) Risk governance of offshore oil and gas operations. Cambridge University Press, Cambridge
11. PSA (2017) Regulations relating to health, safety and the environment in the petroleum activities and at certain onshore facilities (the framework regulations). Last amended 15 Dec 2017
12. PSA (2017) Regulations relating to management and the duty to provide information in the petroleum activities and at certain onshore facilities (the management regulations). Last amended 18 Dec 2017
13. PSA (2017) Regulations relating to design and outfitting of facilities, etc. in the petroleum activities (the facilities regulations). Last amended 18 Dec 2017
14. PSA (2017) Regulations relating to conducting petroleum activities (the activities regulations). Last amended 18 Dec 2017
15. SCR (2015) The offshore installations (offshore safety directive) (safety case etc) regulations 2015

16. HSE (1995) Prevention of fire and explosion, and emergency response regulations. HMSO, London
17. HSE (1995) Offshore installations and pipeline works (management and administration) regulations. HMSO, London
18. HSE (1996) The offshore installations and wells (design and construction, etc.) regulations. HMSO, London
19. HSE (2015) The offshore installations (offshore safety directive) (safety case etc.) regulations 2015—Guidance on regulations
20. ISO (2003) Analysis, design, installation and testing of basic surface safety systems for offshore production platforms. International Standards Organisation, Geneva: ISO10418:2003
21. ISO (2015) Control and Mitigation of Fires and Explosions on Offshore Production Installations—Requirements and Guidelines, International Standards Organisation; Geneva: ISO13702:2015
22. ISO (2010) Requirements and guidelines for emergency response. International Standards Organisation, Geneva: ISO15544:2010
23. ISO (2016) Petroleum and natural gas industries — offshore production installations - major accident hazard management during the design of new installations. International Standards Organisation, Geneva: ISO17776:2016
24. ISO (2014) Safety aspects—guidelines for their inclusion in standards, ISO/IEC guide 51:2014. ISO, Geneva
25. ISO (2009) Risk management - vocabulary, ISO/IEC guide. International Standards Organisation, Geneva: ISO Guideline 73:2009
26. ISO (2018) Risk management—guidelines, Geneva: ISO31000:2018
27. Standard Norway (2010) Risk and emergency preparedness analysis, NORSOK Standard Z-013, Rev.3, 2010
28. IEC (2010) IEC 61508—functional safety of electrical/electronic/programmable electronic safety-related systems. Geneva
29. IEC (2016) IEC 61511—functional safety—safety instrumented systems for the process industry sector. Geneva
30. Norwegian Oil and Gas (2016) 104—norwegian oil and gas recommended guidelines on information security baseline requirements for process control, safety and support ICT systems. Revision 06. 05 Dec 2016
31. Aven T, Vinnem JE (2007) Risk management, with applications from the offshore petroleum industry. Springer, London
32. Woods D (2006) Essential characteristics of resilience. Resilience engineering: concepts and precepts, p 19–30
33. Hollnagel E, Woods D, Leveson N (2006) Resilience engineering: concepts and precepts. Ashgate Publishing, Ltd
34. Roberts KH, Bea R (2001) Must accidents happen? lessons from high-reliability organizations. Acad Manag Exec 2001(15):70–78
35. Leveson N (2012) Engineering a safer world: applying systems thinking to safety, MIT Press
36. Kletz T (2003) Inherently safer design—its scope and future original research article, process safety and environmental protection, 81/ 6, Nov 2003, pp 401–405
37. Lees FP (1996) Loss prevention in the process industries, 2nd edn. Butterworth-Heinemann, Oxford

Chapter 2
Risk Picture—Definitions and Characteristics

2.1 Definition of Risk

2.1.1 What Is Risk?

There is no common definition of risk even among experts in the risk field. In 2015, the Society for Risk Analysis developed a glossary defining relevant terms [1]. When developing the glossary, it was considered unrealistic to come up with one common definition of risk. Instead, the intention was to come up with a list of alternatives. The resulting definitions were as follows:

(a) Risk is the possibility of an unfortunate occurrence
(b) Risk is the potential for realization of unwanted, negative consequences of an event
(c) Risk is exposure to a proposition (e.g., the occurrence of a loss) of which one is uncertain
(d) Risk is the consequences of the activity and associated uncertainties
(e) Risk is uncertainty about and severity of the consequences of an activity with respect to something that humans value
(f) Risk is the occurrence of some specified consequences of the activity and associated uncertainties
(g) Risk is the deviation from a reference value and associated uncertainties.

In ISO [2], risk is defined as the 'effect of uncertainty on objectives'. As emphasized by SRA [1], this definition can be interpreted in different ways, one as a special case of (d) or (g) in the list above, with consequences seen in relation to the objectives.

The Norwegian petroleum safety authorities have since 2015 defined risk as 'the consequences of the activities, with associated uncertainty'. This definition is consistent with (d) in the SRA glossary [1]. This is also the definition being applied in this book.

© Springer-Verlag London Ltd., part of Springer Nature 2020
J.-E. Vinnem and W. Røed, *Offshore Risk Assessment Vol. 1*,
Springer Series in Reliability Engineering,
https://doi.org/10.1007/978-1-4471-7444-8_2

2.1.2 Description of Risk

Since we define risk as 'the consequences of the activities with associated uncertainty', we may ask ourselves which consequences and what kind of uncertainty we are talking about. Before we carry out an activity, for example, development of oil and gas, there may be different outcomes. For most activities there are many different potential outcomes, ranging from great success, better than anticipated, to severe loss, accidents and other severe consequences. Beforehand, we do not know what the real outcome will be out of all the potential outcomes. In other words, there is uncertainty. Risk is the concept of being in a situation with numerous potential outcomes and uncertainty regarding which of the potential outcomes will materialize into the real outcome.

To describe risk, we need tools, often referred to as risk metrics. Examples are probabilities and expectation values, and indexes such as FAR and PLL. We will come back to some risk metrics later in this chapter. It is essential to understand the difference between risk as a concept and a risk metric as a way of 'measuring' or describing the risk. To explain this, we will use an analogy with another concept that is, perhaps, more intuitive than risk: the concept of distance. Distance exists regardless of how it is measured. If two objects are located in two different places, there is distance. In the same way, if an activity is performed with more than one potential outcome, and there is uncertainty as to what will be the real outcome, this means that there is risk. We may ask ourselves: how large is the distance between two objects. Then we need a way to measure or describe the distance. We may use meters, inches, nanometers, feet or light years as the measuring tool, depending on the context. In the same way, we may choose between different risk metrics when we are going to describe risk. Sometimes we use words ('high', 'low' etc.), sometimes we use probabilities, sometimes we use indexes such as FAR values. No matter what choice we make, we will not be able to give a perfect description of risk: there will always be aspects of risk that are not captured in the risk metric that we use to describe risk, and that may influence risk-informed decisions. In the following, one aspect that is lacking in many risk assessments will be explained: the strength of knowledge.

2.1.3 Strength of Knowledge

Strength of knowledge is a concept that many struggle to understand. To explain this concept, we will use a hypothetical risk-related activity. Suppose you are planning to walk from your house to work tomorrow. The outcome of this activity is uncertain—there are numerous potential outcomes ranging from arriving as planned on time to being hit by a car or a meteorite. This means that there is uncertainty associated with the activity. In other words, there is risk. Now, suppose you are particularly worried about the weather. For simplicity, let's consider the weather as a binary situation with only two possible outcomes: rain or no rain. Beforehand, you do not know if the real outcome will be rain or no rain. Hence, there is uncertainty and there is risk. Now, suppose you are going to carry out a risk assessment of the planned activity. Then, how can you assess and describe the risk related to the rain/no rain situation? Suppose you decide to use probability as your risk metric when describing the risk.

When expressing the probability of rain, you will need to base it on some kind of background knowledge. Suppose you look out of the window and see the rain pouring down, and based on this information you consider it more likely to be raining tomorrow than having no rain. After all, you have experienced that when it is raining one day, it tends to continue for some days in the region where you are living. Based on this knowledge, you express a probability of 80% that it will be raining when you are going to work tomorrow. You express $P(rain|K_1) = 0.8$, where K_1 equals your background knowledge.

Suppose you get a phone call from your brother. He is a meteorologist and has access to state-of-the-art models predicting the weather. His models are based on an understanding of the physical phenomena involved when weather changes occur, and relevant experience data from decades are included in the models. You ask your brother what he thinks about his rain/no rain situation. He checks his models, and by coincidence he comes up with the same probability as you. He expresses the probability $P(rain|K_2) = 0.8$, where K_2 is his background knowledge including all the state-of-the-art knowledge included in the weather forecast models.

Both you and your brother expressed the same probability of 80% that it will be raining tomorrow. However, the background knowledge that the probability was based upon was not equal for the two of you. Your own probability was based on poor knowledge while your brother's knowledge was based on state-of-the-art weather prediction models. In general, in a situation where risk analysis is going to provide risk-informed decision support, the 'quality' of the background knowledge is relevant for the decision maker. In this case, one risk-informed decision could be whether or not you should wear a rain coat. Admittedly, the consequence of making a poor decision would not be severe since you only risk becoming wet. However, for other risk situations, such as different kinds of oil and gas related activities, it may be important for the decision maker to understand the strength of the knowledge upon which the probability expressed is based, since this will affect the decision being made. The decision maker may come up with two different decisions for poor and strong knowledge, even if everything else, including probability numbers, was unchanged. This means that the risk metric, for example the probability, does not alone give sufficient decision support. We also need to present information about the strength of the knowledge upon which the risk numbers are based. This can be done by distinguishing between 'weak' and 'strong' background knowledge.

Now, what does it mean that the knowledge strength is strong or weak? Based on [4], it may be argued that the strength of knowledge is related to:

- justification of assumptions,
- access to reliable data and information,
- agreement among experts, and
- understanding of the phenomena involved.

Based on combinations of the above dimensions, we may consider the knowledge strong or weak. There are also other more sophisticated methods of assigning the strength of knowledge. For example, [3] suggests two ways to assess the strength of knowledge. The first method is based on direct grading of the strength of knowledge in line with the scoring used by Flage and Aven [4]. The second approach is based on identifying all the main assumptions on which the probabilistic analysis is based and converting these into uncertainty factors. Another method is provided by Berner and Flage [5], where the assumptions being made in the risk analysis are categorized into six different categories, and where guidance is given on how to treat each of the six situations. In [6], this is taken one step further, by suggesting dedicated risk management (decision) strategies for each of the six settings. Two other examples are provided in [7, 8], showing the practical implications of the method suggested. Finally, a suggestion on how to consider the strength of knowledge in a QRA setting is suggested by Khorsandi and Aven [9].

2.1.4 Consequence Dimensions

2.1.4.1 Personnel Risk

When personnel risk is considered in the case of an offshore installation, only risk for employees (historically usually called second party, but now often called first party) is considered, whereas risk for the public (third party) is not applicable. For risk to personnel, the following may be considered as elements of risk:

- Occupational accidents
- Major accidents
- Transportation accidents
- Diving accidents.

These elements are common for production installations and mobile drilling units. It is stressed that these risk contributions statistically have to be considered separately. The discussion below is mainly concerned with the risk to personnel on production installations, relating to how such risk is commonly regarded.

Transportation from shore is also often considered. There are advantages and disadvantages associated with this approach. One disadvantage is that important variations associated with the installation may be masked by the risk contribution from transportation. It may also be argued that the risk contribution from helicopter transport cannot be significantly influenced by the offshore operations.

In other circumstances, it is very relevant to include the risk contribution from transportation. This occurs if two field development alternatives are being compared, involving significantly different extents of transportation. Another argument is that the risk contribution from helicopter transportation is a significant source of risk for offshore employees, and as such should be included in order to illustrate the total risk exposure. It should be noted that current Norwegian legislation actually requires that the risk from helicopter transportation should be included in the overall risk estimation for offshore personnel.

2.1.4.2 Risk to Environment

The following hazards relating to production installations and associated operations may lead to damage to external environment:

- Leaks and seepages from production equipment on the platform as well as subsea
- Excessive contamination from production water and other releases
- Large spills from blowouts
- Pipeline and riser leaks and ruptures
- Spills from storage tanks
- Accidents to shuttle tankers causing spill.

It is usual that the third, fourth and fifth of these items are considered in relation to offshore installations. If two different transport alternatives are considered, then number six in this list also has to be included. The two first elements are usually considered as 'operational discharges', and are not included in environmental risk assessment.

2.1.4.3 Risk to Assets

Risk to assets is usually considered as non-personnel and non-environment consequences of accidents that may potentially have personnel and/or environment consequences. It may be noted that modelling of risk to assets in many circumstances is relatively weak. The following types of hazards may cause accidental events which have the potential to damage the assets:

- Ignited and unignited leaks of hydrocarbon gas or liquid
- Ignited leaks of other liquids, such as diesel, glycol, jet fuel, etc.
- Fires in electrical systems
- Fires in utility areas, accommodation, etc.
- Crane accidents
- External impacts, such as vessel collision, helicopter crash, etc.
- Extreme environmental loads.

Usually all of these types of accidental events are included in asset risk. However, there may be a need to coordinate with a regularity (or production availability) analysis, if such analysis is carried out.

A regularity analysis considers all upsets which may cause loss of production capacity, both from unplanned and planned maintenance. Some accidental events of the least magnitude, especially the utility systems, may be included in both types of analysis. This is not a problem, as long as any overlap is known, implying that double counting may be removed if a total value is computed.

A summary of how the different risk elements are usually considered in QRA studies for production installations is presented in Table 2.5, which distinguishes between manned and unmanned installations. Risk associated with material handling and diving are usually outside the scope of such studies.

2.1.5 Risk Metrics

Risk may be expressed in several ways, by probability distributions, expected values, single probabilities of specific consequences, etc. In QRAs, expectation valves are commonly used along with qualitative categorizations of strength of knowledge. An expectation value means that potential consequences are assigned and multiplied with the associated probabilities. In other words, the expected loss is calculated by multiplying a numerical value of the consequence with the corresponding probability for each accident sequence i, and summed over all (I) potential accident sequences:

$$R = \sum_i (p_i \cdot C_i) \qquad (2.1)$$

where:

p probability of accidents
C consequence of accidents

This formula expresses risk as an expected consequence. The expression may also be replaced by an integral, if the consequences can be expressed by means of a continuous variable.

It should be noted that the expression of risk as expected consequence is a statistical expression, which often implies that the value in practice may never be observed. When dealing with rare accidents, an average value will have to be established over a long period, with low annual values. If during 40 years we have five major accidents with a total of ten fatalities, this corresponds to an annual average of 0.25 fatalities per year, which obviously can never be observed.

The comment should also be made here that risk as expected consequence gives limited information about the risk picture. Much more information is provided if the distribution is provided in addition to the statistical expected value. We will revert to this in Chap. 16.

The definition in Eq. 2.1 is sometimes called 'statistical risk' or technological risk. Some authors have referred to this expression as 'real risk' or 'objective risk'. These two last terms give misleading impression of interpretation of risk. 'Risk' is always reflecting interpretations and simplifications made by, for instance the analyst, and as such to some extent subjective. It is therefore misleading to give the impression that some expressions are more objective than others.

'Risk aversion' is sometimes included in the calculation of risk (see for instance Eq. 2.9). Risk will be a combination of the probability of an accident, the severity of the consequence, and the aversion associated with the consequence. This is not supported by the authors. It is acknowledged that risk aversion is an important aspect associated with the assessment of risk, in particular relating to the evaluation of risk results. However, risk aversion should not be mixed with technological risk analysis. Risk aversion is a complex phenomenon. It is misleading to give the impression that this complex process may be adequately captured by a single parameter, risk aversion, a.

Further details about risk aversion and ethical adjustment of the risk assessments are presented in [10].

2.1.6 Accident Sequences and Risk Dimensions

When accident consequences are considered, these may be related to personnel, to the environment, and to assets and production capacity. Such consequence dimensions are sometimes called 'dimensions of risk', which are those shown in the list below. Some sub-categories are also presented in the following:

- Personnel risk

  - fatality risk (see Sect. 2.1.7 for definition)
  - impairment risk (see Sect. 2.1.8 for definition)

- Environmental risk (see Sect. 2.1.9 for definition)
- Asset risk (see Sect. 2.1.10 for definitions)

  - material damage risk
  - production delay risk.

It might be considered that fatality risk is a subset of injury risk, and that the latter is the general category. Fatality risk and injury risk are nevertheless quantified in such different ways that it may seem counterproductive to consider these two aspects as one category.

It should be noted that risk to personnel in a QRA context is mainly focused on fatality risk, or aspects that are vital for minimisation of fatality risk. This reflects the focus of the QRA on major accidents, as opposed to occupational accidents as noted in the introduction. When observing historical accidents, however, occupational accidents have played a major role. In Norwegian operations for instance, all fatalities on installations during the last 20 years have been due to occupational accidents. If helicopter accidents are included, there has been two major accidents the last 25 years; one in 1997 with 12 fatalities and one in 2016 with 13 fatalities.

There is no universal definition of the term 'major accident'. One often used interpretation is that 'major accidents' are accidents that have the potential to cause five fatalities or more. The petroleum safety authorities in Norway have defined a major accident as 'an acute incident, such as a major discharge/emission or a fire/ explosion, which immediately or subsequently causes several serious injuries and/ or loss of human life, serious harm to the environment and/or loss of substantial material assets'. This definition includes accidents with only one fatality. In fact, it also includes accidents without fatalities, for example in cases with severe spills to the environment.

Somebody may react to the classification of 'impairment risk' as a sub-category of 'personnel risk'. Impairment risk is discussed in greater depth in Sect. 2.1.8. At this point it is sufficient to note that although the impairment mechanisms are related physical arrangements (such as escape ways), it is indirectly an expression of risk to personnel.

2.1.7 Fatality Risk

Fatality risk assessment uses a number of expressions, such as; platform fatality risk, individual risk, group risk and f–N curve. It should be noted that some of these expressions are calculated in a particular way in the case of offshore installations. The offshore way of expressing risk is the main option chosen, but differences are indicated.

2.1.7.1 Platform Fatality Risk

The calculation of fatality risk starts with calculating the Potential Loss of Life, PLL. Sometimes, this was in the past also called Fatalities Per Platform Year, FPPY. PLL or FPPY may be considered as the fatality risk for the entire installation, if it is calculated for the group consisting of all personnel at the installation. There are two ways to express PLL:

- Observed PLL in terms of accident statistics. PLL = No of fatalities experience in a period (usually per year).
- Fatality risk assessment (through QRA): PLL is calculated according to Eq. 2.2 below.

From the PLL, either Individual Risk (IR) or Group Risk (GR) may be computed. The PLL value can, based on a QRA, be expressed as follows:

$$PLL = \sum_n \sum_j (f_{nj} \cdot c_{nj}) \qquad (2.2)$$

where:

f_{nj} annual frequency of accident scenario (event tree terminal event) n with personnel consequence j

c_{nj} expected number of fatalities for accident scenario (event tree terminal event) n with personnel consequence j

N total number of accident scenarios (event tree terminal event) in all event trees

J total of personnel consequence types, usually immediate, escape, evacuation and rescue fatalities.

The types of personnel consequences which are relevant for analysis of fatality risk may be illustrated as follows:

Immediate fatalities	which occur in the immediate vicinity of the initial accident, or immediately in time.
Escape fatalities	which occur during escape from the place of work prior to or immediately after the initial accident back to a shelter area (temporary refuge).
Evacuation and rescue fatalities	which occur during evacuation from the installation or during rescue from sea and/or evacuation means.

A comment on the use of the expression 'escape fatalities' may be appropriate. Sometimes (for instance in regulations) 'escape' is used as the process of leaving the installation when orderly evacuation is not possible. 'Evacuation' may on the other hand sometimes be used as the expression for the entire process of leaving the workplace until a place of safety is reached. None of these alternative definitions are used in this book, which uses the interpretation stated above.

The annual frequency of an accidental scenario, f_{nj}, may be expressed as follows, if it is assumed that the factors are related (dependent) as shown below:

$$f_{nj} = f_{leak,n} \cdot p_{ign,n} \cdot p_{protfail,n} \cdot p_{escal,n} \cdot n_{nj} \qquad (2.3)$$

where

$f_{leak,n}$ frequency of leak

$p_{ign,n}$ conditional probability of ignition, given the leak

$p_{protfail,n}$ conditional probability of failure of the safety protective systems, such as ESD, blowdown, deluge, passive fire protection, etc, given that ignition has occurred

$p_{escal,n}$ conditional probability of escalation, given ignited leak and failure protective systems responses

n_{nj} fatality contribution of the accident scenario (fraction of scenarios that result in fatalities).

Equation 2.3 reflects the failure of the five main barrier functions: containment, ignition prevention, protection, escalation and fatality prevention; see further discussion in Sect. 2.5.2.

2.1.7.2 Individual Risk

There are principally two options with respect to the expression individual risk, namely:

- FAR (Fatal Accident Rate), or
- AIR (Average Individual Risk).

AIR is also known by other acronyms, such as IR (Individual Risk) or IRPA (Individual Risk Per Annum). The following sections will use AIR.

The FAR value is the number of fatalities in a group per 100 million exposed hours, whereas the AIR value is the average number of fatalities per exposed individual. The following are the equations which define how the individual risk expressions are computed:

$$FAR = \frac{PLL \times 10^8}{Exposed\ hours} = \frac{PLL \times 10^8}{POB_{av} \cdot 8760} \qquad (2.4)$$

$$AIR = \frac{PLL}{Exposed\ individuals} = \frac{PLL}{POB_{av} \cdot \frac{8760}{H}} \qquad (2.5)$$

where:

POB_{av} average annual number of personnel on board
H annual number of offshore hours per individual (on-duty and off-duty hours).

It should be noted that 8760 is the number of hours in one year. The ratio of 8760/H is therefore the number of individuals required to fill one position offshore. Three persons per position is quite common in Norwegian offshore operations, whereby H is 2920 h per year, 1460 on-duty hours and 1460 off-duty hours. If the schedule is two weeks 'on' [the installation]; four weeks 'off', three persons per position is required, and an average of 8.7 periods are spent offshore each year.

From the definitions above it is obvious that AIR and FAR values are closely correlated. The following is the relationship:

$$AIR = H \cdot FAR \times 10^{-8} \qquad (2.6)$$

If H is 2920 h and FAR is 5.0, then AIR equals 0.00015. Thus, it does not matter whether FAR or AIR is calculated. One may be derived from the other, as long as the shift plan is known.

Onshore, H would not be summed over on-duty and off-duty hours, because off-duty hours are not spent in the plant.

FAR and AIR values may be calculated as average values for different groups, for instance the entire crew on an installation, or groups that are associated with specific areas on the installation such as drilling crew or process technicians.

When 'exposed hours' are considered in relation to the definition of the FAR values for offshore operations, this expression may be interpreted in at least two ways:

- On-duty hours (or working hours) are most typically used for occupational accidents, exposure to these are limited to the working hours.
- Total hours on the installation (on-duty plus off-duty hours) are most typically used for major accidents, exposure to these is constant, irrespective of whether a person is working or not, used in Eq. 2.4.
- When helicopter transportation risk is considered, the exposed hours are those spent in the helicopter.

If FAR values from different activities are to be added, then they must have the same basis. This is discussed thoroughly later.

It should be noted that at present, the total number of working hours offshore on production installations on the entire Norwegian Continental Shelf is just below 25 million hours per year. This implies that during a four year period, roughly 100 million hours are accumulated. In practical terms, we can therefore express that the observed FAR value during the last four years, is the number of fatalities during this period. This reflects occupational accidents only.

2.1.7.3 Example: Calculation of FAR Values[1]

For an offshore installation, the following are the main characteristics that are used in order to form the example shown in Table 2.1.

- The average number of persons on the platform is 220.
- Each person has an annual number of 3000 exposure hours offshore.
- Elements of risk are shown in Table 2.1.

Table 2.1 presents frequencies of fatality consequences for five groups of fatalities. The last line sums up the different contributions from occupational accidents, immediate fatalities and evacuation fatalities, implying that fatalities during escape and rescue have been disregarded.

---

[1]Examples are marked with gray shading throughout the book.

Table 2.1 Example, risk contributions

Location/accident type	Average manning	Fatalities per accident				
		1	2–5	6–20	21–100	101–220
Quarters Occupational accidents	140	0.010	0			
Fatalities, evacuation	140	0.001	0			
Process/utility equipment Occupational accidents	80	0.012	0			
Immediate fatalities	80	0.010	0			
Fatalities, evacuation	80	0	0	0.01	0.003	0.0008
Total Occupational accidents	220	0.022	0			
Immediate fatalities	220	0.010	0			
Fatalities, evacuation	220	0.001	0	0.01	0.003	0.0008
Sum all groups	220	0.033	0	0.01	0.003	0.0008

Table 2.2 Example, calculation of PLL, FAR, AIR

Risk values	Average manning	Fatalities per accident				
		1	2–5	6–20	21–100	101–220
Sum frequencies	220	0.033	0	0.01	0.003	0.0008
Geometrical mean consequence		1	3.2	10	44.7	148
PLL contribution		0.033	0	0.1	0.134	0.118
Total PLL	0.386					
FAR value	20.0					
AIR value	0.00058					

The summed results are used as the basis in order to calculate the PLL, FAR and AIR values for the installation, as shown in Table 2.2. The risk values for the installation are the following:

- PLL = 0.386 fatalities per year
- FAR = 20.0 fatalities per 10^8 manhours
- AIR = 0.58×10^{-3} fatalities per year.

Since this is an offshore installation, the FAR value is calculated as an average over all exposure hours, on-duty as well as off-duty. The personnel stay on the installation continuously for 2 weeks at the time, before leaving by helicopter. This implies that the total exposure hours is $220 \times 8760 = 1.93$ million hours.

If this had been an onshore facility, this would have been more complex. There will usually be shift crews as well as day crews, which have different exposure. The day crew is typically for maintenance, modification, etc. The total exposure hours will usually be equal to the on-duty hours. It is not common to calculate FAR values for the sum of on-duty and off-duty hours, because off-duty hours are not influenced by the risk exposure on the facility. The total PLL value will usually have to be calculated for two periods, one period being the time when both shift crew and day crew are present, the other period is the remaining time with only one shift crew present at the time.

2.1.7.4 Group Risk

The most common risk measure is risk to individuals. Experience has shown however, that society is concerned about the effects of accidents on society as a whole. Some measure of risk to society i.e., the total effect of accidents on society (or the affected group), is therefore required. This is what Group Risk (GR) is used to express.

Risk tolerance criteria are usually expressed for individual risk levels. Sometimes it is necessary also to be able to express tolerability for the group risk. A relationship between individual risk and group risk is therefore sometimes necessary.

Group risk is often expressed in terms of an 'f–N' diagram, see example later in this chapter. The derivation below shows how the f–N diagram may be connected to individual risk measures in situations where there are a limited number of people exposed to the risk. The derivation is a generalization of an expression first described by Schofield [11]. The paper by Schofield includes consideration of risk aversion, which is reproduced in the following, although the use of risk aversion in not recommended in most risk calculations.

Let *POB* be the number of personnel on the installation at any one time (not assumed in this illustration to vary).

Let F_N denote the annual frequency with N or more fatalities.

Let f_N denote the annual frequency of exactly N fatalities.

Then it follows immediately:

$$f_N = F_N - F_{N+1}, \ N = 1, \ldots, POB - 1 \tag{2.7}$$

$$f_N = F_N, \ N = POB \tag{2.8}$$

Consider F_N to have the form:

$$F_N = \frac{F_1}{N^b}, \ 1 \le b \le 1.3 \tag{2.9}$$

This equation is valid for all values of b $\ge$ 1.0, but b = 1.3 is considered in relation to its interpretation to be an upper limit [11]. The factor b is usually called the 'aversion factor', which, as noted above, takes account of the fact that it is usually harder for a society to accept an accident with 10 fatalities than 10 accidents with 1 fatality each, even though the frequency of the former is only one tenth of the latter i.e., the expected value is the same. With b = 1.3, F10 = F1/20, whereas F10 = F1/10 with b = 1.0.

It is recommended that risk aversion is considered as a separate factor, then b should be set to 1.0. The general expression is that b may exceed 1.0, but we have advised against using an aversion factor in the calculation of risk. The derivation of equations below is therefore based on the value $b = 1.0$. From Eqs. 2.7, 2.8 and 2.9 it follows that:

$$f_N = F_1 \frac{1}{N(N+1)}, \quad N = 1, \ldots POB - 1 \tag{2.10}$$

$$f_N = \frac{F_1}{N}, \quad N = POB \tag{2.11}$$

Let AIR now be the average individual risk for an average employee on the installation, expressed as the probability of death per annum. Let there be in total K groups of POB persons, where each individual spends H number of hours offshore per annum, such that:

$$H \cdot K = 8760 \tag{2.12}$$

Then, by combination, we have the following:

$$\frac{1}{K \cdot POB} \sum_1^{POB} N \cdot f_N = AIR \tag{2.13}$$

Combination of Eqs. 2.10, 2.11 and 2.13 gives the following:

$$F_1 = K \cdot POB \cdot AIR \cdot \left[1 + \sum_1^{POB-1} \frac{1}{N+1}\right]^{-1} \tag{2.14}$$

By using (2.14) the FN–plot can be determined for any given set of values AIR, POB, K, b. The frequency of fatalities in the range N1 ≤ N ≤ N2 is given by:

$$f(N_1, N_2) = \sum_{N_1}^{N_2} f_N = F_1 \sum_{N_1}^{N_2} \frac{1}{N+1}, \quad N_2 < POB \tag{2.15}$$

$$f(N_1, N_2) = F_1 \cdot \left[\frac{1}{POB} + \sum_{N_1}^{N_2-1} \frac{1}{N(N+1)}\right], \quad N_2 = POB \tag{2.16}$$

Please note that geometric mean may be used in the calculation of PLL, FAR and AIR values (see Sect. 2.1.7.3), but shall not be used in the calculation of values for the f–N curve (see Table 2.3). The actual lower limits of the intervals in Table 2.1 are used directly in the calculation of values for the f–N curve, as shown in Table 2.3.

2.1.7.5 Example: f–N Diagram Transformation

PLL, FAR and AIR values were calculated for the example installation referred to on Page 36. Using the equation for F_1 above, the F_N values may be calculated, and plotted in the diagram. Please note that f–N diagrams commonly use logarithmic scales for both axes. The f–N diagram expresses the frequency of accidents with N fatalities or more, and as such is always a cumulative frequency. This should not be confused with the distinction between f_N and F_N in Eq. 2.7.

Please note that when using the bottom row of Table 2.1 to plot the f–N curve, the values are summed from left towards right. The first value is the frequency of at least 101 fatalities, which is 0.0008 directly from Table 2.1 (X = 101; Y = 0.0008). The second point from the left is frequency of at least 21 fatalities, which is 0.0008 + 0.003 = 0.0038 (X = 21; Y = 0.0038). The third point from the left has coordinates X = 6; Y = 0.0138, whereas the fourth point from the left has coordinates X = 2; Y = 0.0138. The last point from the left has coordinates X = 1; Y = 0.0468. These five pairs of X–values and Y–values (see Table 2.3) give the f–N diagram of Fig. 2.1.

An f–N diagram plotted in a double logarithmic diagram can never have the value zero on any axis, because the logarithm of zero is $\div\infty$. The x–axis usually starts on 1.0, and the y–axis usually starts with 10^{-x}, according to what is the lowest frequency. The curve will always be monotonously falling.

Table 2.3 Example, points for f–N curve in Fig. 2.1

Location/accident type	Fatalities per accident				
	1	2–5	6–20	21–100	101–220
Sum all groups (Table 2.1)	0.033	0	0.01	0.003	0.0008
Frequency of at least N fatalities					
N = 101					0.0008
N = 21				0.0038	
N = 6			0.0138		
N = 2		0.0138			
N = 1	0.0468				

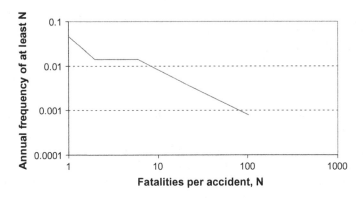

Fig. 2.1 Example f–N diagram (cumulative distribution)

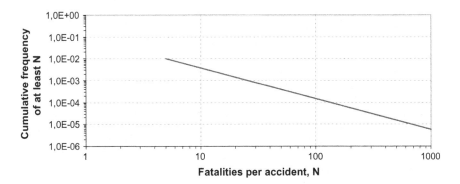

Fig. 2.2 Risk tolerance limit in f–N diagram

2.1.7.6 Example: Calculation of AIR from f–N Diagram

An oil company has a risk tolerance limit for major accident risk on offshore installations as shown in Fig. 2.2. The curve starts with 5 fatalities, and the cumulative frequency is 0.01 for major accidents exceeding 5 fatalities. The aversion factor, b, is 1.4. The installation has an average POB of 80 persons. We want to calculate the AIR value that corresponds to the curve in Fig. 2.2.

Equation (2.13) with the values given above gives 5.4×10^{-4} fatalities per year as the average individual risk for all employees.

2.1.8 Frequency of Impairment

Frequency of impairment is an indirect way to express risk aspects that are vital for the safety of personnel. The aspects for which the impairment frequencies are calculated are usually called 'main safety functions'. 'Main safety functions' are aspects that assist in ensuring the safety of personnel in the event of a major accidenta l event. When frequencies of impairment are calculated, these may be based on physical modelling of responses to accidental loading, and one will thereby avoid the problems of explicitly calculating the consequences of an accident in terms of fatalities.

The frequencies of impairment are usually calculated somewhat differently, reflecting differences in UK and Norwegian legislation, but cover essentially the same overall functions. The wording of the main safety functions is somewhat different in the two countries' legislation. Typically, the following are calculated:

- UK: Impairment of temporary refuge, including the following functions: (according to Safety Case Regulations)

  - Life support safety function
  - Command safety function
  - TR Egress safety function
  - Evacuation safety function.

- Norway: Impairment of several safety functions (according to Facilities regulations):

  - Escalation of accident situations
  - Capacity of main load-bearing structures
  - Rooms of significance to combatting accidents
  - Safe areas
  - Escape routes.

Thus, only one impairment frequency is required to be quantitatively determined under UK legislation, whereas typically five are calculated in Norway. The scope of what is covered is nevertheless virtually the same. The impairment frequency, $f_{imp, i}$, is calculated as follows:

$$f_{imp,i} = \sum_n f_n \cdot p_{imp,n,i} \tag{2.17}$$

where:

$p_{imp,n,i}$ probability of impairment for scenario n with for safety function i
N total number of accident scenarios.

2.1.9 Environment Risk

The environment risk from offshore installations is dominated by the largest spills from blowouts, pipeline leaks or storage leaks. Process leaks, although more frequent, are not normally capable of causing extensive damage to the environment. The quantified risk to the environment is usually expressed as one of the following:

- Expected value of spilled amount.
- Frequency of events with similar consequences for the environment.

Consequence is often measured in restoration time. 'Restoration time' is the time needed for the environment to recover after a spill, to the conditions existing before the spill. This is further discussed in Sect. 6.10. Previously, the expression 'expected spilled amount' has been commonly used. Expected spilled amount per year, Q_{sp}, is expressed as:

$$Q_{sp} = \sum_n f_n \cdot q_n \qquad (2.18)$$

where:

q_n amount spilled for scenario n

The accumulated frequency, $f_{spill\ cons\ i}$, of events with similar consequences (restoration time) is assessed as follows:

$$f_{spill\ cons\ i} = \sum_n f_n \cdot p_{n,i} \qquad (2.19)$$

where:

$p_{n,i}$ probability of environmental consequence i for scenario n.

2.1.10 Asset Risk

The asset risk is comprised of possible damage to equipment and structures, as well as the resulting disruption of production. Risk is expressed similarly for material damage and production delay. Asset risk is usually expressed as either of the following:

- Expected damage to structures and equipment.
- Expected duration of production delay.
- Frequency of events with similar consequences, either in extent of damage or duration of production delay.

Expected value of damage per year (or expected duration of production delay), D, is expressed as:

$$D = \sum_n f_n \cdot d_n \qquad (2.20)$$

where:

d_n extent of damage (duration of delay) for scenario n.

The accumulated frequency, $f_{\text{damage cons } i}$, of events with similar consequences is assessed as follows:

$$f_{\text{damage cons } i} = \sum_n f_n \cdot p_{D,n,i} \qquad (2.21)$$

where:

$p_{D,n,i}$ probability of damage consequence i for scenario n.

The expected value of damage and production delay are, as for the expected amount spilled, entirely artificial values, as the events are rare, and usually large, once they occur.

2.2   Risk Picture, North Sea

An extensive study of the risk levels on the Norwegian Continental Shelf (NCS) was carried out in the late 1990s [12], concerned mainly with accidents in the Norwegian sector of the North Sea, and covered risk to personnel, environment, and assets. The following is a summary of the results for the risk to personnel, limited to fatality risk, updated with the latest available data.

2.2.1   Overview of Fatal Accidents

There have been 86 fatal accidents and 283 fatalities in Norwegian offshore operations since the start of oil and gas operations in 1966 until the end of 2018. This excludes fatalities on shuttle tankers (which are used for transport of crude oil and condensate to shore), but includes fatalities on attendant vessel and other special vessels and barges that engaged in associated oil and gas activities. One fatality on a survey vessel used for geophysical surveys is omitted. Figure 2.3 shows a condensed summary of the statistics.

It should be noted that Fig. 2.3 does not relate the number of accidents to the level of activity. This is done later in this section.

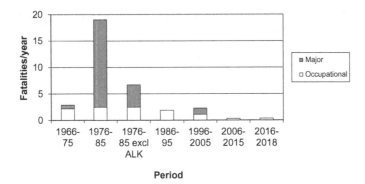

Fig. 2.3 Overview of fatalities, Norwegian Continental Shelf (ALK = Alexander L. Kielland accident)

The frequencies are presented for four ten year periods and the last seven year period, where the following is shown for each interval:

- Average number of fatalities per year from occupational accidents (including diving accidents)
- Average number of fatalities per year from major accidents (including helicopter accidents).

The second period is strongly influenced by the capsize of Alexander L. Kielland in 1980. The average number of fatalities per year is 17.4 if this accident is included, 5.4 fatalities per year if it is excluded (see distinction in the diagram).

It may be observed that the overall frequency has been quite stable during the last decades, when somewhat crudely expressed as average number of fatalities per 10 year periods. This may be further illustrated by the following number of major accidents (including helicopter accidents) on the NCS (see also Fig. 2.4);

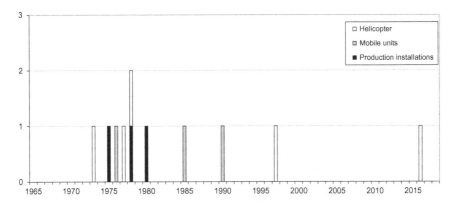

Fig. 2.4 Occurrence of fatal major accidents in Norwegian offshore operations

1966–75:   2 major accidents
1976–85:   6 major accidents
1986–95:   1 major accident
1996–2005: 1 major accident
2006–2018: 1 major accidents

It may be argued that the period 1966–85 was a period with higher frequencies, for occupational as well as major accidents. It is therefore appropriate to discuss the period after 1985 more thoroughly.

2.2.2 Overview of Accidents to Personnel

The total number of fatal accidents in the period 1986–2018 is 33, with 60 fatalities. This includes the Turøy helicopter accident in 2016 with 13 fatalities. The fatal accidents on the Norwegian Continental Shelf have occurred on the following different platform and vessel types:

- Production installations: 9 fatal accidents   10 fatalities
- Mobile installations: 8 fatal accidents   8 fatalities
- Attendant vessels: 9 fatal accidents   10 fatalities
- Crane and pipe-laying vessels: 3 fatal accident   3 fatalities
- Diving: 1 fatal accident   1 fatality
- Helicopter accident (maintenance):   1 fatal accident   3 fatalities
- Helicopter transportation (to shore): 2 fatal accident   25 fatalities

Accidents that have occurred inshore or at-shore are excluded from the values considered, even though in some few cases similar accidents could have occurred at an offshore location. In one case for instance, a lifeboat fell 20 m to the sea, fatally wounding the two onboard. Such an accident could also have occurred offshore (and a few accidents of this type actually have occurred).

The average number of fatal accidents over the year period is 1.0 per year, with 1.8 fatalities per year (1.1 per year if the helicopter accidents are excluded). It should further be noted that except for helicopter accidents, no other fatalities in major accidents have occurred in the period. The jack-up 'West Gamma' capsized under tow in 1990, but all crew members were rescued from the sea. An accident in 1991 involved a helicopter which was being used for maintenance of a flare tip on a fixed installation. The tail rotor touched the flare causing the helicopter to crash, killing the three people onboard. This accident has been classified as an occupational accident. The distribution of fatalities over time is shown in Fig. 2.5.

If only occupational accidents are considered, it is obvious that there has been a reduction during the period. This topic is further discussed in Chap. 13.

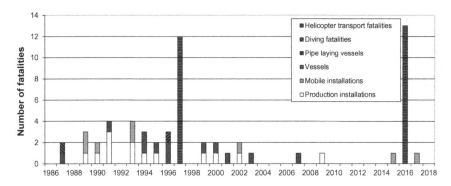

Fig. 2.5 Fatalities on the NCS in the period 1986–2017

Figure 2.6 presents an overview of the main causes of fatal accidents in the Norwegian sector in the period 1986–2017. It is clearly demonstrated that persons being hit or crushed by/between moving or falling objects is by far the most important cause of fatalities, a total of 22 fatalities in the period is due to this category alone.

The other main cause is helicopter crash, two cases where the helicopter crashed in the sea, and one case when a helicopter was used for maintenance purposes, as described previously. Eight fatalities arose from persons falling, either to a lower deck, or in the sea, the latter including the lifeboat maintenance accident where a conventional lifeboat was undergoing maintenance and fell uncontrolled into the water from normal height, due to failure of the prevention mechanism.

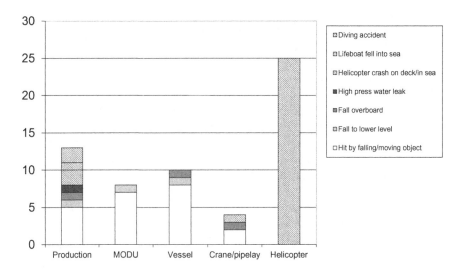

Fig. 2.6 Causes of fatalities on NCS in the period 1986–2018

Table 2.4 Overview of FAR values, average 2002–2011, per working hours

Activity	FAR value
Fixed installations	0.72
Mobile drilling units	1.11
Attendant vessels	4.8
Crane and pipe-laying vessels	8.2
Total for all (2002–2011)	1.21

2.2.3 Fatal Accident Rates

This section presents historic FAR levels for occupational accidents (except helicopter and diving risk), averaged for all personnel on board. These rates are based upon exposure only during working hours i.e., 12 h of exposure per 24 h of offshore stay. These values are calculated as averages over the period 2002–2011. The estimates of FAR levels are presented in Table 2.4. For crane and pipe-laying vessels, the FAR values is based on the last 20 years, 1992–2011, due to lack of fatal accidents the last ten years. No diving accidents have occurred in the period. Diving is therefore not included in any of the FAR values.

The values are limited to occupational accidents, due to the fact that no major accident with fatalities has occurred on offshore installations in the period considered. In general, a risk picture needs to consider more than historical data only, especially for major accidents; see further discussion in Chap. 17.

2.2.4 Trends in Fatality Rates

An important aspect has been to identify possible trends in historic fatality risk level s, in order to identify areas or operations where special efforts may be necessary. Trends are based on activities which takes place on the installations and vessels. Thus fatalities in the Alexander L. Kielland accident and other major hazard accidents are excluded. These trends have been established separately for production installations, mobile drilling units and attendant vessels.

The FAR value is calculated based on fatalities and estimated exposure manhours (in the case of production, these values are available from PSA. For the other activities, the values are mainly derived from activity levels).

FAR values for production and mobile installations are read against the primary vertical axis in Fig. 2.6, and have been reduced a lot from about 250 over the 50 year-period with oil activities. For attendant vessels, the values are read against the secondary (right-hand side) vertical axis of Fig. 2.7, and the FAR values start around 750. The sharp falls in FAR values took place before 1985 and since then the levels have been much more stable.

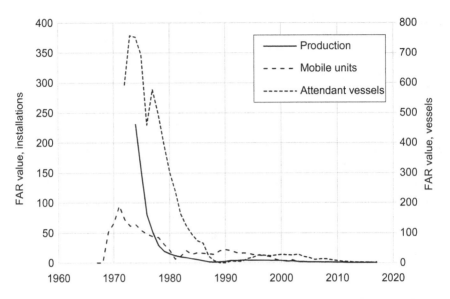

Fig. 2.7 Long term trend in average FAR values for occupational accidents on all offshore installations and attendant vessels

The next illustration looks more closely at the last part of the period, from 1986; see Fig. 2.8. The values include all fatalities that have occurred in the Norwegian sector of the North Sea, and are calculated as rolling ten year average values.

For production installations there is a downward trend over the entire period, although it could be argued that the level was stable until year 2000. The frequency has been falling since year 2000.

For mobile drilling units, there is a generally falling trend over the entire period, perhaps with a small and non-significant increase in the last period due to two occupational accidents in 2015 and 2017, each with one fatality. Since 2000 there has been little or no difference between production installations and mobile units.

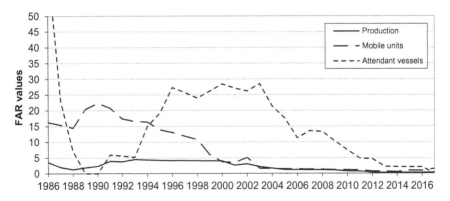

Fig. 2.8 Trend in average FAR values for production installations, mobile units and attendant vessels

The variations are actually the largest for attendant vessel s, but it should be noted that as fewer manhours are worked on these vessels one additional fatality will have a larger influence. Still, attendant vessels is the group with the highest number of fatalities in the period (see Sect. 2.2.3). The number of fatalities fell sharply during the second half of the 1980s, then increased for almost ten years, and has been falling again since year 2000. It is still higher than on offshore installations, although with a small difference during the last 5–year period.

It should be noted that the attendant vessel owners and the oil companies in the Norwegian sector worked together in 1998 and 1999 to improve the safety aspects of attendant vessels. This has obviously had a major long term effect.

Years 2004, 2005 and 2006 were free from fatal accidents completely. It is the first time that no fatalities have occurred in three consecutive years. This occurred again in 2010–2014. In 2015–2018, however there has been two occupational accidents and one major accident with fatalities.

What do these trends imply for the occurrence of fatal accidents in the future? We do not know, but there are indications that we may hope that accidents will be few and rare. This will also depend on actions taken by all parties involved. It should be noted that taking the value calculated for the last year of a period, is actually taking the average value for the preceding 10 years, due to the rolling average calculation. Taking this average may be too optimistic, where there is a clearly increasing trend. Where the trend is close to constant, the average value may be more representative.

The values in this subsection are used in the basis for the evaluation of risk to personnel that is discussed in Chap. 17.

2.2.5 Comparison Offshore–Onshore Activity

A brief comparison has been made between the average offshore fatality risk level and the most appropriate onshore activity, namely oil refineries. This is more relevant now compared to previously, because PSA is responsible for all offshore installations and the main onshore petroleum facilities in Norway. A previous comparison, based on [12], used only refineries in Norway, for the period 1996–2005. For the period 2002–2017 the calculation of FAR value is made for the eight onshore petroleum facilities that PSA has jurisdiction for, refineries, oil and gas terminals and gas plants, including an onshore LNG plant. The comparison is presented in Fig. 2.9. All estimates are based on actual working hours, calculated on the basis of 12 h per day offshore and 7.5 h per day onshore (less for shift personnel). The total number of working hours per year should be about the same per person for both industries.

There have been two fatalities on offshore production installations and one fatality on onshore petroleum facilities in the period 2002–2017. There is large difference in manhours, about 450 million manhours on offshore production installations, versus about 267 million manhours on onshore petroleum

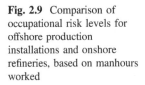

Fig. 2.9 Comparison of occupational risk levels for offshore production installations and onshore refineries, based on manhours worked

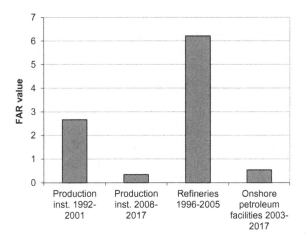

installations. It should be noted that manhours had to be guessed for the period 2002–2005, based on the manhours reported for 2006, for the six facilities in operation and two being constructed.

Figure 2.9 suggests that the fatality risk level may be higher at refineries compared to other onshore petroleum facilities. The RNNP project does not specify the number of working hours for refineries only and, for this reason, the refinery FAR calculation has not been updated since 2005. However, there have been no fatalities on refineries since 2005, indicating that the current FAR value for refineries is substantially lower than the 1996–2005 calculation. The only fatality on onshore plants occurred on a gas terminal in the construction phase. The injury data in RNNP [13] do not confirm that the injury rates at terminals are higher than at other onshore facilities.

2.3 Risk Presentation

In general, QRA results should be presented as detailed as possible. There are usually quite a lot of detailed results available from a QRA, but they are seldom provided for the insight of the reader.

The main objective in the presentation of QRA results is to illustrate relative comparisons of contributions and mechanisms that may show the elements of the risk picture. Thus as many relevant details and illustrations as possible should be presented. Some general principles for presentation of results are stated below (Table 2.5).

The most important, in addition to overall presentations of results, is that contributions to risk are illustrated in a number of ways. The following sections present most of the relevant ways that risk may be presented, including the contribution of the following parameters:

Table 2.5 Risk elements for production installations

Risk element	Manned installation	Unmanned or not normally manned installation	Comments
Accidents on the installation	Occupational accidents	Occupational accidents	
	Major accidents	Major accidents	Including all types, related to HC leaks, external impact and environmental loading
Accidents during transportation	Accidents during helicopter transport from shore	Accidents during helicopter transport from shore	Required to be included for offshore employees as per Norwegian regulations
		Accidents during shuttling between unmanned and manned installation	Usually included for the not normally manned installation
	Accidents during material handling of supplies and transport from shore	Accidents during material handling of supplies and transport from shore	Usually not included
Diving accidents	Not included	Not included	Usually carried out from dedicated vessel, not considered for the fixed installations

- Different types of scenarios
- Types of failures in an MTO (Man, Technology and Organisation) perspective
- Type of activity which contributes to risk
- Barrier failures that led to the actual scenario
- Location where the initiating accident occurred
- Relevant cause (where applicable) of accident initiation.

2.3.1 Fatality Risk

2.3.1.1 Overview

Fatality risk should, as a minimum, be presented using the following parameters, irrespective of whether risk tolerance criteria are formulated for each of them or not:

PLL   PLL is the annual risk exposure of the entire installation, and is thus an important measure of overall risk.

FAR/AIR   FAR or AIR are alternative expressions of individual risk, and are thus
            complementary to the PLL value. FAR and AIR are usually average
            values for groups, such as the entire population and smaller groups.
            Norwegian regulations require that risk is calculated for the most
            exposed persons.
Group risk Usually presented as an f–N diagram (see illustration in the following).

It should be noted PLL may sometimes be expressed for a particular group on an
installation during the execution of a particular task, thus it may occasionally depart
from what is referenced above, the entire population of an installation and a whole
year. It is then important to state clearly what is the reference for the PLL i.e., what
group and the duration of the operation in question.

2.3.1.2 Potential Loss of Life (PLL)

PLL values can be presented with the different contributions, arising from the
different hazards that are applicable, as shown in the following table.

Table 2.6 also shows the relative contributions and the average size of the
accident in terms of fatalities per accident. These PLL values are based upon a study
carried out a number of years ago (numbers not updated).

Figure 2.10 shows the contributions to total PLL from different accident accidental effects:

Table 2.6 PLL contributions from accident types

Hazard category	Annual PLL values	(%)	Fatalities/accident
Blowout	4.1×10^{-3}	27.3	5.3
Process accidents	9.6×10^{-4}	6.4	0.7
Riser, pipeline accidents	5.3×10^{-3}	35.3	4.1
External accidents	2.3×10^{-4}	1.5	1.4
Occupational accidents	2.6×10^{-3}	17.3	1.1
Helicopter accidents	1.4×10^{-3}	9.3	2.8
Total all categories	1.5×10^{-2}	100.0	2.3

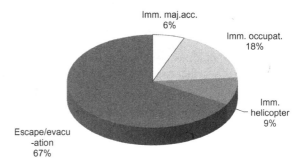

Fig. 2.10 Contributions to
PLL from types of fatalities

- Immediate fatalities

  - occupational accidents
  - major accidents

- Helicopter accidents
- Fatalities during escape and evacuation.

The 'immediate fatalities' are those fatalities that occur as a direct result of the initiating event, either in a major accident or an occupational accident. All fatalities are virtually always 'immediate' in the case of an occupational accident.

2.3.1.3 Average Fatality Risk

Relative contributions to FAR/AIR are often the same as the contributions to PLL. FAR or AIR should be presented in respect of the following categories:

- for personnel located in each main area of the installation
- for each main group of personnel
- contributions to total risk from different accident types
- contributions to total risk from different phases of the accident.

Figure 2.11 shows typical FAR values for different areas of a platform. The minimum that should be presented in this regard is:

- FAR value for personnel spending on-duty and off-duty time in accommodation area.

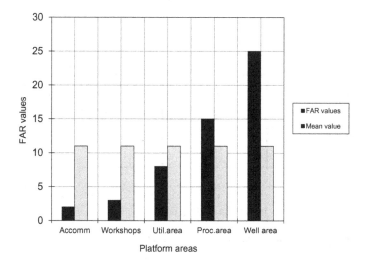

Fig. 2.11 FAR values in different platform areas, and mean platform value

- FAR value for personnel who spend on-duty hours in 'hazardous' areas on the installation (i.e. outside accommodation area) and off-duty time in the accommodation area.

It is often appropriate to use FAR or AIR to compare different concepts or different layout solutions. There are, on the other hand, some situations where it is not appropriate to compare FAR (or AIR) values. Examples of such situations are comparison of:

- Field development alternatives involving different numbers of personnel.
- Field development alternatives that are drastically different, for instance if a new field can be developed with a separate installation or tied into an existing, possibly with extensive modification being required.

PLL has to be used for comparison in these situations, in order to get an understanding of the overall risk, as shown in the following example.

2.3.1.4   Activity Based Variations of Fatality Risk

Different activities have very different FAR levels. Figure 2.12 shows how the average FAR level may vary for a worker who takes part in well intervention as well as process operation. The FAR level during a short helicopter trip is also shown. Further, he is only exposed to the structural failure associated risk level, as well as the possible need for all onboard to evacuate, during sleep in the quarters. Office employees working inside the accommodation block for the entire day would also be exposed to this low risk level. The following points should further be noted:

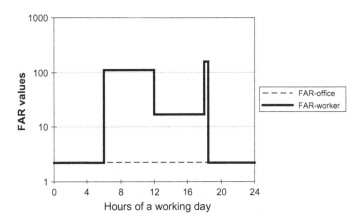

Fig. 2.12 FAR variations depending on activity performed

- The actual values shown are typical, and may vary considerably, depending on the platform.
- The FAR level shown for helicopter transport in the figure is 140 fatalities per 10^8 person flight hours. This used to be the average FAR for the North Sea. It should be noted that according to Chap. 13, the current average FAR for the North Sea is closer to 200.
- The well intervention FAR rate shown relates to moderately hazardous operations. Other well intervention activities may give FAR values almost as high as during helicopter flights. These activities however often have relatively short duration.

2.3.1.5   Adding of FAR Values

It should be noted that FAR values may only be added if they apply to the same time period. The FAR values shown in Fig. 2.12 cannot be added, but the total FAR for a 24 h period may be calculated in the following way:

$$FAR_{tot} = \sum FAR_i \cdot t_i \tag{2.22}$$

where

FAR$_{tot}$ total FAR value over a working day
FAR$_i$ FAR value for period i
t$_i$ duration of period i.

The calculation of the daily average FAR for the platform worker experiencing the risk regime shown in Fig. 2.12 is shown in the Table 2.7.

The PLL) for one such worker over a total working day is 0.89×10^{-5} fatalities (per day), which is shown to correspond with a daily average FAR value of 36.9.

Table 2.7 Calculation of daily average FAR for offshore worker (example)

Activity	FAR	Duration (h)	PLL (10^{-8})
Sleeping	2.28	6	14
Well intervention	114.2	6	685
Process operation	17.12	6	103
Helicopter transport	140	0.5	70
Sleeping	2.28	5.5	13
Total value	36.9	24	885

2.3.1.6   Example: Comparison of Field Development Schemes

There is a fixed installation which has been producing for some few years on a field, producing mainly oil and some associated gas. A second part of the field has been found, also with oil and associated gas, and two options exist for development of this part of the field:

• Subsea wells tied back to the existing fixed installation.
• A Floating Production Storage Offloading (FPSO) unit.

The comparison in these circumstances is focused on the additional PLL that might occur due to production from the second part of the field, including the modification phase and the remaining production period. If subsea completions are to be used and the existing installation is to house production from the second part of the field, then the following aspects need to be considered:

• Increased fatality risk during the modification phase.
• Increased fatality risk during the production phase, due to new equipment.
• Increased number of personnel onboard during modification and during production.

The total value of increased PLL in this case is compared to the total PLL for the FPSO. The possible fatality risk to workers in the construction yards is disregarded in both cases. This would, if included, add more risk for the FPSO alternative, because considerably higher yard manhours would be required, in order to construct a new hull. Figure 2.13 shows the calculated PLL for the two alternative development scenarios.

The FPSO alternative is seen to add 33% more PLL to the field development and residual production phase, mainly due to the higher number of personnel exposed to risk offshore.

This example assumes that the extra personnel needed during the modification phase can live on the installation itself i.e, that one hundred extra

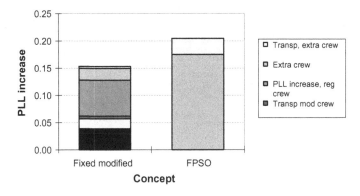

Fig. 2.13 Comparison of PLL increase for two different field development concepts

beds are available. Some installations have this capacity, but not all. The alternatives would be the following:

- Connecting a flotel to the fixed installation by bridge, in order to increase the accommodation capacity, or
- Shuttling extra personnel to the platform each day either from onshore or an adjacent platform.

These alternatives have significant implications for risk, as the next example shows.

2.3.1.7   Example: Flotel Versus Helicopter Shuttling

When offshore modifications or tie-ins are being carried out, it is sometimes not possible to accommodate all personnel required for the work on the installation itself. It is then a choice between shuttling the personnel to shore or to another installation or providing a bridge-connected flotel beside the platform. Figure 2.14 is provided to illustrate the difference in fatality risk (PLL). It is an add-on to Fig. 2.13, when it is assumed that the extra personnel during the modification cannot be accommodated on the fixed installation. The personnel will either have to be shuttled to shore or accommodated on a bridge-connected flotel. The resulting total life cycle PLL values for the two alternatives are shown in Fig. 2.14.

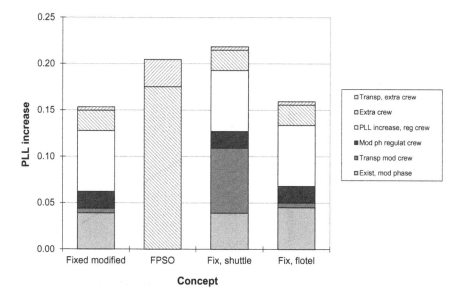

Fig. 2.14 Comparison of development schemes with shuttling and flotel

The life cycle risk is increased by virtually 50% if all personnel have to be shuttled to shore each day. Such an extent of shuttling is too extensive to be the normal solution, but is used in order to illustrate the most extreme case. As shown, this is the highest PLL value in Fig. 2.14, whereas the alternative with flotel for provision of extra beds is almost as low as the alternative using the fixed installation alone for all purposes. The flotel solution will usually be quite expensive, but will probably have to chosen, because it is unlikely that the shuttling of all personnel will be permitted.

2.3.2 Group Risk

The group (or societal) risk may be presented by an f–N curve for fatalities. The f–N function was introduced in Sect. 2.1.7.3, where the possibility to include risk aversion also was discussed. Risk aversion is initially omitted from the discussion in the following, but the importance of aversion is also illustrated. Figure 2.15 shows the usual shape of the f–N diagram, which is just a particular way to draw a cumulative function.

It should be noted that the f–N diagram usually requires both axes to be drawn with a logarithmic scale, because the range of values is usually several orders of magnitude on both axes. Typically, fatalities can range from 1 to the maximum present on board at any one time. On large installations this may reach several hundred persons. Figure 2.16 shows how the same diagram would look, if linear scaling were used on both axes. An actual example from a platform QRA is presented in Fig. 2.17.

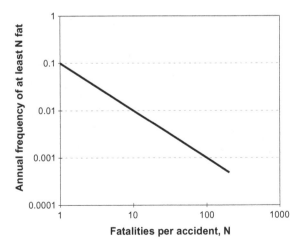

Fig. 2.15 f–N diagram, double logarithmic scaling

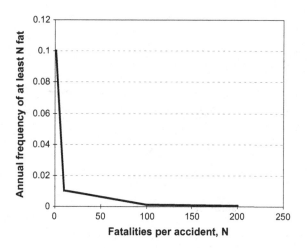

Fig. 2.16 f–N diagram, linear scaling

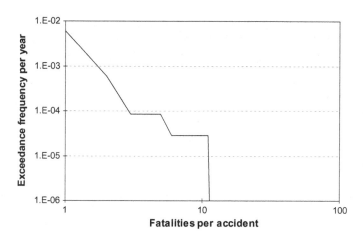

Fig. 2.17 f–N curve for offshore installations, all fatalities

The two next diagrams (Figs. 2.18 and 2.19) show important aspects of f–N diagrams. The first diagram illustrates the importance of risk aversion, if this is included. The curves in Fig. 2.18 have been produced for the following case:

- POB: 25
- Upper tolerability limit, AIR: 1×10^{-3} per year
- Number of hours exposed offshore: 3200 h per annum (50% on-duty, 50% off-duty).

The two curves are developed in order to present the variation dependent on the extent of risk aversion, ranging from no aversion ($b = 1.0$), to maximum aversion

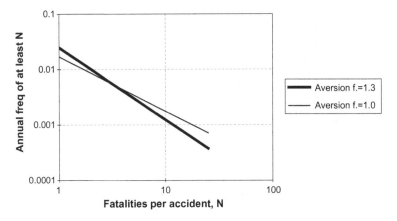

Fig. 2.18 Variation of group risk curve as a function of aversion factor

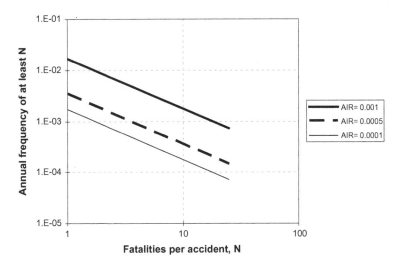

Fig. 2.19 Group risk curve for different values of AIR limit

($b = 1.3$) (see Sect. 2.1.7.4). The difference between the two cases may seem insignificant but for accidents with at least 20 fatalities, the difference is a factor of 1.8 (0.0009 vs. 0.0005 per year).

Figure 2.19 has been developed in order to show how differences in upper tolerability limit for AIR (thereby also FAR) values affect the f–N diagram in situations where finite numbers of people are subjected to risk. No risk aversion has been included in the presentation.

It is seen from the diagram that the values are linearly dependent on the value of the AIR (or FAR implicitly). This is also understood from Eq. 2.14.

Table 2.8 Annual impairment frequency, escape ways

Hazard category	Annual impairment frequency	(%)
Blowout	7.3×10^{-5}	48.5
Process accidents	8.2×10^{-6}	5.4
Riser, pipeline accidents	1.97×10^{-5}	13.0
External accidents	5.0×10^{-5}	33.1
Total all categories	1.51×10^{-4}	100.0

2.3.3 Impairment Risk

Impairment risk is usually related to impairment of so-called 'Main Safety Functions'. There are often three to five main safety functions defined, each of which has a separate impairment frequency. These frequencies should therefore be presented individually. Typical contributions to annual frequency of impairment of escape ways for a wellhead platform are shown in Table 2.8.

Impairment of escape ways is here presented per hazard category, as required by Norwegian regulations. Also the total value is shown; this is not required by Norwegian regulations. Impairment of escape ways is sometimes expressed on a 'per area' basis.

2.3.4 Risk to Environment

Table 2.9 shows one way to express results for risk to the environment. The table must be accompanied by a definition of the consequence categories. These are often based on the effect on the coastline bearing in mind the following aspects:

- Amount of oil reaching the shore
- Length of coastline affected by spill
- Extent of areas of special environmental value (including areas with particular value) that are affected.

Sometimes, the expected value of the spilled amount per year (at source or that reaching shore) is given as a measure of risk. This would be a value such as 0.56 tons per year, if the amounts corresponding to the consequence categories in

Table 2.9 Annual frequency of environmental damage, for categories of spill effects

Environmental consequence category	Corresponding amount spilled (tons)	Annual frequency
Minor effect	10	3.4×10^{-3}
Moderate effect	500	8.6×10^{-4}
Major effect	10,000	9.7×10^{-6}

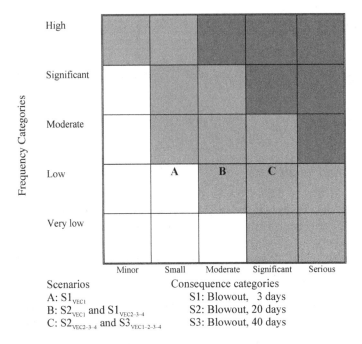

Fig. 2.20 Risk matrix with results plotted

Table 2.9 are used. It is clearly seen that such a value is virtually meaningless for prediction purposes. There will never be anything like 0.56 tons spilled in any one year, if we consider major accidents. Either the amount is nil, or a substantial amount. The expected value is virtually useless when the probabilities are very low, and the consequences are high values. This applies to prediction of future spills. For some other purposes, like comparison of concepts or systems, the expected values may be quite useful.

Another way to present risk to the environment is by the use of a matrix presentation, as shown in Fig. 2.20, which uses the following consequence categories:

Minor damage: <2 years
Small damage: 2–5 years
Moderate damage: 5–10 years
Significant damage: 10–20 years
Serious damage: >20 years

There are also other alternative result presentations that may be chosen. The blowout scenarios S1; S2; S3; imply different durations, as shown in the diagram. There are four Valued Ecological Components (VECs) considered, these are denoted VEC1; VEC2; VEC3; VEC4. There are different frequencies for each consequence category for each VEC; this is indicated in the matrix by the results falling in boxes, A, B and C.

2.3.5 Asset Risk

There are normally two dimensions of asset risk that are presented separately; Material damage risk and Production delay (deferred production) risk. In actual situations the production delay often dominates material damage if both are converted to monetary values. Table 2.10 presents an example of material damage risk contributions for a wellhead platform. The risk of production delay may be presented in a number of ways:

- Expected value i.e, expected delay per year due to accidents.
- Frequencies of consequences of different magnitude, similar to the presentation for material damage above (see Eq. 2.21).
- Exceedance diagram showing the accumulated frequency of delays of a certain duration or longer.

There are often four or five categories presented, the following are used in the WOAD® database [14], and could be considered a 'standard' to some extent:

Total	Total loss of the unit including constructive total loss from an insurance point of view. The platform may be repaired and put into operation again.
Severe damage	Severe damage to one or more modules of the unit; large/medium damage to load-bearing structures; major damage to essential equipment.
Significant damage	Significant/serious damage to module and local area of the unit; damage to several essential equipments; significant damage to single essential equipment; minor damage to load-bearing structures.
Minor damage	Damage to several non-essential equipments; minor damage to single essential equipment; damage to non-load-bearing structures.
Insignificant damage	Insignificant or no damage; damage to part(s) of essential equipment; damage to towline, thrusters, generators and drives.

Table 2.10 Annual frequency, material damage

Hazard category	Annual damage frequency		% total loss
	Partial loss	Total loss	
Blowout	1.07×10^{-3}	2.61×10^{-4}	79.8
Process accidents	2.06×10^{-3}	5.76×10^{-6}	1.8
Riser, pipeline accidents	1.62×10^{-4}	1.04×10^{-5}	3.2
External accidents	9.62×10^{-3}	5.0×10^{-5}	15.3
Total all categories	1.29×10^{-2}	3.27×10^{-4}	100.0

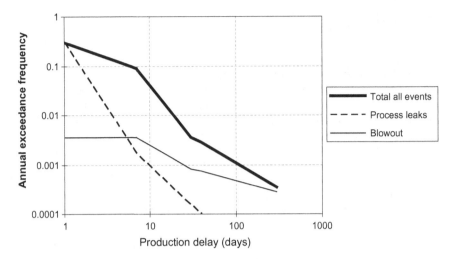

Fig. 2.21 Production delay curve in an exceedance fashion

Figure 2.21 shows an exceedance diagram for production delay. Three curves are shown, the total, and the two most important contributions, blowout and process accidents. There are also other contributions which are not shown. It should be noted that the expected value for the exceedance curve, can be expressed as:

- 1.5 days of production delay per year.
- Equivalent of 0.40% reduction of production availability.

About 60% of the contribution to production delay comes from short duration events, but there is also substantial contribution from events of longer duration and rare occurrences. A more informative presentation of values is therefore as follows:

- On average 0.85 days per year of short duration delays (up to one week).
- 1% probability per year of long duration damage; on average, 66 days delay.

2.3.6 Load Distribution Functions

The exceedance diagram is similar to the f–N diagram for fatalities, shown in Fig. 2.17. Figure 2.14 presents the annual exceedance frequency for collisions with a North Sea wellhead platform. This is similar to the presentation of production delay, as shown in Fig. 2.21. There are four curves shown for the three contributions from merchant vessels, shuttle tankers and supply vessels, and the total frequency.

It may be argued that the load distribution functions are not risk expressions, but that they present intermediate results that are used in the further risk calculations. This may be the case, but sometimes these loads correspond to what is seen as a

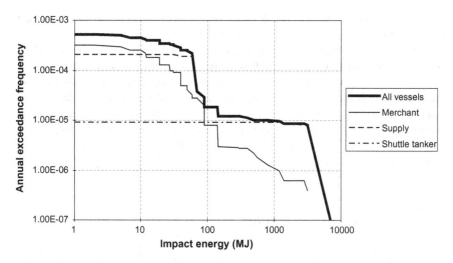

Fig. 2.22 Typical exceedance frequency for collision risk

design requirement or feature, such that some persons may be interested in these results as a risk output. For instance for collision loads, this may be the case. These curves may be used as input to structural design against collision loads (Fig. 2.22).

2.4   Uncertainties

2.4.1   Basis for Uncertainty Consideration

Risk quantification is often characterised by a mixture of the classical statistical approach and the use of subjective probabilities. Most professionals are trained in the former approach, where the probability of end events is considered to be independent of the analyst, and as a quantity characterising the object being studied. The classical concept of probability implies that the results of the risk analyses are calculations (estimations) of these 'true' probabilities.

The alternative is the use of subjective, where the concept of probability is used to express the analyst's degree of belief. There is still significant resistance among risk analysts to the idea that their results are not 'objective' risk results, but rather subjective values. Most risk analysts would, however, accept that there are some elements of their work which are subjective values. For example, subjectively assessed conditional probabilities are commonly used for some of the nodes of the event trees, typically where simplifications of complex physical phenomena are introduced.

The approach adopted in this book is the use of subjective probabilities, whereby the risk values are considered to be expressions of the uncertainty related to whether accidents will occur or not. The implication of this consideration is that uncertainties shall not be quantified in QRA studies, because the risk assessment in itself is an expression of uncertainty. This does not, however, imply that the subject of uncertainty is without interest. It will be important for the analysts to be aware of what is influencing the extent of subjectivism in the analysis, in order to focus on results that minimise the inherent uncertainty. We will therefore consider aspects and factors that are important for the extent of subjectivism in an analysis. The difference is, however, that no attempt is made to quantify these elements of uncertainty. A more thorough discussion of these aspects is presented in 'Foundations of Risk Analysis' [15].

2.4.2 Influence of Uncertainty

There will always be uncertainty as to whether certain events will occur or not, what the immediate effects will be, and what the consequences for personnel, environment, or assets may be. This uncertainty reflects the insufficient information and knowledge available for the analysis, in relation to technical solutions, operations, and maintenance philosophies, logistic premises etc. The uncertainty will be reduced as the field development project progresses. But there will always be some uncertainty about what may be the outcome of accidental events, even when the installation has been installed and put in operation.

The uncertainties are expressed by the probabilities that are assigned. There is as such no other expression of uncertainty. But it is nevertheless important to consider and reflect on what are the sources of uncertainty.

As emphasized in Sect. 2.1.3, the strength of knowledge that the risk assessment is based upon can be categorized, for example, by considering to what extent the assumptions can be justified, whether relevant and reliable data has been used, understanding of the phenomena being studied, and to what extent there is agreement among experts in the field.

It is important to consider how risk is calculated in order to understand the influence of uncertainties. The calculation of event sequences (see further discussion of event sequences in Chap. 4) from an initiating event to a final situation may be illustrated as follows:

Causes ⟹ Initiating events ⟹ Physical accidental loads ⟹ Physical consequences ⟹ Damage

Historically, the causes of events have often been omitted in QRA studies. For example, the causes of a leak of hydrocarbons may not be addressed particularly. This is discussed in detail in Sect. 2.5.1. One example of risk calculations relating to an event sequence may be as follows:

Event		Physical accidental loads		Physical consequences		Damage
Leak	⇨	Fire load, .. kW	⇨	Fire loads on escape ways	⇨	Fatalities

The extent of assumptions that have to be made will usually increase as one gets further into the accident sequence, and more and more uncertainty is introduced. There are more sources of uncertainty associated with calculation of fatality risk compared to physical accidental loads or consequences. This should also be considered when choosing the risk parameters to be used in decision-making (see discussion in Sect. 14.13).

The way to treat uncertainties in the analysis should be defined prior to performing this evaluation. It is recommended here that subjective probabilities are used. This implies that sensitivity studies should be carried out in order to illustrate the criticality of assumptions and data used in the analysis.

The NORSOK Z–013 standard expresses that the 'best estimate' risk levels from the risk analysis, rather than the optimistic or pessimistic results, should be used as basis for decision-making. This is based on a classical statistical approach, and some interpretation is needed. The implication of this requirement is that expected values should be used, rather than alternative values.

Where the analyst considers that a particular evaluation, or calculation, is particularly uncertain, it is common practice to aim to 'err' on the conservative side. This is considered good practice, but care should be taken to ensure that the conservatism is not exaggerated. For instance, if a maximum blast load is calculated as 1.2 bar, then we may be certain about what effects of fragments on personnel may be (disregarding other effects in this example) and consider conservatively that 50% of the persons present may be injured by fragments. The conservatism in calculating the fraction of persons injured is OK, but we should not apply conservatism on all the factors leading up to the frequency of blast loads from such explosions.

2.4.3   Calculation Based on Observations

There is one situation where it may be appropriable to consider variability in a statistical sense. This may be illustrated with the following example. Let us consider three offshore installations that have been in operation for many years. The number of hydrocarbon (HC) leaks (all leaks, also the smallest) that have occurred over the years, is shown in Table 2.11.

Let us first of all calculate the average frequencies for the three installations individually, as in Table 2.12. The average number of HC leaks per year is 50% higher on installation 'A' compared to installation 'B', whereas the number of leaks per million manhours are quite similar for these two installations. The average frequencies for installation 'C' are considerably lower. The question may be whether these differences are statistically significant?

Table 2.11 Experience data for installation, manhours and HC leaks

Installation	HC leaks during 15 year period	Manhours during 15 year period
A	100	80 mill
B	65	50 mill
C	25	100 mill

Table 2.12 Average number of leaks for example

Installation	Average number of leaks per year	Average number of leaks per million manhours
A	6.67	1.25
B	4.33	1.30
C	1.67	0.25

Based on an assumed Poisson distribution, we may now calculate 90% prediction intervals for the three installations. This implies the intervals that we can compare next year's occurrences with, in order to conclude whether there are significant improvements or increases.

The interpretation of statistical significance in this context is different from that used in the classical interpretation of risk assessment. A 90% prediction interval means that if all sources and mechanisms of risk are unchanged in the future from what is was in the past, the future observations (here number of leaks) will fall within the interval given with an assigned probability of 90%. If future observations fall outside the interval, we have strong evidence for concluding that that conditions have changed to an extent that risk is influenced. This is referred to as 90% confidence level.

It should be noted that this approach is different from the confidence intervals used in the classical approach to risk assessment, although the approaches appear to be the same. The basis for the confidence interval is an assumption that 'true' values exist, generated by averages of properties of an infinite thought-constructed population of similar situations, which is not part of our assumptions here (Table 2.13).

If there is one HC leak on Installation A next year, this will represent a statistically significant reduction. On the other hand, even if there are no leaks at all on Installation B, there is insufficient data to conclude there is no significant reduction.

If we, as a last illustration, calculate the average of Installations A and B, then we have a larger database, and the basis for conclusions about significant changes is broader, as shown in Table 2.14.

Table 2.13 Average number of leaks for example

Installation	Average number of leaks per year	Prediction interval (number of leaks per year)
A	6.67	2–11
B	4.33	0–8
C	1.67	0–5

Table 2.14 Average number of leaks for Installations 'A' and 'B'

Installation	Average number of leaks per installation per year	Prediction interval (number of leaks per year)
Average of A and B	5.5	2.5–8.5
Average per 1 million manhours	1.27	0.58–1.96

If next year there is one HC leak on each of the Installations 'A' and 'B', then there is a significant reduction. Moreover, if there are three leaks on installation 'C' and 6 million manhours worked, then this corresponds to 0.5 leaks per million manhours, which is a statistically significant reduction, compared to the average of 'A' and 'B'.

We are now able to compare prediction intervals for these three installations. It will be seen that all three prediction intervals overlap to some extent, implying that possibly the differences in average number of leaks per million manhours are due to statistical variations. Let us therefore consider 80% confidence level in Table 2.15. Now the prediction intervals of installations A and B do still overlap. For Installation C, there is still a slight overlap with Installation A, and differences may be due to statistical variations. With 70% confidence level there would not have been any overlap. This would imply that there is about 70% probability that the average FAR level on Installation A is higher than on Installation C. Finally, the Fig. 2.23 shows the prediction intervals with 90 and 80% confidence levels.

Table 2.15 Prediction intervals for installations A, B and C, 80% confidence

Installation	Prediction intervals (80%)	
	Lower limit	Upper limit
A	0.38	1.69
B	0.00	2.10
C	0.00	0.45

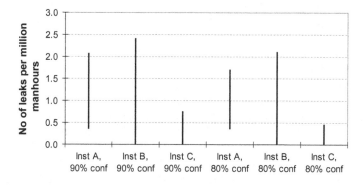

Fig. 2.23 Comparison of prediction intervals for different confidence levels

2.5 Basic Risk Modelling Concepts

This chapter has introduced some essential concepts related to calculation and presentation of risk. Some further concepts are needed, for use in the modelling of hazards and risks. 'Defence in depth', 'barriers', 'root causes' and 'risk influencing factors' are discussed in the following. Further concepts and terms are given in the Glossary section (see Page xxx).

2.5.1 Defence in Depth

Defence in depth is a term that is closely associated with accident prevention in complex industries. This principle has been introduced in different ways by several authors in the past. Haddon's ten accident prevention strategies [16], and the principles of the investigation logical tree, Management, Oversight and Risk Tree (MORT), reflect the same basic principles:

- Contain energy at source, AS WELL AS
- Stop flow of energy between source and target, AS WELL AS
- Protect targets against energy.

Professor Reason [17] has focused on organisational causes which may result in a similar breakdown of defences, for instance by the TRIPOD model. Kjellén [18] provides an in-depth discussion of accident models and principles.

If we adopt the 'energy flow' concept according to Haddon and MORT, barriers are the instruments that may be used to contain energy, stop energy flows and protect targets against energy. This interpretation implies that 'barriers' should be regarded as physical 'fences'. There are, on the other hand, many authors and experts who regularly refer to 'organisational barriers'. We will in this book maintain the principle that barriers are those actions or functions that may control (change) the flow of energy of some kind. This ties in with the term 'barrier function', which is introduced in Sect. 2.5.2.

An illustration of the defence in depth is perhaps most easily achieved through starting with a situation where this ability is completely lacking. The use of family cars on single lane roads (one lane in each direction) is a system without any defences in depth. If the operator (driver) loses control over the car, then other cars may easily be hit, possibly with severe (fatal) consequences. We have over the years made better cars, in the sense that there are zones that are specially designed in order to absorb energy during a collision. This implies that we have the driver to ensure that control over the energy is not lost, and we have some barriers on other cars, in order to protect personnel inside in case of a head-on collision. But we do not have any barriers in order to stop energy flow, if control is lost over a car.

So what can be done in order to provide in-depth defences? The solution to put a physical barrier in the middle of the road was started in Sweden and has gained wider application, at least in Scandinavia. The barrier between the lanes will stop the uncontrolled energy of a car that loses control and enters the opposite lane with head-on traffic. So we have a very effective barrier against head-on collisions. In theory we could also install some detection equipment in order to try to detect an unwanted development, before control is lost. This is not in widespread use. If we, for instance, could have a detector that discovers when the driver is about to fall asleep, we would improve the situation considerably. All these measures taken together would at the end of the day imply that defence in depth is available against head-on accidents.

The approach to barriers is most commonly used in relation to fire and explosion hazard, and may easily be illustrated in relation to this hazard. An overview of barrier systems and elements is presented in the following.

2.5.2 Barriers

The terminology proposed by a working group from 'Working together for safety' [19] is used, involving the following levels:

- Barrier function
- Barrier system
- Barrier element
- Barrier influencing factor.

The differences between these levels may be explained as follows:

Barrier function	The task or role of a barrier.
Barrier system/element	Technical, operational and organisational measures or solutions involved in the realisation of a barrier function.
Risk/performance influencing factors	Factors identified as having significance for barrier functions and the ability of barrier elements to function as intended.

The above definitions are based on the 2017 version of the PSA barrier memorandum. The wording in these definitions is simplified compared to the 2004 Norwegian Oil and Gas Association version, although the content/understanding is not significantly changed. The term 'barrier' is as such not given a precise definition, but is used in a general and imprecise sense, covering all aspects. The main emphasis in relation to barriers is often on barriers against leaks in the process area, comprising the following barrier functions:

- Barrier function designed to maintain integrity of the process system (covered largely by reporting of leaks as an event based indicator)
- Barrier function designed to prevent ignition

- Barrier function designed to reduce cloud and spill size
- Barrier function designed to prevent escalation
- Barrier function designed to prevent fatalities.

The barrier function may for instance be 'prevention of ignition', which may be divided in sub-functions; gas detection; electrical isolation as well as equipment explosion protection. One of the barrier elements in the gas detection sub-function is a gas detector; the process area operator may be another example. If we consider the process operator as the barrier element, there may be several barrier influencing factors, such as working environment; competence; awareness and safety culture. The different barriers consist of a number of coordinated barrier systems and elements.

The PSA regulations require the following aspects of barrier performance to be addressed:

- Reliability/availability
- Effectiveness/capacity
- Robustness (antonym vulnerability).

The reliability/a vailability is the only aspect of performance which varies significantly during operations, effectiveness/capacity and robustness are mainly influenced during engineering and design. Slow degradation may over a long time, on the other hand, change these values.

The following are aspects that influence reliability and availability of technical barrier systems:

- Preventive and corrective maintenance
- Inspection and test programmes
- Management and administrative aspects.

Figure 2.24 shows a simple block diagram which outlines the main barrier functions with respect to prevention of fatalities through fire and/or explosion caused by loss of hydrocarbon containment.

The barrier functions listed for hydrocarbon leaks in the process area, are also applicable to blowouts and leaks from risers and pipelines. For the blowout hazard, the integrity barriers are well control barriers. Well control barriers are outside the scope of the discussion in this chapter.

Fig. 2.24 Barrier functions for hydrocarbon leaks

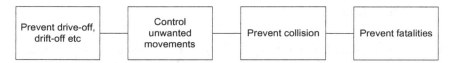

Fig. 2.25 Barrier functions for shuttle tanker station-keeping failure due to DP-system failure

For marine and structural accidents, there are fewer barriers. The corresponding barrier functions may be:

- Barrier function designed to maintain structural integrity and marine control
- Barrier function designed to prevent escalation of initiating failure
- Barrier function designed to prevent total loss
- Barrier function designed to prevent fatalities.

Figure 2.25 shows a similar diagram for the loss of station-keeping by DP-operated shuttle tankers in tandem off-loading (see further discussion in Sect. 11.4) with respect to prevention of fatalities due to collision between the shuttle tanker and an FPSO.

The relationship between barrier function, barrier elements, failures of barrier elements and risk influencing factors is illustrated in Fig. 2.26. The function is to detect a valve in the wrong position, for which purpose there may be several barrier systems or elements. These may have failures, as indicted by two basic failure events in the fault tree. Risk influencing factors are shown as influences for the failures of the barrier elements. The diagram is from the BORA approach (see Sect. 6.2.6.1).

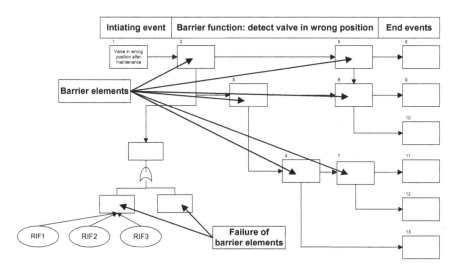

Fig. 2.26 Relationship between barrier function, elements and risk influencing factors

2.5.3 Root Causes

A root cause is, according to the TapRoot® system, the most basic cause that can reasonably be identified that management has control over, in order to fix, and when fixed, will prevent or significantly reduce the likelihood of the problem's reoccurrence [20].

Root causes are the most essential element of accident and incident investigations, because they are essential in order to prevent or reduce the likelihood of reoccurrence. It may be claimed that identification of root causes is often a weak element in the internal investigations made by the industry.

'Immediate causes' are often the main focus in internal investigations. Consider for instance the following illustrations from an investigation of a gas leak on an installation in the North Sea:

Immediate cause: Operation of the wrong manual isolation valve by process personnel.

Underlying causes: 'Best practice' for actual operation not described in manuals. Valves not labelled.

Attitude on shift is not to use instructions for routine tasks.

Are these underlying causes (as claimed by the investigation team) root causes? No, they cannot be considered root causes, because they are not the most basic causes that management can control. Possible root causes may have been 'too high time pressure', 'implicit acceptance by management that it is acceptable not to follow procedures', or simply 'bad safety culture'.

2.5.4 Risk/Performance Influencing Factors

Risk influencing factors (sometimes called performance influencing factors) are factors that influence the performance of barrier systems and elements. Consider as an example the manual gas detection function performed by personnel performing manual inspection in the hydrocarbon processing areas. Factors that will influence the ability of such personnel to detect possible gas leaks are as follows:

- Procedures for manual inspections
- Organisation of work, work patterns
- Training of plant operators
- Experience of plant operators
- Motivation of plant operators
- etc.

What we here consider as influencing factors will often be considered as 'barriers' according to some of the definitions that are used.

It may be discussed whether 'causal factors' and 'influencing factors' are synonymous expressions, and to some extent they are. It may be argued that 'influencing factors' is a wider term than 'causal factors', but little emphasis is placed on this. Causal factors as well as influencing factors offer good opportunities to identify risk reducing measures, which may have a significant effect on the risk level.

References

1. SRA (2015) Society for risk analysis. SRA glossary developed by Committee of Foundations of Risk Analysis
2. ISO (2009) Risk management—vocabulary, ISO/IEC guide. International Standards Organisation, Geneva, ISO Guideline 73:2002
3. Aven T (2013) Practical implications of the new risk perspectives. Reliab Eng Syst Safety 115:136–145. Wiley
4. Flage R, Aven T (2009) Expressing and communicating uncertainty in relation to quantitative risk analysis. Reliab Risk Anal Theory Appl 2(2):9–18
5. Berner CL, Flage R (2016) Strengthening quantitative risk assessments by systematic treatment of uncertain assumptions. Reliab Eng Syst Safety 151:46–59. Wiley
6. Berner CL, Flage R (2017) Creating risk management strategies based on uncertain assumptions and aspects from assumption-based planning. Reliab Eng Syst Safety 167:10–19. Wiley
7. Tuft V, Wagnhild B, Pedersen LM, Sandøy M, Aven T (2015) Uncertainty and strength of knowledge in QRAs. In: Proceedings of ESREL 2015, Zurich, 7–10 Oct 2015
8. Tuft V, Wagnhild B, Slyngstad O (2018) A practical approach to risk assessments from design to operation of offshore oil and gas installations. Knowledge in risk assessment and management, 1st edn. Wiley
9. Khorsandi J, Aven T (2017) Incorporating assumption deviation risk in quantitative risk assessments: a semi-quantitative approach. Reliab Eng Syst Safety 163:22–32. Wiley
10. Aven T, Vinnem JE (2007) Risk management, with applications from the offshore petroleum industry. Springer, London
11. Schofield SL (1993) A framework for offshore risk criteria. Safety Analysis Unit, Offshore Safety Division, HSE, Liverpool
12. Vinnem JE, Vinnem JE (1998) Risk levels on the Norwegian Continental shelf. Report No.: 19708–03. Preventor, Bryne, Norway
13. PSA (2018) Trends in risk levels. Main Report 2018 (in Norwegian only), Petroleum Safety Authority
14. DNV (1998) WOAD, Worldwide Offshore Accident Database. Høvik
15. Aven T (2003) Foundations of risk analysis—a knowledge and decision oriented perspective. Wiley, New York
16. Haddon W (1980) The best strategies for reducing damage from hazards of all kinds. Hazard prevention, 8–12
17. Reason J (1997) Managing the risks of organizational accidents. Ashgate, Hampshire
18. Kjellén U (2000) Prevention of accidents through experience feedback. Taylor and Francis, London
19. Norwegian Oil and Gas (2004) Working together for safety. http://samarbeidforsikkerhet.no
20. Paradies M, Unger L (2000) TapRoot, The system for root cause analysis, problem investigation, and proactive improvement. System Improvements Inc., Knoxville

Chapter 3
Risk Assessment Process and Main Elements

3.1 Selection of Risk Assessment Approach

It is prudent that the selection of risk assessment approach should reflect the technical and operational challenges that the facilities are faced with. The ISO standard [1] suggests that the risk assessment methods being used should depend upon factors such as:

- size and complexity
- what are the credible major accident hazards
- the severity of the consequences
- the degree of uncertainty
- the level of risk
- the number of people exposed, as well as
- the proximity of environmentally sensitive areas.

The ISO 17776 standard also emphasizes that the risk assessment approach can vary depending on the scale of the installation and the project phase during which the risk assessment is conducted. For example, it is suggested that for simple platforms such as wellhead platforms, checklists based on previous risk assessments for similar installations and operations can give a consistent approach to major accident safety. It is also mentioned that, if an earlier design is repeated, the evaluations for the original design can be used if certain criteria are met. For complex platforms, however, the re-use of major accident safety evaluations is less relevant.

At present there is ongoing work in the Norwegian oil and gas industry to detail out how the above principles can be applied in practice in order to re-use some of the decision support obtained for previous concepts and still obtain dimensioning loads. At present, this work has not been finalised, and thus, it is not included in this book.

© Springer-Verlag London Ltd., part of Springer Nature 2020
J.-E. Vinnem and W. Røed, *Offshore Risk Assessment Vol. 1*,
Springer Series in Reliability Engineering,
https://doi.org/10.1007/978-1-4471-7444-8_3

3.2 Quantitative or Qualitative Risk Assessment?

The purpose of risk assessment is primarily to support decision-making, including decisions on risk reducing measures in the context of a structured, systematic and documented process. The documentation requirements for the safety case under UK legislation are in this respect the most explicit, when they require documentation of the outcome of the decision-making process for risk reduction measures based on a risk assessment.

This overall purpose is often forgotten, in the sense that companies may think that the purpose of risk assessment is to document that the risk level is tolerable. Even worse, a risk assessment may sometimes be conducted in order to demonstrate that it is acceptable to deviate from regulatory requirements or common industry practice. This is what is referred to as 'misuse of risk analysis' (see Sect. 3.14).

The next question is to what extent the risk assessment needs to be quantitative. It is sometimes argued that qualitative risk assessment is better, because the numbers are often rather uncertain. In the authors' opinion, any risk assessment, quantitative or qualitative, should address uncertainties and reflect on the strength of knowledge on which the evaluations are based. Then the choice between quantitative or qualitative techniques is a question of what is fit for purpose and how decision support will be achieved.

There are usually several studies that are used as input to decisions about risk reducing measures, the majority of these are normally quantitative. Many of them will analyse economic consequences, and will definitely be quantitative. It is therefore often essential that the risk assessment studies are quantitative, in order to have the same precision level as the other studies that are used as input to decision-making.

The authors' opinion is further that quantification improves the precision when a study is carried out. A qualitative study will discuss various factors, but will often not perform a detailed trade-off between the factors. When quantification is needed, such a trade-off is needed as part of the quantification, and a more precise answer is produced.

The proper attention to evaluation of uncertainty and evaluation of model sensitivity is extremely crucial in quantitative studies (see Chap. 16). This applies to quantitative risk assessment, but it certainly also applies to other studies that are part of the basis for the decision-making about risk reduction measures. Uncertainties are actually more often under-communicated in other studies of investment costs, operating cost, production capability and so on.

The opponents of quantification argue that the quantification is not to be trusted. Used properly with due consideration of uncertainties, it is the author's clear view that quantitative risk assessment is more trustworthy than qualitative risk assessment. Qualitative risk assessments are easier to manipulate to conclude with whatever some decision-makers will prefer, and such risk assessments will as argued above not be capable of making trade-offs between opposing factors.

Some risk assessments are used in order to establish design accidental loads (see Sect. 19.2.1.2), such as the structural resistance to impact and/or hat loads. It is not possible to understand how qualitative risk assessments can be used in such cases.

But it should also be realised that there are some examples of use of quantitative risk assessments that are as far from trustworthy as more or less possible. One such case is presented in Sect. 3.13, and some other examples are mentioned in Sect. 3.14.

One final overall aspect of quantification may be added; the best use of such studies is often to use "quantitative studies in a qualitative manner". Put differently, the quantification is not the goal itself, but just a means to achieve better decision-making. The numeric values can be used to demonstrate what are the important aspects and evaluations. This is the important output, not the actual numerical values.

3.3 Risk Assessment Approach

There has been considerable focus in the past few years on models for risk assessment in various industries, not the least the offshore oil and gas industry. The most commonly used approach is the ISO 31000 standard: Risk management—guidelines [2]. The same approach has also been adopted in the NORSOK Z–013 standard: Risk and emergency preparedness analysis [3], as well as by the petroleum regulations in Norway, issued by PSA.

The main elements of the process for risk assessment according to ISO 31000 are presented in Fig. 3.1, referred to as the risk management process. This process is general and can be applied to any risk-informed decision in the organization. It can be used for strategic decisions made by top management, to treat economic risk in the finance department, as well as for treating operational risk when running an offshore installation in the operations phase. What varies between these application areas are two things in particular: the scope and the risk analysis method being used. If the scope is to treat risk related to overruns in a project, a project risk analysis method will be the natural choice of risk analysis method. If the scope is to perform a work task in the operations phase, a safe job analysis or HAZID may be used as the risk analysis method. And if the scope is to treat major accident risk, QRA may be used as the risk analysis method. Since QRA and major accident risk is the main scope of this book, the risk management process will in the following be described based on the treatment of major accident risk as the scope and QRA as the risk analysis method.

The core of the process: (a) scope, context, criteria, (b) risk assessment, and (c) risk treatment, is consistent with common practice for many years in the offshore petroleum industry. The elements on each side, communication and consultation, and monitoring and review, have also played a role in the offshore petroleum industry as they were introduced in the previous version of the ISO 31000 standard

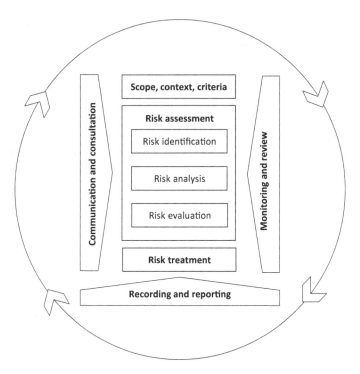

Fig. 3.1 The risk management process described in ISO 31000

in 2009 [4]. The purpose of communication and consultation is to assist both internal and external stakeholders in understanding risk and the basis for supporting decisions (Standards Norway, 2010). Communication seeks to promote awareness and understanding of risk, whereas consultation involves obtaining feedback and information to support decision-making. Both of these aspects are needed when QRA results are being implemented in an organization. Monitoring and review include periodic review of the risk management process and its outcomes, in order to improve the risk management process in the future [3]. In a QRA context, this includes learning from previous development projects and how QRAs can be used to provide decision support.

The element at the bottom of the figure, recording and reporting, was introduced in the 2018 version of ISO 31000 [2]. As emphasized in the NORSOK Z–013 standard, recording and reporting should be documented to provide decision support, communication and improvement of risk management as well as interaction with other stakeholders [2].

Both the NORSOK Z–013 standard [2], and the ISO 17776 standard [1], include figures explaining their own content relative to the ISO 31000 standard. Both of them cover all elements except risk treatment (even though recording and reporting is not explicitly mentioned, since this element was included in ISO 31000 after the two standards were published). Each of the main elements of this process is outlined in Sects. 3.4–3.10.

3.4 Scope, Context, Criteria

Since the risk management process is general and can be applied to any risk-informed decision, it can be applied to many different contexts. A success criterion for any risk assessment is to be explicit on what the purpose of the risk assessment is. In other words, it is crucial to understand which risk-informed decisions the risk assessment is going to support.

Establishing the context covers all activities carried out and all measures implemented prior to or as a part of the initiating phase of a risk assessment process, with the intention of ensuring that the risk assessment process to be performed is:

- Suitable with respect to its intended objectives and purpose
- Executed with a suitable scope and level of quality
- Tailored to the facility, system(s) and operations of interest
- Tailored to the required and available level of detail.

Focus on these aspects has often been lacking, especially before the requirement was formulated as the ISO 31000 standard. In particular, authorities have often claimed that risk analysis is unsuitable for its purpose. This is probably a relevant criticism for many risk assessment processes.

Selecting an analytical approach that is suitable for its purpose is dependent on the description and consideration of the context. In particular, the objectives of the risk assessments, the required timing and the relevant stakeholders need to be taken into account, when defining the scope of the risk assessment process.

The following example illustrates how this can fail dramatically. An oil company needed to decide whether to install a lift on a production installation in order to improve working conditions. It was a difficult decision and management and workforce were at odds with respect to which decision to take.

They agreed to perform a quantitative risk assessment. This was specified as the scope of work for a risk analysis contractor that decided to carry out a standard QRA approach for two alternatives, with and without the lift. However, they did not consider what input data would be available for the study before they selected the analytical approach.

When the contractor searched for the relevant incident data, they found only one event which with some goodwill could be considered to be relevant, although not all would agree that even that one incident was relevant. The contractor performed a standard QRA of the two alternatives, without any particular consideration of the uncertainties or robustness of the conclusions, which was appropriate considering the sparse data available. When this was discovered, the client did not approve of the report at all, the study had to be scrapped completely, and decision-making had to be made without input from risk assessment.

Ethics in risk assessment is another topic related to the establishment of the context. Recent experience in Norway has demonstrated the importance of this topic, and Sect. 3.13 therefore discusses ethics in risk assessment.

3.5   Hazard Identification

A comprehensive and thorough identification and recording of hazards is critical, as a hazard that is not identified at this stage would be excluded from further assessment. Well planned and comprehensive hazard identification (HAZID) is therefore a critical and important basis for the other elements of the risk assessment process. The objectives of hazard identification are:

(a) Identification of hazards associated with the defined systems and of the sources of these hazards, as well as events or sets of circumstances that may cause the hazards and their potential consequences.
(b) Generation of a comprehensive list of hazards based on those events and circumstances that might lead to possible unwanted consequences within the scope for the risk and emergency preparedness assessment process.
(c) Identification of possible risk reducing measures.

It is often claimed, especially by authorities, that too little emphasis is placed on the identification of hazards. This may result in only well-known hazards being identified, and nothing beyond those that are well known, in which case the hazard identification has failed completely.

It is actually more important to search for unknown threats (sometimes referred to as 'unknown unknowns'), i.e. those hazards that nobody has any experience of. Some 15 years ago two people died because of the narcotic effect of an unignited hydrocarbon gas release inside a concrete shaft on an installation in the UK sector. At the time, this was an unknown threat—nobody had been similarly exposed to hydrocarbon gas at such a concentration before—and the lethal effect was not generally known in the offshore industry. Today this is no longer an unknown threat.

However, a warning about the inefficient use of hazard identification should also be given. It is not necessary to 'reinvent the wheel' each time! On one hand, we should guard against 'Xerox engineering' or more specifically 'Xerox HAZID', i.e. just copying from a previous study. On the other hand, one should not spend too much time on the identification of hazards from separators, compressors, heat exchangers, and so on because this is very well known by experienced risk analysts. Sometimes, authorities imply that 'the wheel needs to be reinvented' for each new project; however, this is misleading and it will not lead to the most efficient result.

The intelligent use of resources reuses previous HAZIDs for well-known hazards and systems, focuses the most extensive effort on those parts of the system that are novel or unconventional, and aims to identify unknown threats.

Further information may be found in Sect. 15.1 and in the NORSOK Z–013 standard.

3.6 Risk Analysis

Figure 3.2 is a conventional diagram showing the main elements of risk analysis: analysis of initiating events, analysis of consequences and establishment of the risk picture. The diagram is in line with ISO 31000, although the risk treatment and recording and reporting parts are not highlighted. It can be considered as a diagram 'zooming in' to the risk assessment part of Fig. 3.1.

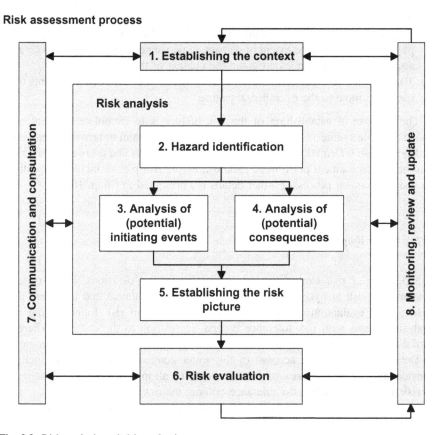

Fig. 3.2 Risk analysis and risk evaluation processes

The analysis of initiating events is the analysis and identification of potential causes of initiating events, in other words, this aims to assess the probability or frequency of initiating events occurring. This is further discussed in Sect. 14.3. It is extremely important to consider the purpose of the study, when the analysis of initiating events is planned. If the purpose is to search for risk reducing measures and quantify their effects, then an analysis based on failure statistics, is usually insufficient and unsuitable for the purpose.

The term 'analysis of potential consequences' is used in a wide sense, covering the entire accident sequence or sequences that may be the outcome if an initiating event should occur. As the objective and scope of a risk assessment may vary, the way to analyse potential consequences may range from detailed modelling (using extensive event-trees including a comprehensive assessment of the various branches) to coarse judgemental assessment (by e.g. extrapolation from experimental studies or from the available data). The analysis of potential consequences may therefore be qualitative, semi-quantitative or quantitative, depending on the context. The following are the objectives of consequence analysis:

(a) To analyse potential event sequences that may develop following the occurrence of an initiating event.
(b) To determine the influence of the performance of barriers, the magnitude of the physical effects and the extent of damage to personnel, environment and assets, according to what is relevant given the context of the assessment.
(c) To assess the possible outcomes of identified and relevant initiating events that may contribute to the overall risk picture.

The purpose of establishing of the risk picture is to formulate a useful and understandable synthesis of the risk assessment, with the aim to provide useful and understandable information to the relevant decision-makers and users about the risk and the risk assessment performed. Establishing the risk picture includes reporting the risk assessment process. Further details are presented in Chap. 16.

3.7 Risk Evaluation

The purpose of risk evaluation is to assist in making decisions, based on the outcomes of risk analysis, namely which risks need treatment and treatment priorities. Risk evaluation involves comparing the level of risk found during the analysis process with risk tolerance criteria, in relation to the context. Where a choice is to be made between options, this will depend on an organisation's context.

Decisions should take account of the wider context of the risk and include consideration of the tolerance of the risks borne by all involved parties. If the level of risk does not meet risk the tolerance criteria, the risk should be treated.

In some circumstances, the risk evaluation may lead to a decision to undertake further analysis. It may also lead to a decision not to treat the risk in any way other than maintaining existing risk controls. Risk evaluation for personnel is discussed in more depth in Chap. 17.

3.8 Risk Treatment

Risk treatment involves selecting one or more options for addressing risks, and implementing those options. It may involve a cyclical process of assessing a risk treatment plan, deciding that residual risk levels are not tolerable, generating a new risk treatment plan, and assessing the effect of that treatment until a level of residual risk is reached which is one within which the organisation can tolerate based on the risk criteria.

Risk treatment options are not necessarily mutually exclusive or appropriate in all circumstances. The options include the following:

(a) Avoiding the risk by deciding not to start or continue with the activity that gives rise to the risk
(b) Seeking an opportunity by deciding to start or continue with an activity likely to create or maintain the risk
(c) Changing the likelihood
(d) Changing the consequences
(e) Sharing the risk with another party or parties (including insurance)
(f) Retaining the risk, either by choice or by default.

Risk treatment is outside the scope of this book, and it is not considered any further here.

3.9 Monitoring and Review

The purpose of monitoring, review and updating the risk assessment is to monitor the established context, with respect to its validity due to the decisions made, new knowledge about the system or operation being analysed or other factors that may jeopardise the validity of the context. The results from scoping or framing studies, performed after the context was updated, or those from studies or assessments performed as a part of the risk assessment process may also require the context to be updated.

The purpose is further to update the context throughout the risk assessment process if and when required, and to assure that the risk assessment process and its various elements are executed based on an updated context.

A risk analysis is in general only valid as a basis for decision-making as long as the basis for the analysis (its methods, models, input data, assumptions, limitations, etc.) is assessed to be valid. Any deviation from the basis for analysis should therefore initiate an assessment of the deviation with respects to its effect on the risk and/or the validity of the analysis and its results, provided that the analysis is intended to be used as a basis for future decisions. When updating an analysis (or using an analysis as basis for sensitivity studies) all basis for the analysis should be reviewed and documented.

Monitoring is discussed in more depth in Chaps. 22 and 23. Further details are also presented in NORSOK Z–013.

3.10 Communication and Consultation

The objective is to involve the relevant internal and external stakeholders (relative to the operator), at the right time and with the appropriate level of involvement throughout the entire risk assessment process, as a measure to improve the quality of the risk assessment process and its ability to be tailored and suitable for its intended purposes. Experience transfer from personnel with operational knowledge from the practical utilisation of critical equipment and systems is of importance to establish a high level of safety and predictable risk assessment outcomes.

Effective internal and external communication and consultation is carried out to ensure that those affected by the hazards and those accountable for managing the risk understand the established context on which the results are calculated and evaluations are made, the risk picture, and the reason why particular priorities may be needed in the risk treatment.

Communication and consultation is often a weak element in the risk assessment process, see for instance the case study presented in Sect. 3.13. This case demonstrates the complete failure of communication and consultation, not because it was not performed, but because it was far from open and honest.

At the other end of the scale are the risk assessment studies for the abandonment of the Frigg installations in the North Sea [5]. In this case, it was decided, mainly based on risk assessment that three large concrete installations had to be left on site permanently due to the high probability of loosing at least one of them during a possible refloating and transporting to shore. The communication and consultation process with all stakeholders, including authorities in the UK and Norway as well as fishermen's organisations and NGOs in two countries was conducted in a very open, honest and forthcoming manner with an extremely good atmosphere. This consultation process, which is mandatory according to OSPAR and national regulations, was so successful that a unanimous decision was reached to leave the installation in place without any disagreement.

Regulations may have requirements with respect to what communication and consultation process is needed for the public.

3.11 Recording and Reporting

As emphasized in the NORSOK Z–013 standard, recording and reporting should be documented to provide decision support, communication and improvement of risk management as well as interaction with other stakeholders [3]. For any risk assessment, it is crucial to document all steps. A good risk analysis report is capable of explaining to the reader all choices made during the risk assessment. Examples that should be explained include the scope and which risk-informed decisions the risk assessment is going to support, the choice of risk analysis method, how the hazards have been identified, how risk has been analysed, as well as the risk results. For the latter, it is particularly important that it is presented in such a way that the reader can follow the arguments. Also assumptions need to be presented, as well as the strength of knowledge upon which all the evaluations have been based.

3.12 Who Are These Requirements Applicable for?

The context of risk analysis studies in the offshore petroleum industry is often such that there is a client (typically a project or a unit within an oil company) and a risk analysis contractor. The client prepares the scope of work, a few potential contractors bid for the contract, upon which one is chosen to perform the work.

It is often assumed that the requirements in the risk analysis and assessment process are mainly applicable to the contractor, who has the responsibility for selecting the appropriate tools and data and performing the analytical work.

However, the client plays an important part in this process, and thus the requirements of risk assessments are equally applicable to it as they are to the contractor in the client—contractor relationship outlined here.

It is essential that the requirements in the risk assessment process are carefully considered by the client, for instance, when the scope of work is prepared.

3.13 Ethics in Risk Assessment

The important ethical aspects of risk assessment are discussed based on a recent case study from Norway. An extensive summary of the case is presented initially, as a basis for the general discussion about ethical challenges.

3.13.1 Case Study

The case study relates to an LNG plant that began operations in 2011, which is located on the west coast of Norway, in an urban area outside the city of Stavanger, about 4 km from Stavanger Airport. The LNG plant is an onshore facility and is as such outside the scope of this book. It is nevertheless included because it demonstrates some of the ethical challenges of risk assessment. The case study is described in more detail in Ref. [6].

The part of this case of interest to the present discussion is the early planning process, used to select the location of the plant and consider its risk exposure to local communities. The actual site of the plant used to be a refinery jetty area: the refinery was decommissioned around 2000. In addition to the LNG plant the following facilities are also found in the Risavika bay area:

- Gasoline depot
- Offshore supply base with extensive shipping traffic
- Container terminal (new)
- Passenger cruise/ferry terminal (new).

The gas supply to the plant comes via a subsea pipeline from a large gas processing plant at some 50 km away. LNG export from the LNG plant is by sea and by road. Annual production is 300,000 tons of LNG, but capacity can be increased to 600,000 tons if market conditions call for such an increase. The LNG plant has the following main components, as shown in Fig. 3.3:

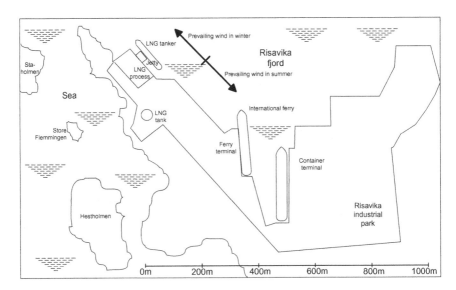

Fig. 3.3 LNG plant and surroundings

1. Pipeline landfall
2. Gas reception facilities
3. Pre-treatment system
4. LNG production
5. LNG tank
6. Export facilities/jetty.

The LNG plant was built by a company consisting of 50% private ownership and 50% ownership by a group of local communities; however, the private ownership was terminated due to the economy of the project. The site is owned by a company where local communities (some identical to the plant owners) have about 49% ownership.

Local politicians are heavily involved as non-executive directors of all the companies involved. Several newspaper articles at the time raised questions about the politicians' behaviour, for example whether it would meet the standards of Transparency International.

It is worth noting that all these companies acted as if they were private enterprises, even though local authority ownership was 100%. This must be seen as a severe deficiency in the applicable regulations or democratic process, namely that companies dominated or completely owned by local governments on behalf of local communities are allowed to act as if they are private companies without any restrictions or obligations to the public, thus completely avoiding the 'open government' policies that apply to the national as well as local governments in Norway.

3.13.2   Overview of Risk Studies and Risk-Informed Decision-Making

During the early planning and engineering phases, the following risk studies were carried out:

1. Several preliminary hazard analysis (PHA) studies (referred to in Norwegian as 'ROS' studies, risk and vulnerability studies, but essentially the same as PHA)
2. Initial QRA study
3. QRA study in detailed engineering.

The use of risk analysis studies in risk-informed decision-making for the LNG plant in the early project phases is outlined below, focusing on the following aspects:

- Choice between alternative locations
- Risk analysis results in relation to risk tolerance criteria
- Risk communication to the public
- Use of risk analysis in design process.

3.13.3   Choice Between Alternative Locations

Prior to the selection of the final location of the LNG plant, a decision-making process was carried out in which, as far as it is known, two alternative locations were considered. This process was carried out entirely without any public knowledge or insight, in spite of the extensive involvement of local councils as owners of the company responsible for the development. From what has been revealed indirectly from this process, the decision seems to have been made without any consideration of safety aspects. Market conditions and project cost seem to have been the only factors taken into account. The two alternatives are nevertheless known.

It should be noted that the two locations are quite different, especially from a safety point of view:

- Location 1: Risavika (chosen site)

    - Close to urban area
    - Several other public and industrial facilities quite close, including an international ferry terminal
    - Jetty located in the mouth of a long bay with extensive passing traffic of large ships
    - Unfavourable direction of prevailing wind with respect to densely populated areas, such as the international ferry terminal.

- Location 2: Mekjarvik

    - Scattered houses, not an urban area
    - Local topology favourable with respect to restricting the spreading of a gas cloud and blowing it over open sea
    - Easy to separate marine traffic from quay location
    - Favourable direction of prevailing wind with respect to densely populated areas.

In summary: no risk analysis was performed at that stage and no risk evaluations seem to have been made in the process of choosing the location of the plant.

A comparison may be made with the location of petroleum facilities that fall under Norwegian petroleum law. Risk evaluations are compulsory activities in all stages of planning and preparation, according to the PSA management regulations. Rigid requirements ensure public knowledge and involvement in the evaluation of premises and decision-making at certain stages.

The LNG plant falls under the jurisdiction of the Norwegian Fire and Explosion Prevention Act, which based on the case study, has severe deficiencies with respect to ensuring that safety is integrated into early planning and screening processes.

In the present case, risk evaluations would probably have shown that Location 2 was significantly better from the point of view of major hazard risk, perhaps to the extent that major hazard risk would have prevented Location 1 from being accepted.

3.13.4   Risk Analysis in Tolerability Evaluations

The main presentation of risk in the initial QRA study presented the expected risk level in relation to the upper and lower limits of tolerability, as included in Fig. 3.4. A list of release scenarios contributing to the curve was also presented.

When an updated analysis was presented, the risk results were seen to have increased as shown in Fig. 3.4. This further illustrates one of the problems with the initial QRA study, namely a lack of robustness. The updated risk level shows that the frequency of accidents with at least 100 fatalities increased by a factor of 56 compared with the results from the initial QRA study. It should be noted that the initial QRA study was performed before engineering studies had started, whereas the updated study carried out by the engineering contractor reflected all the engineering details. This difference might be expected, given that the initial study was a coarse study. However, this should have been recognised and due allowance made for uncertainties.

The operator had adopted the risk tolerance limits proposed by HSE for the study in question. The results were interpreted as 'negligible' in relation to these risk tolerance limits, which were intended for use with respect to existing facilities.

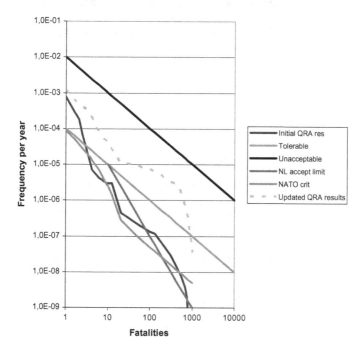

Fig. 3.4 Updated risk results for LNG plant and alternative risk tolerance criteria (Reprinted from Reliability Engineering and System Safety, Vol 95/6, Vinnem, Risk analysis and risk acceptance criteria in the planning processes of hazardous facilities—A case of an LNG plant in an urban area, 9 pp., 2012, with permission from Elsevier)

Stricter limits, where risk aversion has been incorporated, have been stated for new facilities [7]. However, other risk tolerance limits may be used for a broader illustration of the tolerability of risk, as shown in Fig. 3.4. This is discussed later.

3.13.5 Risk Communication with the Public

The communication of risk results to the public was one of the main problems of risk communication in the present case. The initial risk results were not available until one year after the public hearing of the planning proposal. For the input to the public hearing, the enterprise conducted coarse PHA-type studies, where risks were presented in a risk matrix (see example in Fig. 3.5).

The most severe consequence category was 'critical', which was translated as 'more than one fatality'. Risk presentation of this nature is highly unsuitable for the characterisation of third party risk exposure from major hazards. Furthermore, the analysis was limited to risk to employees on the LNG plant, although these risk results were used by the enterprise in the compulsory communication to the neighbours to present their risk exposure. Several complaints were made to the enterprise and all involved authorities to the effect that employee risk was completely unrepresentative for the risk to the public, but none of the parties acknowledged that this was so. The reason is difficult to comprehend, the most obvious interpretation being a severe lack of competence.

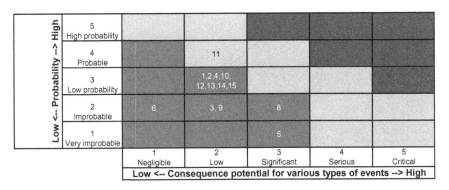

Fig. 3.5 Risk picture for the LNG plant and export facilities (translated from the PHA study) (Reprinted from Reliability Engineering and System Safety, Vol 95/6, Vinnem, Risk analysis and risk acceptance criteria in the planning processes of hazardous facilities—A case of an LNG plant in an urban area, 9 pp., 2012, with permission from Elsevier)

3.13.6   Use of Risk Analysis in the Design Process

In several LNG plant studies, it was assumed that in the event of the impact of a passing vessel on an LNG tanker loading at the jetty, the gas release would ignite immediately, presumably by sparks generated by the collision itself. No explanation was provided for how the ignition of very heavy and extremely cold gas could occur physically. In fact, it is very hard to foresee how it could be caused in this way. However, the implication of this assumption was that it was unnecessary to consider in the studies any spreading of the gas cloud due to wind and heating of the liquefied gas, with obvious consequences for the scenarios the public might be exposed to. Such a critical assumption should at least have been subjected to a sensitivity study in order to illustrate how changes in the assumption would affect the results, and the robustness of the assumption discussed. None of this, however, was provided in any of these studies.

In the case of the LNG plant the risk analysis was used to 'prove' that it was safe enough not to follow the US practice for safety zones for LNG plants. This analysis rested on arbitrary assumptions about leak rates and durations and an optimistic assumption that all safety systems would always function as intended.

The result of this approach was that a safety zone regarded by many as drastically insufficient was adopted. Confirmation of its insufficiency came from the updated QRA study that, in spite of several weaknesses, confirmed that the risk level for the public had increased considerably, by a factor of more than 50 for accidents with at least 100 fatalities (see Fig. 3.4).

The local authorities responsible for the planning permits raised no objections. In this case, specialist authorities (the Directorate for Civil Protection and Emergency Planning, DSB) also had no objections to this use of risk analysis. They thus took a view contrary to that of the PSA who in November 2007 issued a letter to the industry [8], warning about the malpractice of using risk analysis to demonstrate the acceptability of deviations from accepted practice and regulatory requirements (see further discussion in Sect. 3.14). The approach taken by the DSB may be interpreted as a lack of competence or a lack of professional maturity.

The consultant performing the initial QRA used unrecognised software. This was clearly identified by an independent verification of the risk analysis, but no remedial actions were taken by the authority responsible for commissioning the review. Recognised software was used in the updated QRA study, but not software suitable for a detailed engineering stage. At that stage, one would expect more detailed modelling based on computational fluid dynamics. Good practice would be to use several models in order to illustrate robustness.

As stated above, in the case of the LNG plant the calculation of the safety zone was based on a specific leak scenario with a short duration of only 30 s. Such a brief duration assumes that all relevant safety systems (barriers) function exactly as intended with no delay at all. Leak statistics from onshore and offshore petroleum facilities have demonstrated clearly that gas leak cases rarely occur exactly as 'planned'. Delays in detection, or the failure of automatic isolation to the effect that manual isolation is required is more the norm than the exception.

The approach taken in this case may be characterised as liberal ('non-conservative'). No sensitivity studies were conducted or the question of possible variations addressed, and the safety zone must therefore be considered to be an unrealistic minimum case of 125 m radius. It may be mentioned that US authorities do not accept the use of risk analysis in the definition of safety zones but stipulate a safety zone radius of up to about 1600 m.

A robust definition of safety zones should probably be based on a leak with a duration of several minutes, which should reflect scenarios where one or a few barriers fail, at least partially, or where delayed response occurs. Of similar importance would be the performance of sensitivity studies to assess the effect of changes in assumptions and premises in relation to the dimensions of safety zones.

In the case of the LNG plant, the initial QRA study made assumptions about the explosion strength of the closest buildings outside the safety zone, but the local authorities did not ensure that this was followed up in the construction of the building in question. A contributory factor was the poor documentation of the initial QRA study, making identification of what should apply far from clear.

3.13.7 Cause of Deficiencies

The summary of the case study presented above shows several serious failures of competence on the part of the enterprise as well as the local and national authorities. Some would argue that the failures are so conspicuous that the parties involved cannot have absorbed any relevant professional experience from the past 20 years.

Current Norwegian legislation is aimed at fulfilling the EU Seveso II directive in a prudent manner and ensuring that the public is not compelled to accept an unreasonable level of risk exposure. In the present case the public was obliged to do just this.

Several societal control systems have been designed to prevent third parties from being subjected to significant risk levels without being consulted and to ensure that their views are taken seriously. All these control systems failed in the present case, with the result that the public was forced to accept exposure to risk considered to be unreasonable by most of the neighbours. A complicating factor was that the highest exposure would apply to passengers at the international ferry terminal. For obvious reasons, these passengers have no spokespersons and their voices were never heard in the process. The owner of the land on which this terminal has been built is also the owner of the LNG plant and is therefore an inappropriate person to represent the passengers in the terminal in an objective way.

One of the most surprising and worrying aspects of the case is that all authorities (local and national) failed to realise that the obligations under current laws and regulations to allow the neighbours a substantial say in the matter had not been observed, because the premises for the work were wrong. One of the arguments in the present case centred on the assumption by the authorities that they could not all be wrong and thus those who objected (including the present author) had to be wrong, apparently because the majority opinion reflected the opposite point of view.

However, there can be no uncertainty about the fact that qualitative analysis for evaluation of third party risk exposure and determination of safety zones is grossly insufficient. The practice in the Netherlands [9], is a further confirmation of this. Neither can there be any uncertainty about the unrepresentativity of employee risk used to inform the public about their risk exposure. The lack of satisfactory involvement of the public thus becomes a serious failure of societal control systems.

It is a paradox that, in an industry where it is essential that authorities take seriously the obligation to protect the public as a third party, the enterprise as well the authorities in this case all failed completely to assume this responsibility, whereas in the petroleum industry the need for the careful consideration of third party risk exposure is less pressing, yet companies and authorities all take this responsibility seriously.

3.13.8   Main Ethical Challenges

The description above documented the serious failures and deficiencies on the part of the enterprise and authorities. This is a separate discussion. When it comes to ethical challenges, the focus is on risk analysis contractors.

If you are a risk analysis contractor and you are prevented by the client from carrying out a risk assessment with scope of work in compliance with applicable the regulations, do you have a responsibility to rectify this situation (or inform authorities), or is it just fine to accept the client's instructions? This is an ethical challenge that the contractor for the LNG plant failed to meet in a responsible manner. The result was an unethical risk analysis study, which the client used as if it were in accordance with the relevant regulations.

Further, is it ethical for you as a risk analysis contractor, in order to please the client, to hide a crucial assumption that may change the results dramatically so well in an appendix, that the typical reader would never discover it? The question mark is for rhetorical purposes, it is obvious what the answer is. However, this is precisely what was done in the LNG case with the assumption that all leaks from the tankers would ignite immediately. Not only is it a wrong assumption, but it was hidden and not open for discussion and evaluation.

The first unethical aspect may have been made unwillingly based on the client's insistence and the contractor's insufficient stamina. However, the second aspect gives the impression of having been done deliberately, which would be even more serious.

This second aspect illustrates the extreme importance of the assumptions and premises usually adopted in substantial numbers during the execution of a detailed risk analysis. There is a vast opportunity to manipulate these assumptions and premises in an unethical manner. It is therefore extremely important that all risk analysis contractors are aware of this aspect, and take all reasonable steps in order to avoid such actions.

It does not make life any easier when you consider that many risk analysis practitioners are young people with limited experience. A client that wishes to influence strongly which assumptions to make, may thus have an easy task.

It is also considered to be unlikely that a scope of work not in compliance with applicable regulations would be accepted if risk analysis contractor personnel are well aware of their ethical obligations. It is therefore important for the education of risk analysis specialists and for training in work practices for risk analysis contractors that such issues are focused on.

Awareness is important, but practical guidance about what the principles imply in practice is crucial. A quick search on the web for the leading Norwegian risk analysis contractors shows that only one company makes any public announcements about its ethical values. This company displays the following information (translated from Norwegian):

- Integrity

  - We shall deliver what we promise
  - We shall follow acts and regulations
  - We shall be honest and open with our customers, and we shall maintain an honest and open dialogue internally in the company.

It is commendable to have such principles openly communicated to the public and customers. More companies should have the same. However, of equal importance is that it is focused upon and debated in the companies how such principles are interpreted in practice, especially in relation to the dilemmas facing employees in their daily work.

3.14 Misuse of Risk Analysis

When the use of risk analysis started in the Norwegian oil and gas sector in 1981, there were several cases that NPD had to censure. After some 25 years, the PSA had to reiterate their warnings about misuse of risk analysis in November 2007 [8]. PSA issued a letter to all oil companies, engineering contractors and suppliers of risk analysis services, with the following title:

- About unacceptable use of risk calculations in order to deviate from requirements in the regulations for health, environment and safety.

In this letter, PSA summarises the purpose of carrying out risk analysis for installations on NCS, as follows:

- Identify what are the most important contribution to risk for personnel, environment and assets
- Identify new and improved solutions
- Analyse and evaluate the risk level associated with specific technical solutions and activities (individually as well as accumulated).

PSA further states that they during a period prior to November 2007 had seen several instances where the offshore industry has used risk calculations as part of the basis for deviating from the specific requirements in the regulations. This practice or interpretation of the regulatory framework is not consistent with the principles of health, environment and safety in the Framework Regulations Chapter III in general and the Framework Regulations Section 9 about the principles for risk reduction in particular. Since this malpractice had occurred several times, they preferred to issue guidelines to the entire industry, in order to remove misunderstandings and assist the industry in the understanding of the requirements and intentions with the regulations.

There are two specific situations that represent misuse of risk analysis that PSA have focused on:

- Unacceptable use of risk calculations as an argument for deviating from specific requirements in the regulations relating to health, environment and safety
- Unacceptable use of risk calculations as an argument for choosing solutions that give inferior safety level compared to the established minimum level.

As an example of the specific requirements in the regulations, the requirement to have fire partitions (H–0 as minimum) between main areas on an installation is mentioned. This requirement is applicable even if a risk analysis could demonstrate that no fire scenario had an annual frequency of at least 1×10^{-4}.

Adherence to accepted norms in the industry is used as an example of possibly choosing solutions that give inferior safety level.

It is further emphasised in the letter that challenging of accepted solutions is not prohibited, as long as the objective is further improvement of the safety level. It is only when the arrow is pointed in the opposite direction, deterioration of the safety level, often with the underlying ambition to save money, that it becomes misuse of risk analysis.

3.15 Risk Reduction Priorities

In the first set of unified regulations for NCS [10], the following order of priority for risk reduction was stated:

(a) Probability reducing measures, with the following order of priority:

   (aa) measures which reduce the probability for a hazardous situation to occur, (ab) measures which reduce the probability for a hazardous situation to develop into an accidental event.

(b) Consequence reducing measures, with the following order of priority:

(ba) measures relating to the design of the installation, to load-bearing structures and passive fire protection,

(bb) measures relating to safety and support systems, and active fire protection,

(bc) measures relating to contingency equipment and contingency organisation.

This prioritised list has never been repeated in later regulations, but it is still considered to be important rules; firstly that accident prevention is to be preferred over mitigation of accident consequences and secondly, passive protective systems are to be preferred over active protective systems.

If these principles are implemented during the Front End Engineering and Design (FEED) phase, this may imply an installation concept which is robust with respect to major accident risk.

However, there are challenges with such an approach. One of these challenges is related to documentation of risk reduction, and the requirements for such documentation. It may be that consequence risk reduction is much easier to document compared with probability risk reduction, as the following example may illustrate.

Some years ago a new floating production installation with flexible risers was being studied in a FEED study. The major hazard risk contribution from flexible risers is often one of the highest. The project team had two options with respect to major accident risk reduction from flexible risers:

- Change flexible riser material from multilayer Coflexip type to titan risers
- Install Subsea Safety Isolation Valve (SSIV) on the flowlines/pipelines.

Both options involved considerable investment costs. The SSIV option in addition may have significant operational costs, especially if replacement is required. Titan was therefore considered to be a better solution, provided it could be documented that there was a significant reduction in failure probability. Such documentation was not available, and not easy to obtain, realising this would be new technology.

The project decided at the end that installation of SSIVs was preferable, because of the uncertainty about the documentation of the probability reduction implied by the titan as a material for flexible risers. This illustrates how documentation requirements to some extent lead to what by many would be considered as non-optimal solutions.

3.16 Norwegian and UK Approaches Suitable as Models?

The National Commission into the Macondo accident recommended that new US regulations could be modelled based on UK and Norwegian legislation [11]. It may therefore be valuable to discuss to what extent these two approaches are suitable as models for other countries.

It may first of all be noted that the NORSOK standard Z–013, Risk and emergency preparedness analysis [3], is the most detailed standard for risk analysis, considerably more detailed than ISO 17776 [1] (See also Sects. 3.1 and 3.2).

The author has for many years argued that the key document in the UK legislation, the Safety case [12], is an improvement over the Norwegian system, which actually is lacking a document into which all risk assessment arguments may be integrated.

The Norwegian regulations have an implicit higher safety level, which in some respects call for more robust solutions and improved margins for the safety of personnel.

But the Norwegian system is not at all perfect. The two main principles of Norwegian risk regulations in many societal activities and industries—including offshore petroleum exploration and production—are risk-based regulations and internal control. The case study summarised in Sect. 3.13.1 has demonstrated clearly that risk-based regulation and internal control is a fragile combination, which is a unique Norwegian combination. There is no other country that has a similar combination. This ultimately implies that companies can set their own risk tolerance criteria without interference from authorities, and perform the risk assessment without needing to seek any approval from authorities. How bad this may go is illustrated by the case study in Sect. 3.13.1. UK authorities have published recommended risk tolerability criteria. Holland has gone even further, and published failure data to be used and approved models for physical effects [9].

Norwegian legislation has focus on risk tolerance criteria as well as ALARP—as low as reasonably practicable. But the industry has neglected the ALARP approach for a long time, as least when it comes to main principles for major accident prevention.

Another main principle in the Norwegian petroleum legislation is the requirement that companies shall focus on 'continuous improvement', in their new projects and during operations of existing installations. This implies that they should strive for reduced likelihood of and consequences of accidents. This may be functional in some respects, but certainly not on a broad basis, and not very actively with respect to major accident hazards.

This may also be illustrated through reference to risk tolerance criteria, which as mentioned above, the industry is free to define by itself. Section 17.6 documents that revision of risk tolerance criteria virtually never occurs, and certainly not the last 20 years. When ALARP is not used extensively, this implies that the risk for major accidents has virtually been unchanged during the past 15–20 years. Section 17.6 suggests how improved risk tolerance criteria could be developed.

Finally, the supervisory approach adopted by PSA has been under debate for some time. It started with recommendations from an expert group in 2013 [13], which recommended that PSA adjusted their supervisory practices as a precautionary step, in view of the recent changes in the composition of the stake holders and actors. Several other committees and reports voiced the same recommendations in 2017 and 2018, culminating with the strong recommendation from the Auditor General to the Parliament in Norway in their report in January 2019 [14]. PSA was criticised by the Auditor General for not choosing supervisory practices which ensure that the industry complies with the regulations, based on a detailed assessment of four recent cases in the Norwegian petroleum sector (Goliat, Songa Endurance, Mongstad and Nyhamna). It was revealed through the case studies that PSA in several cases failed to discover that the industry sometimes did not comply with the injunctions given by the authorities, to an extent which could jeopardize the high safety level implied by the regulations. PSA has stated in early 2019 that they will change their supervisory practices.

There are thus various aspects of UK and Norwegian legislation that may be models for other countries, but they will need to consider carefully which aspects to copy or not.

References

1. ISO (2016) ISO 17776 petroleum and natural gas industries—Offshore production installations—Major accident hazard management during the design of new installations
2. ISO (2018) ISO/IEC 31000 Risk management—Guidelines
3. Standard Norway (2010) Risk and emergency preparedness analysis, NORSOK Standard Z-013
4. ISO (2009) ISO/IEC 31000 Risk management—Principles and guidelines
5. Kierans L, Vinnem JE, Bayly D, Decosemaeker P, Stemland E, Eriksen T (2004) Risk assessment of platform decommissioning and removal. In: Presented at SPE international conference on health, safety, and environment in oil and gas exploration and production, 29–31 March 2004, Calgary, Alberta, Canada
6. Vinnem JE (2010) Risk analysis in the planning processes of hazardous facilities—a case of an LNG plant in an urban area. Reliab Eng Syst Saf 95(2010):662–670
7. HSE (2008) Guidance on 'as low as reasonably practicable' (ALARP) decisions in control of major accident hazards, 10 June 2008. www.hse.gov.uk/comah/circular/perm12.htm
8. PSA (2007) About unacceptable use of risk calculations in order to deviate from requirements in the HES legislation (in Norwegian only), letter 26.11.2007 to oil companies and contractors
9. Uijt de Haag P, Gooijer L, Frijns PJMG (2008) Quantitative risk calculation for land use decisions: the validity and the need for unification. In: Kao, Zio, Ho (eds) Proceedings of PSAM9, international conference on probabilistic safety assessment and management, 18–23 May 2008, Hong Kong, China
10. NPD (1992) Regulations relating to explosion and fire protection of installations in the petroleum activities. Norwegian Petroleum Directorate, Stavanger, February 1992
11. Commission (2011) National commission on the BP deepwater horizon oil spill and offshore drilling. Report to the President, January 2011
12. HSE (2015) The offshore installations (offshore safety directive) (safety case etc.) regulations 2015, 2015 No. 398, Health and Safety Executive. HMSO, London

13. Supervisory practice and HES regulations in the Norwegian petroleum sector (in Norwegian only), report by an expert group to the Ministry of Labour, submitted 27.8.2013 according to ta mandate dated 31.10.2012 https://www.regjeringen.no/globalassets/upload/ad/temadokumenter/arbeidsmiljo_og_sikkerhet/utvalgsrapport_hms-regelverk_endelig_2008_2013.pdf
14. The Auditor General's assessment of the Petroleum Safety Authority's practice in the supervision of health, environment and safety (in Norwegian only), Document 3:6 (2018–2019), Auditor general, 15.1.2019

Chapter 4
Lessons from Major Accidents

4.1 Overview

Experience from major accidents in the past is an important source of information to prevent the occurrence of similar accidents in the future. There have been a number of major accidents in worldwide operations in the second half of the twentieth century. There was a positive trend in the period 1980–2000, with fewer and fewer major accidents. The trend has changed after year 2000, where the number of major accidents starts to increase.

Some of the catastrophes that had a strong effect on the offshore industry occurred several decades ago. This has the drawback that the important experience gained from these occurrences may be forgotten and not be brought forward to future generations. We have therefore seen it as important to document the experience from major accidents in the past, in a concise but still comprehensive manner. The information should be useful for modelling risks to offshore installations, but also as background information to explain why some of the requirements were implemented and why they may be important.

The main focus area is on hostile environments in North-west Europe, North Atlantic, South China Sea as well as deep water operations elsewhere. But also some of the major accidents in other areas have been included, where they are considered to give important lessons or messages for the future.

Also some accidents that had major accident potential and near-misses which could have developed into major accidents are covered, if they have important messages for the prevention of similar occurrences in the future. These are all from the North Sea, mainly from the Norwegian Continental Shelf.

For each accident we will describe the main sequence of events, as well as focus on barriers, those that failed as well as those that functioned. The lessons learned are, in addition, spelled out explicitly. For some of these cases, the information is available in great depth, especially if an official commission was appointed, or if a public inquiry took place. In other circumstances, comprehensive investigations did

© Springer-Verlag London Ltd., part of Springer Nature 2020
J.-E. Vinnem and W. Røed, *Offshore Risk Assessment Vol. 1*,
Springer Series in Reliability Engineering,
https://doi.org/10.1007/978-1-4471-7444-8_4

not take place, and the available information is more limited. The Macondo blowout is discussed in a separate Chap. 5 (Table 4.1).

The events are grouped according to the type of event that initiated the sequence of events, and secondly in a chronological sequence. The following are the groups of hazards used:

- Blowouts
- Hydrocarbon leaks on installations, leading to fire and/or explosion
- Hydrocarbon leaks from pipelines/risers, leading to fire/explosion
- Marine and structural failures, possibly leading to total loss
- Other accidents.

The terminology relating to barriers is that which was introduced in Sect. 2.5.2. The following classification is used:

Table 4.1 Overview of major accidents and near-misses discussed

Hazard area	Blowout	Process leak	Pipeline/ riser leak	Marine accident	Other accidents
UK	Ocean Oddesey, 1989	Brent A, 1988 Piper A, 1988		Gryphon, 2011	
Norway	Ekofisk B 1977 West Vanguard 1985 2-4-14, 1989 Snorre A, 2004 Gullfaks C, 2010		Ekofisk A, 1975 Jotun, 2004	Deep Sea Driller, 1976 Alexander L. Kielland, 1980 West Gamma, 1990 Norne Shuttle Tanker, 2000 Ocean Vanguard, 2004	
Canada				Ocean Ranger, 1982	
Brazil	Enchova, 1984 Frade, 2011	Cidade de São Mateus, 2015		P–36, 2001 P–34, 2004	
South China Sea	Seacrest, 1989				Penglai 19-3 Oil Spills
Americas/ US	Ixtoc, 1979 Macondo, 2010	Abkatun, 2015 & 2016			Exxon Valdez oil spill, 1989
Other areas	Temsah, 2004 Montara, 2009 Usumacinta, 2007		Mumbai High North, 2005	Bohai II, 1979 Glomar Java Sea, 1983	

Loss of containment, hydrocarbons

HL1. Barrier function designed to maintain the integrity of the process system
HL2. Barrier function designed to prevent ignition
HL3. Barrier function designed to reduce cloud and spill size
HL4. Barrier function designed to prevent escalation
HL5. Barrier function designed to prevent fatalities.

Loss of structural capability

STR1. Barrier function designed to maintain structural integrity and marine control
STR2. Barrier function designed to prevent escalation of initiating failure
STR3. Barrier function designed to prevent total loss
STR4. Barrier function designed to prevent fatalities.

These barrier functions are addressed in the following discussion of incidents and accidents in general. The discussions are not focused on each barrier function specifically, but the discussions may be related to these barrier functions in general.

There are a number of accidents, some of them very severe, that are not included in this section. This is because the relevant national authorities do not publish investigation reports. We also have contacted operators to seek information, but usually without result. Those events that we would have like to include are the following:

- Capsizing and sinking of Kolskaya jack-up during tow, (Russia, 2011)
- Several fatal accidents in Azerbaijan petroleum operations, 2015 & 2016.

4.2 Ekofisk B Blowout

The description of this accident is based on a report on evacuation means [1].

4.2.1 Event Sequence

A blowout occurred on 23rd April 1977, on the steel jacket wellhead platform Ekofisk Bravo, during a workover on a production well. The BOP was not in place on the platform, and could not be reassembled correctly in time. All personnel were evacuated without injuries by means of evacuation capsules and one dinghy, and were subsequently rescued by a supply vessel. The well was mechanically capped 7 days after the blowout. The capping was performed by well control specialists from the USA. The oil spill was approximately 20,000 m^3, although no oil ever reached shore. Production on the platform was stopped for six weeks to allow clean-up operations. There was virtually no material damage to the platform.

The Ekofisk Bravo blowout is the only blowout in the Norwegian sector where a substantial amount of oil was spilled into the sea.

4.2.2 Barrier Performance

The barriers that failed were well control barriers, which are outside the scope of this discussion. It could be noted, though, that the barriers in question were operational barriers. No other barriers except well control barriers failed.

The following barrier functions performed as required in order to prevent the blowout from developing into a catastrophe with extensive consequences for personnel:

- Ignition prevention
- Fatality prevention.

It is noteworthy that ignition was prevented in this case. The statistics at the time when this blowout occurred indicated that one in three blowouts in average would ignite. It would not have been surprising if an ignition had occurred, either due to ongoing operations or due to equipment in the drilling area.

The Ekofisk B blowout is one of very few major accidents on the Norwegian Continental Shelf which did not involve fatalities. The following barrier systems performed successfully:

- Conventional evacuation means were used successfully without any injuries in excellent weather conditions.
- Successful rescue operations (from evacuation means) were conducted by a supply vessel in good weather conditions.

The following is a summary of the barrier performance:

HL1	Barrier function designed to maintain the integrity of the well system	Failure
HL2	Barrier function designed to prevent ignition	Success
HL3	Barrier function designed to reduce cloud and spill size	Not applicable
HL4	Barrier function designed to prevent escalation	Success
HL5	Barrier function designed to prevent fatalities	Success

4.2.3 Lessons Learned for Design

There are no lessons from this accident which are related to design.

4.2.4   Lessons Learned for Operation

The Ekofisk B blowout did not involve fatalities, and the material damage was limited to spills that had to be cleaned up. It is therefore important to extract the lessons which may be learned from the accident (see above for the evacuation):

- Mechanical capping of a blowout on a platform is possible, although it takes time.

4.3   IXTOC Blowout

This section is based on information posted on the web by the Hazardous Materials Response Division, Office of Response and Restoration, National Ocean Service, National Oceanic and Atmospheric Administration.

4.3.1   Event Sequence

On 3rd June 1979, the IXTOC I exploration well blew out in the Bahia de Campeche, some 850 km south of Texas in the Gulf of Mexico. The IXTOC I was being drilled by the Sedco 135, a semi-submersible drilling rig on lease to Petroleos Mexicanos (PEMEX). A loss of drilling mud circulation caused the blowout to occur. The oil and gas blowing out of the well ignited, causing the platform to catch fire. The fire caused the derrick to collapse into the wellhead area hindering any immediate attempts to control the blowout.

PEMEX hired blowout control experts and other spill control experts including Red Adair, Martech International of Houston, and a diving company. The Martech response included 50 personnel on site, the remotely operated vehicle TREC, and the submersible Pioneer I. The TREC attempted to find a safe approach to the Blowout Preventer (BOP). The approach was complicated by poor visibility and debris on the seafloor including derrick wreckage and 3,000 m of drill pipe. Divers were eventually able to reach and activate the BOP, but the pressure of the oil and gas caused the valves to begin rupturing. The BOP was reopened to prevent destroying it.

Two relief wells were drilled to relieve pressure from the well and to allow response personnel to cap it. Norwegian experts were contracted to bring in skimming equipment and containment booms, and to begin cleanup of the spilled oil. The IXTOC I well continued to spill oil at a rate of 10,000–30,000 barrels per day until it was finally capped on 23rd March 1980. The total volume spilled has been estimated at 3,522,400 barrels.

Prevailing northerly currents in the western Gulf of Mexico carried spilled oil towards the USA. A 100 km by 110 km patch of sheen containing a 90 m by 150 m patch of heavy crude moved towards the Texas coast. On 6th August 1979, tarballs from the spill impacted a 17 mile stretch of Texas beach. As of 1st September, all of the south Texas coast had been impacted by oil. A storm lasting from 13–15 September removed the majority of the oil.

In the initial stages of the spill, an estimated 30,000 barrels of oil per day were flowing from the well. In July 1979 the pumping of mud into the well reduced the flow to 20,000 barrels per day, and early in August the pumping of nearly 100,000 steel, iron, and lead balls into the well reduced the flow to 10,000 barrels per day. Mexican authorities also drilled two relief wells into the main well to lower the pressure of the blowout.

PEMEX claimed that half of the released oil burned when it reached the surface, a third of it evaporated, and the rest was contained or dispersed.

PEMEX contracted an aviation contractor to spray chemical dispersant on the oil. Almost 500 aerial missions were flown, treating 1,100 square miles of oil slick. Dispersants were not used in the U.S. area of the spill because of the dispersant's inability to treat weathered oil. Eventually the OSC requested that Mexico stop using dispersants north of 25° N.

In Texas, the emphasis was on coastal countermeasures protecting the bays and lagoons formed by the Barrier Islands. Impact of oil on the Barrier Island beaches was ranked as second in importance to protecting the inlets to the bays and lagoons. This was done with the placement of skimmers and booms. Efforts were concentrated on some selected locations. Economically and environmentally sensitive barrier island beaches were cleaned daily. Labourers used rakes and shovels to clean beaches rather than heavier equipment which removed too much sand.

Ultimately, 71,500 barrels of oil impacted 162 miles of US beaches, and over 10,000 cubic yards of oiled material were removed.

On 8th August 1979, the United States Fish and Wildlife Service (USFWS) began training volunteers to handle oiled birds and implemented beach patrols on South Padre Island. Bird cleaning stations were set up by the USFWS on Mustang and South Padre Islands. An overall decrease in bird population densities due to movement from their regular habitats along the oiled shoreline may account for the fact that only a few dead, oiled birds were ever found. After the beaches were cleaned, population densities increased, but not to expected levels. Contamination of food supplies caused many birds to leave their habitats for the duration of the spill. A total of 1421 birds were recovered with oiled feathers or feet.

The US government had two months to prepare for the expected impact of the IXTOC I oil on the Texas shoreline. During this time the government realised the importance of coastline mapping with regard to oil sensitivity. This led to a mapping project which resulted in the first Environmental Sensitivity Index (ESI). Placement of containment booms and other response equipment was done after study of the environmental sensitivity as reported in the ESI.

The IXTOC I well blowout was an unusual situation with regard to responsibility for, coordination of, and control and cleanup of the spilled oil. The U.S.

government publicly requested compensation from Mexico for damages associated with the spill without first entering into negotiations with the Mexican Government. Mexico denied being financially responsible for damages incurred, and refused to help pay cleanup expenses to the USA.

Officials reported that tourism along the Texas beaches dropped by 60% during the course of the spill. The cleanup cost was 12.5 million USD.

The semi-submersible installation was extensively damaged; substructure, derrick, drilling machinery, and box girders were completely destroyed. The installation was declared a total loss and was later sunk in deep water in the Gulf of Mexico.

4.3.2 Barrier Performance

In this case also the barrier failures related to well control measures. Loss of mud circulation is an operational failure of well control. The loss of well control occurred while tripping out of hole, which is one of the most hazardous conditions relating to well drilling.

Another barrier function that failed was ignition prevention. It is not known what the cause of ignition was.

The 63 workers on the rig were successfully evacuated. The barrier function to prevent fatalities performed successfully.

The well control operation took over nine months to complete and involved the drilling of two relief wells.

The following is a summary of the barrier performance:

HL1	Barrier function designed to maintain the integrity of the well system	Failure
HL2	Barrier function designed to prevent ignition	Failure
HL3	Barrier function designed to reduce cloud and spill size	Not applicable
HL4	Barrier function designed to prevent escalation	Failure
HL5	Barrier function designed to prevent fatalities	Success

4.3.3 Lessons Learned

Lessons learned for design and operation is unknown for this accident.

4.4 Enchova Blowout

The description of this accident is based on a report on evacuation means [1].

4.4.1 Event Sequence

The Brazilian fixed jacket production platform *Enchova 1* suffered a blowout and fire on 16th August 1984. The platform was producing 40,000 barrels per day oil and 1.5×10^6 m^3 per day gas through 10 wells. Gas released during drilling caught fire.

There were two fires. The first fire was under control when an oil leak apparently caused the blast. The fire was extinguished the following day. The platform's drilling equipment was gutted but the rest of the platform remained intact.

The crew began to evacuate, but 36 were killed when a lifeboat was lowered. One falls lashing was not properly released, and then broke causing a shock load, which broke the bow hook. The lifeboat hung vertically until the stern support broke and it then fell 10 m into the sea. Six other crew were killed when jumping 30–40 m into the sea.

207 survivors were rescued from the platform and lifeboats by helicopters. A total of 42 people were killed, all from the two incidents during evacuation.

4.4.2 Barrier Performance

The causes of the blowout are outside the scope of the present discussion. The causes of the apparent subsequent oil leak and the ignitions are unknown. The only barrier function failures that may be observed are the following:

• Prevention of loss of containment
• Prevention of fatalities (through evacuation).

The failure to prevent fatalities was through human failure associated with launching of lifeboats, which demonstrates the vulnerability of the conventional lifeboats launching mechanism.

There is no information about barrier successes known for this event. The following is a summary of the barrier performance:

HL1	Barrier function designed to maintain the integrity of well system	Failure
HL2	Barrier function designed to prevent ignition	Failure
HL3	Barrier function designed to reduce cloud and spill size	Failure
HL4	Barrier function designed to prevent escalation	Success
HL5	Barrier function designed to prevent fatalities	Failure

4.4.3   Lessons Learned for Design

This event demonstrates the criticality of the use of conventional lifeboats for evacuation purposes. There are also other accidents where hooks have been broken due to various reasons, thus exposing the users to extreme risks.

4.4.4   Lessons Learned for Operation

The causes of the failure to operate the lifeboat properly are unknown in this case, and the following are speculative causes. Lacking competence is an obvious possible explanation of the failure to operate the release mechanism correctly; panic may be another potential explanation.

4.5   West Vanguard Gas Blowout

The presentation of this accident is based on the official investigation report published [2], and the author's visit to the installation shortly after.

4.5.1   Event Sequence

The semi-submersible mobile drilling unit West Vanguard experienced a shallow gas blowout on 6th October 1985 while conducting exploration drilling in the Haltenbanken area in the Norwegian Sea.

Drilling of a 12¼ inch pilot hole had commenced earlier the same day, 6th October 1985, at 13:00 h, with the marine riser connected, but no BOP installed, according to normal practice. Just before 21:00 h the bit entered a thin gas layer 236 m below the sea bed. There was an influx of gas into the well bore, followed by a second gas influx about one hour later. The third influx caused a gas blowout, which occurred at 23:00 h.

When the drilling crew realised what was happening, they started pumping heavy mud and opened the diverter valve to deviate the flow of gas away from the drill floor. Just a few minutes' erosion in the bends of the diverter caused these to leak and the gas entered the cellar deck from below. Attempts to release the marine riser wellhead coupling on the seabed were not successful, due to the perceived ignition hazard in all areas on the platform.

Ignition probably occurred in the engine room just before 23:20 h, setting off a strong explosion, subsequent fire, and further explosions. The damage to the engine

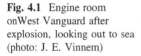
Fig. 4.1 Engine room onWest Vanguard after explosion, looking out to sea (photo: J. E. Vinnem)

room is visible in Fig. 4.1. It can be seen that the outer wall of the engine room is completely blown away due to the blast load.

Lifeboat 2 was being launched when the explosion occurred, Lifeboat 1 followed immediately thereafter. The platform manager tried to release four anchor lines just after the initial explosion although only three lines were released. The stability manager and the platform manager subsequently escaped by climbing down one of the forward columns, wearing survival suits.

One of these two managed to swim about 100 m to the standby vessel; the other had to be picked up by the FRC from the standby vessel, as he was drifting away from the vessel.

The personnel in the two lifeboats were transferred to the standby vessel after 1½ to 2 h, there being no hurry in good weather conditions. The standby vessel gave priority in the first hours to the search for one missing person, who was never found.

4.5.2 Barrier Performance

Barrier failures in this accident are several, the initiating failures are related to well control during drilling through shallow zones. This is outside the scope of the present discussion, but it may be noted that drilling through shallow zones is now usually done without a marine riser, if a BOP is not installed.

The diverter failure is also associated with well control measures in a shallow gas blowout scenario, the details are outside the scope of this discussion.

The second failure of a barrier function is the failure to prevent ignition of the gas cloud around the installation. The most likely cause of ignition was one of the diesel generators in the engine room, in spite of the flame arrestors installed on the diesel engine.

The barrier function to prevent fatalities during evacuation was successful, with one exception. One person was never found after the accident, it was suspected that the person could have been blown overboard in the initial explosion.

Barriers to prevent escalation of the initial explosions and fire were also successful. The engine room was completely burned out, but this was the only room that was completely destroyed.

The rescue operations were also successful; all personnel from two lifeboats were rescued successfully in good weather conditions, even though it was dark. The rescue of two persons wearing survival suits from the sea was reported as a success, but somewhat more critical, at least for one of the two who started to drift away from the standby vessel. The following is a summary of barrier performance:

HL1	Barrier function designed to maintain the integrity of the well system	Failure
HL2	Barrier function designed to prevent ignition	Failure
HL3	Barrier function designed to reduce cloud and spill size	Not applicable
HL4	Barrier function designed to prevent escalation	Failure
HL5	Barrier function designed to prevent fatalities	Mainly success Limited failure

4.5.3 Lessons Learned for Design

The lessons learned were in this case particularly related to well control and operations, as discussed in the following subsection.

4.5.4 Lessons Learned for Operation

An official investigation was also conducted by a commission of enquiry and an official report published, [2]. The focus was primarily placed on drilling technology and management. However, from a risk analysis point of view, there were several lessons to be learned, as below:

- Launching of conventional lifeboats was shown to be successful in good weather conditions in spite of the serious and dramatic accidental conditions.
- It was shown to be important to rescue personnel in the sea rapidly, even though they may be uninjured, wearing survival suits and focusing their efforts on survival.

- The rescue of personnel from undamaged lifeboats proved to be a standard procedure when the weather conditions were good in spite of the darkness (probably quite a lot of light available from the fire).
- The problems caused by noise from the accident had been underestimated.

4.6 Ocean Odyssey Burning Blowout

There is documentation available from the investigation of the Ocean Odyssey accident [3]. The report is focused on the events that led to the occurrence of the gas blowout and the death of one of the radio operators. Fire and explosion loads are not discussed explicitly, and some of the information has needed to be inferred rather than observed.

4.6.1 Event Sequence

The semi-submersible drilling rig Ocean Odyssey suffered a serious fire as a result of a subsea blowout on 22nd September 1988. The rig was drilling in the Fulmar area of the North Sea, some 160 km from Aberdeen, Scotland.

The rig was drilling a reservoir with abnormally high gas pressures and the well drilling programme was designed accordingly with special equipment installed. For nearly two weeks before the blowout there had been no drilling as gas levels on the rig had been consistently high and the well was just being kept under control. It was suggested in the inquiry that ineffective gas monitors had disguised the severity of the situation. At 12.00, on reaching a drilling depth of 4,900 m, the drilling took a kickback. According to the company, annular preventers were closed and heavier mud was being circulated down the drill pipe and back through the choke line. It is thought that the choke line developed a leak; gas flowed to the surface and exploded underneath the rig, possibly also damaging the hydraulic BOP control system.

The first event (reported to be an explosion) came from the mud processing module, suggesting that gas had somehow been ignited as it was dissolved out of the mud. A second event (also reported to be an explosion) occurred beneath the surface of the water, shown by a large bubble of gas, indicating the beginning of the blowout. A fire followed the blowout and swept up from the moon pool to affect the cellar deck and the mud pump room. The accommodation module also suffered severe damage. The fire burned for 10 h. Collapsed drill pipe on the drill floor showed that the temperature there had reached 500 °C, but intact drill collars showed it had not reached 700 °C.

The 67 men on board were put on alert when the kick was taken and all but 10 essential personnel were in lifeboats in anticipation of any problems while the kick was dealt with. It was reported afterwards that no one knew what to do when in the lifeboats, no orders were given and no checks carried out, and that there was total confusion about the evacuation process. In fact it was suggested in the inquiry that the captain of the rig did not recognise the screaming noise of gas being vented as a blowout.

Three lifeboats were launched, and only one person was injured while freeing the jammed lifeboat release system. The standby vessel Notts Forest picked up 38 survivors in total. Its fast rescue craft (only one was available) first recovered eight of these survivors from the sea after they were forced to jump from the burning rig. It then took in tow a lifeboat which was drifting towards the flames. These survivors were eventually transferred to MODU Sedreth 701, which was drilling nearby. The anchor handling tug, the British Fulmar, came from 2 miles away to pick up 28 men from the other two lifeboats.

A radio operator died in the incident. Having been in a lifeboat, he was ordered back to the radio room by the captain. He had not been trained to use the breathing equipment, which might have saved him from dying of smoke inhalation. A communication problem seemed to exist, and he probably did not understand or hear the final command to abandon the rig.

Stadive, a semi-submersible emergency support vessel, was brought onto the scene to help control and extinguish the fire. Anchor lines were later cut with explosive charges and the rig was towed clear of the well.

It was suggested in the inquiry that the safety supervisor had no formal safety training and no authority with which to carry out his job. The crew had not been required to have survival training certificates before working on the rig.

4.6.2 Barrier Performance

Well control barrier failures are outside the scope of the present discussion. The barrier function to prevent fatalities did not succeed completely, but would have succeeded if an untrained person had not been ordered out of the lifeboat. This appears to be a human failure.

The rescue operations were conducted without failure. Also the operations to salvage the installation were partly successful. The installation was declared a 'constructive total loss', but was later repaired and modified into a floating rocket launching facility, called 'Sea Launcher'. This is the barrier performance:

HL1	Barrier function designed to maintain the integrity of the well system	Failure
HL2	Barrier function designed to prevent ignition	Failure
HL3	Barrier function designed to reduce cloud and spill size	Not applicable
HL4	Barrier function designed to prevent escalation	Success (?)
HL5	Barrier function designed to prevent fatalities	Mainly success Limited failure

4.6.3   Lessons Learned for Design

No relevant lessons for design were concluded from this event.

4.6.4   Lessons Learned for Operation

There is no official reporting of this accident, the available documentation is from the investigation carried out by the Sheriff Principal of Grampian, Highlands and Islands [3]. This investigation is focused on legal aspects, rather than identification of lessons to be learned in a HES management context. The focus was primarily placed on drilling technology and management. However, from a risk analysis point of view there were several lessons to be learned, as below:

- Launching of three conventional lifeboats was shown to be successful in good weather conditions in spite of the serious and dramatic accidental conditions. One person received a back injury during attempts to free a jammed hook from the lifeboat.
- One lifeboat needed to be towed away, in order to prevent it from drifting back into the flames, the inferred condition being that engine power was not available.
- No fatalities occurred during evacuation in good weather conditions, the only fatality was caused due to failure to evacuate from the accommodation, due to smoke.
- The rescue of personnel from undamaged lifeboats proved to be a straightforward procedure, when the weather and light conditions were good.
- The noise from the accident was also a complicating factor.

4.7   Treasure Saga 2/4-14 Underground Blowout

The underground blowout near the Albueskjell field in 1989 may be called the 'forgotten accident'. No official investigation report is publically available, and the main source for this summary is a recent Master's thesis [4], the paper by Aadnøy and Bakøy [5] and a textbook in Modern Well Design [6].

4.7.1   Event Sequence

Saga Petroleum started to drill a wildcat exploration well, 2/4-14, with the semi-submersible mobile drilling unit Treasure Saga in October 1988, 9 km (northeast) of the Albueskjell field in the southern Norwegian North Sea, with a 68 m water depth. Drilling operations went smoothly until they reached the $8\frac{1}{2}''$ section at a depth of 4,713 m near the top of the reservoir, where the pore pressure increased rapidly from 1.65 to 2.11 s.g. The maximum gas reading was 68%.

A high pressure kick occurred at a 4,734 m depth, with a 6.5 m^3 influx, and the well was shut in. A cement plug was installed in the wellbore. During the subsequent coiled tubing operation, the plug failed, and in combination with equipment malfunction, this caused the loss of the control with the well.

The shear rams were activated at 16:40 on 20th January 1989. The wellhead pressure subsequently reached more than 700 bar. Altogether, 38 of the 75 people on board were evacuated by helicopter to the nearby Ekofisk centre, while the remaining personnel worked to stabilise the well. Another 10 people were evacuated later that evening.

An attempt to kill the well with heavy mud was performed the following day through the kill line. The flexible hose ruptured at the gooseneck, and the well blew through the kill line for one minute before the fail-safe valves closed and the well was shut in. All communication with the well was then lost, and Treasure Saga was pulled away to make room for another installation.

Treasure Saga then subsequently started drilling a relief well (2/4-15s) on 31st January, 11 days after the kick, and about 1,200 m from the 2/4-14 well location. The jack-up Neddrill Trigon was mobilised, modified and approved for killing operations at the end of April. Wellhead pressure measurements showed that the shut-in pressure had decreased to 210 bar, but no spill into the environment was observed.

The fishing inside the wellbore for the coiled tubing started on 12th May. Indications of flow inside the well were discovered, a packer was blown out when they tried to install it, and the underground blowout was confirmed through logs and shallow seismic. The flowrate was 2,900 m^3 per day (verification impossible), with ruptures of the $9\frac{5}{8}''$ and $13\frac{3}{8}''$ casings most likely. The well started to collapse during further fishing operations, and the casing ruptured at a depth of around 870 m in mid October 1989. Fishing operations were therefore stopped permanently in order not to destroy the well further, which could have resulted in a seabed blowout with significant environment spill. From this point, the successful killing of the underground blowout was fully dependent on the relief well.

The relief well drilling from Treasure Saga took a long time, much longer than planned, mainly due to three sidetracks that had to be performed. The drilling crew was ready for the final section (8–10 m distance between the two wellbores) of the well to be drilled in early December 1989. Communication between the two wellbores was immediately established. The 2/4-14 well was then declared dead and stable on 13th December 1989. The fishing and stabilising of the well then continued for another three months before the well could finally be abandoned in a safe state.

4.7.2   Barrier Performance

Since the incident was limited to an underground blowout, most of the surface barrier functions were not relevant. There were no injuries during the well operations on either of the rigs, but one person was killed on Treasure Saga in connection with the handling of drill pipes on the drill floor. The performance of the main barrier functions is summarised as follows:

HL1	Barrier function designed to maintain the integrity of the well system	Failure
HL2	Barrier function designed to prevent ignition	Not applicable
HL3	Barrier function designed to reduce cloud and spill size	Not applicable
HL4	Barrier function designed to prevent escalation	Not applicable
HL5	Barrier function designed to prevent fatalities	Not applicable

4.7.3   Lessons Learned for Well Operations

This accident does not seem to have been easily preventable. There are few distinct errors as the main causes, somewhat in contrast to several recent blowouts and well incidents. However, knowledge on these aspects may be limited due to the absence of an investigation report in the public domain.

However, there is some anecdotal information on some overlooked signs that occurred a few hours prior to the initial kick, which could have warned about instability in the well. These were not detected from the logs until long after the blowout.

The failure of the initial cement plug is an unwanted incident, but it can-not be completely prevented. The well had high pressure as well as high temperature, which at the time may have been new and unconventional. It was the first high pressure high temperature (HPHT) well drilled in the Norwegian sector. Three more HPHT wells were drilled in the same reservoir, including the relief well. These three additional HPHT wells were drilled successfully. Aadnøy [6] describes that the theories and models developed for the relief well drilling has been used for relief well drilling since 1989.

The failure of the flexible hose in the kill system was also an unwanted incident that contributed to the negative consequences. The causes of this rupture are unknown. There is some anecdotal information on operational errors that were made during the attempted control actions.

The casings were partially eroded because of hydrocarbon induced stress cracking, due to using high strength material, which is no longer used in casing designs.

4.8   Temsah Burning Blowout

10th August 2004, the Global Santa Fe operated Adriatic IV jack-up MODU was
on location over the Temsah gas production platform, operated by Petrobel, off Port
Said, Egypt in the Mediterranean Sea. The rig was drilling a gas well when a
blowout occurred during drilling operations [7]. Apparently, there was an explosion
followed by fire which was initially contained on the jack-up. The fire then spread
to the production platform where it continued to rage for over a week before being
brought under control. More than 150 workers on the jack-up and platform were
evacuated with no casualties, due in part to the prior recommendation that production activities be ceased as a precautionary measure. Both installations were
total failures. Requests for information have been submitted to ENI Norway (now
Vår Energi), but no information has been made available about the accident.

4.9   Snorre Alpha Subsea Gas Blowout

4.9.1   Event Sequence

An uncontrolled subsea gas blowout occurred on Snorre Alpha platform in the
Norwegian North Sea on 28th November 2004. The gas plume came up around the
installation and gas was detected on the lower deck of the installation, but the
blowout did not ignited. The blowout was killed by pumping heavy mud into the
well within a few hours. The accident was investigated by the operator Statoil[1]
(now Equinor) as well as PSA [8]; the latter is the main source of this section.

Snorre Alpha (SNA) is a steel hull Tension Leg Platform (TLP), in operation
from 1992, and has had several operators, starting with Saga Petroleum until 2000.
Norsk Hydro took over in 2000, and Statoil took over the operatorship from 1st
January 2003. SNA is an integrated drilling, production and accommodation platform, with 42 wells immediately below the installation and two subsea templates
with production and injection wells. The total annual production at the time was
around 200,000 bbls of oil per day.

The P–31 well was drilled as an observation well in 1994. There were several
problems during the drilling operations, which also resulted in two to three small
holes in the $9\frac{5}{8}''$ casing, a so-called scab-liner (a pipe with smaller diameter used for
repair) was installed and the maximum allowable operating pressure in the well was
reduced. The well performed satisfactory until 2001 when several problems
occurred, also at the end of 2003 there were problems and a plug was installed
above the reservoir section. The operation planned to be carried out in 2004 was to
drill a new well through the same well slot. The planning of the new well started in

---

[1]Many companies are changing names currently through mergers, etc. We have selected to use the
name at the time of the event.

the spring of 2004, and the well operations on SNA started on 16th November 2004. A new drilling contractor, Odfjell Drilling, took over the well drilling contract on the installation from 1st November 2004. They also took over 80% of the personnel from the previous drilling contractor, according to legislative requirements.

Swabbing (an unwanted piston effect in a well when pipe sections are retrieved) was observed several times during retrieval of production string parts from the well in the period up to the blowout. Mud was circulated through the well each time in accordance with normal practice, and the well was observed for any influxes (of oil or gas), which were not observed. However, there were several losses of mud to the formation observed throughout the afternoon, and the BOP annular preventer was closed once.

A reverse circulation was attempted around 18:00, but increase in mud return was observed and the BOP was closed again, which also caused a significant pressure build up in the well.

An emergency platform management meeting was called around 19:00, and they took the decision to manually shut down the production around 19:30, main power generation and water injection were not stopped. At that time the muster alarm was released and onshore operations and authorities were notified about the situation.

Gas was detected below drill floor at 19:42, based on gas leaking gradually through the BOP. Working pressure in the hydraulics was increased in order to stop this leak.

Precautionary evacuation of non-essential personnel was carried out by helicopter to nearby installations by helicopters; the manning level was thus reduced from 216 to 75 people, the remaining personnel being drilling crew and emergency management.

Several gas alarms were observed at 21:20, and personnel detected that the sea around the installation was 'cooking' with gas. This caused manual shutdown of the installation to be initiated, main power generation was stopped and emergency power generators started. Emergency evacuation (by free fall lifeboats) of remaining persons was considered, but not activated.

The main power generation was restarted around midnight, and additional 40 people were evacuated by helicopter shortly after. The well was observed throughout the night, and preparations for the final well killing operation were made. The final bullheading of mud down through the drill string started at 09:15 on 29th November, at 10:22 zero pressure reading was recorded in the drill string as well as in the annular space outside. At that time the only remaining mud onboard was less than 10 m^3, implying that if this attempt had been unsuccessful, full evacuation was the only option left.

It was realised that gas had leaked through the formation, which was confirmed later by several craters that were found on the seabed under the platform. The most critical hazard in this context is associated with the anchoring structure for the tethers in each corner of the platform. If the foundation below one of the four anchoring structures had failed, the anchor's holding capacity might have been suddenly reduced to zero, thus reducing the tension in the tethers in one corner to zero. The dynamics of the situation could have caused capsizing of the installation, which in the worst case scenario could have collapsed and sunk on top of the 42 wellheads immediately below. The likelihood of additional blowouts from wells with failing safety valves would have been very high. Any such blowout would have required drilling of relief wells, as the access to the wellhead would have been completely prevented by the sunken wreck of the installation. The stopping of the blowouts would have been very time consuming, possibly over several years, depending on the number of blowouts to be stopped.

There was also an explosion and fire hazard on the installation when gas was detected in several places on the lower deck. This detection lasted for less than 30 min. There was additional fear of gas on the installation around midnight, when the wind changed direction and was briefly reduced to zero wind speed.

4.9.2 Barrier Performance

The PSA investigation found 28 deviations from the regulations that had contributed to the subsea blowout, and also a number of improvement suggestions. The deviations found indirectly cover a number of barrier functions that failed. The full list of deviations is as follows:

1. The method chosen for internal audit did not reveal the failure to comply with steering documentation.
2. Milestones in the well planning not according to steering documentation.
3. Planning of well operation with inadequate well barriers when puncturing the $2\frac{7}{8}''$ tail pipe.
4. Consequences of changes in the planning not sufficiently analysed.
5. Inadequate experience transfer in relation to well integrity.
6. Planning with insufficient well barriers during cutting of scab-liner.
7. Risk not evaluated in relation to pulling of scab-liner
8. Planned pulling of scab-liner through BOP with inadequate well barriers.
9. Inadequate management involvement in relation to giving priority to independent verification of well planning.
10. Deficient approval procedures.
11. Signatures for control, verification and approval not in accordance with steering documentation.
12. Meeting to evaluate overall risk during planning is cancelled.
13. Deficient experience transfer from previous incidents on SNA.
14. Performance of the puncturing of the tail pipe with substandard secondary barriers.
15. Execution of puncturing operation not stopped with inadequate barriers.
16. Deficient handling of deviations from procedures.
17. Unclear procedures for drilling and completion.

18. Inadequate approval of HAZOPs, also reflecting the fact that they were conducted after the approval of the intervention program.
19. HAZOPs conducted onshore not communicated to the drilling crew.
20. Technical competence not involved in overall risk evaluation.
21. Focus on risk removed from the intervention program.
22. Scabliner punctured, cut and retrieved without testing of secondary barrier.
23. Deficient preparation for possible well control situations.
24. Inadequate well barriers when pulling scab-liner through BOP.
25. Inadequate risk assessment in relation to possible swabbing.
26. Kelly kock valve blocked in open position when the well control situation developed.
27. Delayed personnel status when muster alarm sounded.
28. Deficient logging of actions during the emergency management.

Several of these deviations imply inadequate planning, change management and execution of well operations. It has been focused in the aftermath that the operations management did not follow procedures which contributed to the occurrence of the blowout. However, the management's failure to follow procedures also implied that they kept a significant crew on the installation for more than 12 h in order to kill the blowout, when compliance the procedures would have implied that they should all have left SNA on 28th November, which would have resulted in a blowout with unknown duration and consequences. The platform manager was severely criticised internally for the failure to evacuate all personnel as soon as possible, was not allowed to continue as platform manager, and resigned from his employment in the company within short time.

The barrier functions that had been successfully completed included the emergency management, the mustering of personnel and the notification and normalisation of the event. The following is a summary of the barrier performance:

HL1	Barrier function designed to maintain integrity of the well system	Failure
HL2	Barrier function designed to prevent ignition	Success
HL3	Barrier function designed to reduce cloud and spill size	Success
HL4	Barrier function designed to prevent escalation	Success
HL5	Barrier function designed to prevent fatalities	Success

4.9.3 Lessons Learned for Well Operations

The summary of deviations above from the PSA investigation documents a number of lessons for planning and execution of drilling and well intervention operations, and is not repeated here.

Statoil (now Equinor) commissioned a research institution (Studio Apertura, NTNU) to carry out an in-depth analysis of the root causes for the SNA subsea gas blowout [9]. The following is a summary of some of the main conclusions from that study.

The operations on SNA may be characterized as an event-driven way of working, marked by 'fire fighting' and the inability to work systematically and long term. This attitude has had negative consequences for risk awareness and training in putting things into context. An aging technical condition has put a high pressure on the management and did make it difficult to find time to perform all the functions they were required to perform. The activity level was also very high in 2004, and had been like that for a long time.

The SNA organisation had changed from Saga Petroleum to Norsk Hydro in 2000, and subsequently to Statoil in 2003 as a consequence, the management was given increased responsibilities, and organisational changes and new procedures consumed unproportionally large parts of their time. The change of drilling contractor added even more to this overload.

An overall impression of repeated deviations from steering documentation reflects the fact that many employees admitted that their knowledge of steering documentation was inadequate. They complained that it was very extensive, complex, unsystematic and difficult to navigate in. Far from all had received adequate and required training and it was impossible to find the time to get familiar with steering documentation.

The platform organisation was not sufficiently well integrated into the Statoil (now Equinor) organisation, and the knowledge of competence groups onshore was low amongst offshore personnel.

The climate for critical objections and afterthought has not been good, management should be trusted, and the overriding goal was to keep production going in an uninterrupted manner through ability to improvise, with basis in very good insight locally. Risk assessment has not been a regular activity and risk assessment competence was low. Risk awareness was also low.

The management and supervision was characterized by frequent interruptions and changes as a result of equipment failures and operational problems. The planning of well operations has been mainly carried out by the onshore organisation, with little involvement by offshore personnel. Plans and programs have often come very late, mainly due to lack of relevant resources.

4.10   Usumacinta Blowout

4.10.1   Event Sequence

The mobile drilling unit Usumacinta was preparing for final drilling and the completion of one well on the small three-well production platform KAB-101, operated by Pemex, located 18 km offshore, north of Frontera, north-east in Mexico, in the eastern part of the Gulf of Mexico. On 23rd October 2007 while the

Usumacinta rig was transitioning from a rig-move state to a production state, the rig shifted in its position close to the fixed wellhead platform, and settled in a way that led to the breaking off of the production tree on one of the wells, Well-121.

A limited initial leak occurred, but the well was sealed for about one hour by the subsurface safety valve (SSSV), until the safety valve started to leak, which required the evacuation of Usumacinta. During evacuation and rescue operations 22 people died.

Pemex commissioned Battelle (Columbus, Ohio) to carry out a root cause analysis of the events leading to the multiple fatality outcome [10], the report from which is an excellent source (878 pages long) of learning from this accident.

All personnel onboard arrived safely on the water in one Whittaker capsule and one Watercraft lifeboat. After landing on the water in good condition, the survival crafts became unsafe through the personnel's opening of hatches, and to some extent their abandonment of the survival crafts. One of the survival crafts was stable for 1.5 h before hatches were opened, personnel were unbuckled from seat belts and they started to move about, and the craft capsized. The fatalities occurred to those who abandoned the survival crafts as well as those that remained in the crafts or reboarded them. The apparent cause of death was drowning.

The blowout has often been reported as an ignited blowout. This is true if the entire duration (50 days) of the blowout is considered, but it is not true for the phase of the accident when evacuation was decided and performed. The blowout was ignited by a vessel involved in the combatment operations, the first fire was on 13th November 2007, and the second fire on 20th November. The second fire lasted until early December, 2007 before it was extinguished.

4.10.2 Barrier Performance

When barriers are considered, two sets of causes and concerns need to be addressed, relating to the leak from the well and to the fatalities during the evacuation and rescue phases.

The barrier failures that contributed to the initial and continued leak from Well 121 are the following:

- Site planning, preparation and management aspects contributed to the settling of the jack-up rig due to weather impacts.
- The root cause analysis report is inconclusive about the failure of well barriers, but observes that the flow from the well provided only a small fraction of the production capacity, implying that the failure was not total.
- Unspecified problems relating to communication between the formation and the annular space are also reported, but it is unclear if this was a contributing factor.

When it comes to the barrier failures that contributed to the loss of life during evacuation and rescue operations, the following are mentioned in the report:

- Both survival crafts apparently capsized due to the inadvertent opening of hatches and/or unrestricted personnel movement inside them.
- A lack of training apparently caused unsafe behaviour by personnel inside the crafts, involving opening hatches and abandoning the survival crafts altogether, at least temporarily.
- Management and organisational errors relating to planning emergency responses and managing emergency responses overall and locally inside the survival crafts caused personnel to behave in an unsafe manner during the emergency response phase.

The following is a summary of the barrier performance:

HL1	Barrier function designed to maintain integrity of the well system	Failure
HL2	Barrier function designed to prevent ignition	Success
HL3	Barrier function designed to reduce cloud and spill size	Failure
HL4	Barrier function designed to prevent escalation	Failure
HL5	Barrier function designed to prevent fatalities	Success

4.10.3   Lessons Learned for Design

The main lesson learned for design relates to the improvement of the SSSV, referred to as the 'storm valve'. The reliability and availability of such valves is important. The corresponding valve in the UK and Norwegian sectors is usually called the surface controlled subsurface isolation valve. It is more than 20 years since these valves were in focus in order to improve their reliability and availability significantly.

Further, it seems that no emergency power was available due to hydrocarbons in the atmosphere, as it would be in the northwest European waters.

4.10.4   Lessons Learned for Operation

The root cause analysis performed by Batelle [10] identified a number of root causes addressing operational issues, and made a number of recommendations relating to these issues.

A number of lessons address those learned from the Alexander Kielland accident (see Sect. 4.25). The following summarises the recommendations from the root cause analysis:

- Site surveys need to be regularly updated, if seafloor operations are performed or seabed is unstable.

- Risk assessment should be carried out for the placement of jack-up rigs if changes have severe effects.
- Emergency response plans need to address the specific risk to the rig and the platform when relevant, such as when a loss of well control occurs in severe weather conditions.
- Living quarters should be well protected in order to be able to serve as a 'safe area' during an emergency, as in the North Sea.
- It seems that no survival training was given to offshore personnel in order to give all platform and contractor personnel training in survival practices.

4.11   Montara Blowout

4.11.1   Event Sequence

The Montara blowout of the H1 well occurred on 21st August, 2009 on the Montara wellhead platform, 250 km off the northwest coast of Australia. The operator was PTT Exploration and Production Public Company Limited (PTTEP). Drilling operations were carried out from the jack-up platform West Atlas, operated by Seadrill AS. The blowout was not stopped until 3rd November 2009 by drilling one relief well. A comprehensive investigation report was issued, which is the main source of the present summary [11].

Once the blowout had occurred, and the crew realised that they did not have time skid the derrick (with BOP) back over the rig, all personnel were promptly evacuated, thus without injury to anybody. This was the proper action, considering the risk of ignition. In fact, the blowout did not ignite until just a few days prior to stopping, on 1st November. The source of ignition is unknown, but it happened during a period when the relief well was circulated with seawater, because no drilling mud was available.

The drilling of the relief well started on 14th September, having to be made from a position over 2 km from the Montara platform due to the gas cloud explosion hazard. Drilling took 22 days, and was completed on 6th October, but the actual killing operations took about four weeks, because five killing attempts were needed before successfully killing the well. Owing to restrictions on the location of the jack-up rig drilling the relief well, the angles available for the interception of wellbores were very unfavourable, and this caused the need to make repetitive attempts. The calculation of the amount spilled is uncertain, but the mass of oil spilled into the sea is given as 4,000–30,000 tons.

4.11.2 Barrier Performance

The main barrier performance issues were well related, as no injuries occurred as a result of the blowout. The performance of the barriers was, however, far from satisfactory. The following are the main barrier failures:

- The $9\frac{3}{8}''$ cemented casing shoe had not been pressure-tested in accordance with the company's procedures despite having had many problems with the shoe cementing. However, personnel did not realise what the problems were, despite having been shown in the daily drilling report. The Inquiry states that the problems were not complicated or unsolvable, and the potential remedies were well known and not costly.
- Two secondary barriers were programmed for installation, but only one was ever installed. The $9\frac{3}{8}''$ pressure-containing anti-corrosion cap (PCCC) was not tested, and in fact not even in accordance with good oilfield practice, as these caps are never intended to be a proper barrier.
- Key PTTEP personnel onboard were under the impression that the fluid inside the casing was overbalanced to pore pressure and would therefore act as an additional barrier (even though the fluid was not monitored and overbalanced significantly to pore pressure in accordance with good practice in order to be regarded as a proper barrier).
- These failures persisted over several months in 2009, while drilling Well H1 was suspended from April until August
- When West Atlas returned in August 2009, it was discovered that the $13\frac{3}{8}''$ PCCC had never been installed, which had resulted in the corrosion of the threads of the $13\frac{3}{8}''$ casing and this, in turn, led to the removal of the $9\frac{5}{8}''$ PCCC in order to clean the threads. This was viewed by key personnel as a mere change of sequence that simply involved bringing forward the time of the removal of the $9\frac{5}{8}''$ PCCC, and a positive decision was made not to reinstall the $9\frac{5}{8}''$ PCCC. This meant that the H1 Well would have been exposed to the air without any secondary well control barrier in place for some 4–5 days, with sole reliance on an untested primary barrier (the cemented $9\frac{5}{8}''$ casing shoe) that had been the subject of significant problems during its installation.
- After the $9\frac{5}{8}''$ PCCC had been removed, the H1 Well was left in an unprotected state (and relying on an untested primary barrier) while the rig proceeded to complete other planned activities as part of batch drilling operations at the Montara WHP. The blowout in the H1 Well occurred 15 h later.

The following is a summary of the barrier performance:

HL1	Barrier function designed to maintain the integrity of the well system	Failure (multiple)
HL2	Barrier function designed to prevent ignition	Success

(continued)

(continued)

HL3	Barrier function designed to reduce cloud and spill size	Not relevant
HL4	Barrier function designed to prevent escalation	Success
HL5	Barrier function designed to prevent fatalities	Success

4.11.3 Lessons Learned for Well Drilling

There are clear similarities between the Montara accident and the Macondo accidents (see Chap. 5). The Petroleum Safety Authority in Norway has also issued a report [12] on the follow-up of recent accidents, where Macondo and Montara are covered together. Failing to utilise common well barriers according to established practice is a common element. Some of the lessons identified by the Inquiry Commission are:

- Management plan for H1 well drilling was in adequate.
- Senior personnel had insufficient experience of batch drilling.
- Steering documentation was partly ambiguous, and this was manifested in different personnel making different interpretations of the steering documentation.
- Inadequate understanding of the procedures by the personnel on the rig.
- Management personnel offshore and onshore failed to make important decisions based on the sequence of events.
- Defective recording and documenting of the interactions between onshore and the rig.
- Defective communication between PTTEP and the rig operator on the West Atlas rig.
- The investigation by the Commission revealed that the culture in PTTEP was to cut corners and 'get the job done', without proper hazard identification and risk assessment.
- The interactions between the company and regulator had also become too comfortable.

The Commission states that "the blowout was not a reflection of one unfortunate incident, or of bad luck. What happened with the H1 Well was an accident waiting to happen; the company's systems and processes were so deficient and its key personnel so lacking in basic competence, that the blowout can properly be said to have been an event waiting to occur. Indeed, during the course of its public hearing, the Inquiry discovered that not one of the five Montara wells currently complies with the company's well construction standards."

4.12   Gullfaks C Well Incident

The severe well kick on Gullfaks C on 19th May 2010 occurred less than one month after the Macondo blowout, and received a lot of attention due to this, but not the least because it was seen to demonstrate that the operator had not learned the necessary lessons after the Snorre Alpha subsea blowout in 2004 (see Sect. 4.9). Statoil (now Equinor) investigated the incident [13], and was required to commission a study of the failure to learn from previous events [14], both of these report form the basis for the following summary.

4.12.1   Event Sequence

Well 34/10-C-06 on Gullfaks C was drilled in managed pressure drilling mode to a total depth of 4,800 m. During the final circulation and reservoir section hole cleaning on 19th May 2010, a hole occurred in the $13\frac{3}{8}''$ casing, with subsequent loss of drilling mud to the formation. The casing was a common well barrier element, and the hole in the casing implied loss of both well barriers. Loss of back pressure lead to influx from the exposed reservoirs into the well, until solids or cuttings packed off the well by the $9\frac{5}{8}''$ liner shoe. The pack-off limited further influx of hydrocarbons into the well.

Both the crew on the platform and the onshore organisation were struggling to understand and handle the complex situation during the first twenty-four hours. The well control operation continued for almost two months before the well barriers were reinstated.

The incident caused a gas release on the platform; the production on the platform was shut down for almost two months.

4.12.2   Barrier Performance

The hole in the $13\frac{3}{8}''$ casing and thereby the loss of a common well barrier element was caused by insufficient technical integrity of the casing. Another cause that allowed a leakage in the $13\frac{3}{8}''$ casing to develop into a hole, was lack of monitoring and follow-up of the pressure in the annulus outside the $13\frac{3}{8}''$ casing. This pressure had increased during the weeks before the incident, but the increase was not noticed.

A contributing cause to the difficulties in handling the subsequent well control situation was that the managed pressure drilling operation was carried out with insufficient margin between the pore and fracture pressures.

The risk assessment carried out in the planning phase was insufficient, and resulted erroneously in accepting the use of $13\frac{3}{8}''$ casing with insufficient technical integrity, and the lacking follow-up and monitoring of pressure in the annulus outside the casing.

There was further inadequate evaluation of risk during the execution of the managed pressure drilling operations, as well as insufficient transfer of experience relating to pressure control during managed pressure drilling from similar operations in Well C-01 in 2009.

The following is a summary of the barrier performance:

HL1	Barrier function designed to maintain integrity of the well system	Failure
HL2	Barrier function designed to prevent ignition	Success
HL3	Barrier function designed to reduce cloud and spill size	Not relevant
HL4	Barrier function designed to prevent escalation	Success
HL5	Barrier function designed to prevent fatalities	Success

4.12.3   Lessons Learned for Well Operations

The following are the main observations from the study commissioned by Statoil of the causes of failure to learn from previous events [14] as mentioned above.

Statoil (now Equinor) and the petroleum division of Norsk Hydro were merged from 2007, with two very different documentation systems and management cultures. The resulting documentation system was to large extent Statoil's procedures implemented into the previous administrative system from Norsk Hydro, which probably has created quite some confusion, and may have been one of the contributing causes of many silent deviations.

A complete management reshuffling in the spring of 2009 completely changed out the local management on Gullfaks C, and an experience transfer loss was created. The attitude on board is stated to have been to loyal, with insufficient critical evaluation and challenges of management. Risk evaluations were inadequate, and often unsystematic and undocumented.

The inability to learn across the organisation is partly explained by the failure to empower the formal management in the organisation, too much power has been retained in the informal leadership. This is probably an area where the two cultures prior to the merger were distinctly different. The procedures for learning from investigations are considered to be inadequate.

It was found that the top management had too high belief in the excellence of their steering documentation, and that incidents and accidents to a large extent were seen as failure of the organisation to follow the 'perfect' system, combined with too little openness in the organisation for identification of improvement potentials.

4.13   Frade Underground Blowout

The source of the text in this section is the investigation by Agencia Nacional de Petroleo, Gas Naturale e Biocombustiveis [15].

4.13.1   Event Sequence

On 7th November 2011, Chevron Brazil Upstream Frade Ltda. took a kick as it reached the upper section of reservoir N560 in a sidetrack well. The reservoir was over-pressured due to the injection of water by the operator itself of the Frade field, which had been in production since 2009. The well was drilled from the second-generation semi-submersible mobile drilling unit, Sedco 706, built in 1976 and operated by Transocean. The water depth at the site was 1,184 m.

The BOP was activated upon kick detection. There were subsequently several indications that an underground blowout had resulted, but Chevron did not recognise any of these indications. A sheen on the sea surface seen from the Roncador field which was immediately to the east of the Frade field, was one such indication.

Chevron conducted three unsuccessful attempts to kill the well on 8th November, followed by three attempts to bullhead into the live well. The circulation after the third bullhead attempt on 9th November showed traces of oil in the return, indicating that the well was still flowing.

Wild Well Control Inc. was contacted by Chevron on 10th November, and arrived in Chevron offices on 12th November. They arrived offshore on the day after, and immediately initiated a dynamic kill procedure, followed by a new log that indicated that the influx from the well had been stopped.

No one was injured during these operations. The spill was calculated to be 3,700 bbls, corresponding to around 500 tons, and was as such a relatively small spill. The potential flow of an underground blowout may however be indicated by the 2/4-14 underground blowout in the Norwegian North Sea in January 1989 (see Sect. 4.7).

4.13.2   Barrier Performance

Chevron made several errors while planning and carrying out well operations, as discussed in Sect. 4.13.3. Several of these errors are similar to those made when drilling the Macondo well in 2010. The following is a summary of the barrier performance:

HL1	Barrier function designed to maintain the integrity of the well system	Failure
HL2	Barrier function designed to prevent ignition	Success
HL3	Barrier function designed to reduce cloud and spill size	Success
HL4	Barrier function designed to prevent escalation	Success
HL5	Barrier function designed to prevent fatalities	Success

4.13.3   Lessons Learned for Well Operations

The local geology and fluid dynamics were not interpreted correctly in spite of drilling 62 wells in the block in question. The main problem seems to have been calculating the effect of the water injection. A too low mud weight was calculated in spite of the calculations performed by Chevron, which gave the correct range.

The test results from drilling other wells were disregarded when planning the new well. If correct data had been used, the well could not have been planned the way it was, and the kick would not have occurred.

The uncertainty involved in planning the well program reflected a development well, not an appraisal well as it was. Thus, several errors or flawed interpretations were made during the well planning phase. These errors and flawed interpretations made it possible for the kick to occur.

The risk analysis of well drilling in accordance with Brazilian regulations and internal procedures was not carried out properly, similar to several other recent incidents and accidents relating to well problems.

Errors or flawed interpretations were also made during the execution of the well programme. These point to a lack of the systematic management of changes during the execution of the well programme. In fact, this was a common root cause with the Macondo accident (see Chap. 5). It was detected that the last cement plug was set 175 m below the required depth, but the effects of this failure were never considered. This error opened up the possibility for the underground blowout to occur.

The setting of the $13\frac{3}{8}''$ casing shoe only 600 m below the seabed increased the risk of an underground blowout with the weakness of the formation below the casing shoe. The casing design thus lacked the expected robustness that could have prevented the underground blowout.

It also took the crew two days to realise that they had suffered an underground blowout in spite of different indications that all pointed in the same direction. It seems that the operator took a long time to take the underground blowout seriously, trying for some time to ignore the severity of the situation.

The attempts by the operator to control the well were significantly inadequate for the actual well situation. The well was not controlled until the Wild Well Control people arrived on the rig. It thus took six days to kill the blowout, which could have been stopped in less than half the time with proactive interpretations and the immediate calling in of Wild Well Control.

4.14   Endeavour Burning Blowout

On Monday 16 January at 4.30 to 5 a.m., Chevron's KS Endeavour drilling rig burst into flames, approximately 6 miles off the coast of Nigeria. Two workers were reported missing, never found. The cause is not confirmed in the public domain, but early reports indicate that the explosion was partly the result of a failed blow out preventer (BOP). The Nigerian state oil company, NNPC, speculated that Chevron's drillers lost control of gas pressure when equipment failure led to a "gas-kick".

Chevron Nigeria Limited (CNL) reported its plans and preparations to drill a relief well, and its investigation into the cause of the incident. CNL confirmed that 152 workers on the shallow water rig and associated barge were safely evacuated from the incident, which occurred about 10 km from shore. Two employees who received medical care for burns have been released from the hospital.

4.15   Uncontrolled Well Leak on Elgin Platform

Shortly after midday on Sunday 25 March 2012 a major gas release occurred on the Total E&P UK Ltd. (Total) Elgin offshore wellhead platform, from well G4. Personnel on the Elgin production platform and the adjacent Rowan Viking jack-up drilling rig were all subsequently evacuated without injury.

On Tuesday 15 May 2012, day 51 of the release, well G4 was successfully 'killed' and the gas release stopped with mud pumped via the West Phoenix drilling rig.

4.16   Brent A Explosion

The description of this incident is based on the Preventor Report [16].

4.16.1   Event Sequence

A gas leak from a ruptured flange gasket in the gas compression module occurred on the steel jacket production platform Brent Alpha on 5th July 1988. Within a few seconds an ignition occurred, causing an explosion that blew off lightweight cladding on two walls, and moved a door in a fire wall more than 30 m. Two persons were injured in the explosion, but not seriously. A relatively short duration fire followed, but this caused only limited and superficial damage. The accident was over in less than an hour.

A thorough investigation by the operating company revealed that the deluge system had been activated by the gas leak, probably by the smoke detectors. It was further concluded that the most likely ignition source was deluge water ingress into a light fitting, which was no longer explosion 'proof' thus allowing a short circuit and a spark.

The maximum overpressure was back calculated to be in the range 0.3–0.4 bar. Today's knowledge implies that the overpressure may have been reduced somewhat by the deluge, but with the relatively small size of the module, this effect would probably not have been large. The subsequent fire lasted for only about 45 min, with not very extensive damage to the gas compression module.

4.16.2 Barrier Performance

There are several barrier failures that were demonstrated by this event:

- Loss of containment occurred from a ruptured flange. It is not known why this flange ruptured.
- The barrier to prevent ignition also failed, probably through failure of Ex-rated equipment. This ignition is actually one of the few in the North Sea where equipment failure is relatively certain to have caused ignition. It is not known why the equipment failed, but it would appear to be reduced reliability of the Ex protection.

The following barriers performed successfully in the gas leak event on Brent Alpha:

- The barrier function to limit the amount of released hydrocarbons performed successfully.
- The barrier function to prevent escalation of explosion and fire performed successfully.
- The barrier function to prevent fatalities performed successfully.

The incident started in a similar way to the Piper Alpha accident (see Sect. 4.17). A comparison of barrier performance in the two events is presented in Sect. 8.1.1. The following is a summary of the barrier performance:

HL1	Barrier function designed to maintain the integrity of the process system	Failure
HL2	Barrier function designed to prevent ignition	Failure
HL3	Barrier function designed to reduce cloud and spill size	Success

(continued)

(continued)

| HL4 | Barrier function designed to prevent escalation | Success |
| HL5 | Barrier function designed to prevent fatalities | Success |

4.16.3 Lessons Learned for Design

The following lessons may be drawn for design purposes from this event:

- There is an important observation to be made about the platform layout. On Brent A lpha the gas compression module was on top of the platform, furthest away from the accommodation, with the least chance of leading to escalation to neighbour areas.
- The heavy door thrown over 30 m demonstrates the escalation potential. Imagine the same door with the same force deep down in the gas compression module of Piper Alpha, and ruptured pipes are not at all unrealistic (see description in Sect. 4.17.1).
- The damage during both the initial explosion and the subsequent fire was considerably more limited than would have been expected.

4.16.4 Lessons Learned for Operation

This incident started in a surprisingly similar way to the events on Piper Alpha the following day, but with a dramatically different result! (See also Sect. 7.1.1.) It was therefore considered to be an interesting incident to include in this overview. Some of the lessons that may be learned from this incident with respect to risk analysis are the following:

- The importance of the availability of fire water to cool the facilities and prevent escalation is probably the most important lesson.
- The deluge system was shown to have both positive and negative effects. As noted above, it has been assumed to be the cause of ignition, but may also have marginally reduced the maximum overpressure from the explosion. It was definitely useful in cooling equipment and limiting the damage.
- One of the two injured persons was actually running towards the scene of the leak to isolate it, when the ignition occurred. Experience from offshore operations indicates that this is quite common behaviour, in spite of company policy which requires personnel to go directly to the shelter area or muster stations. On the other hand, process operators are often quite keen to establish whether a gas alarm is real ('confirmed') or false, which makes the behaviour seen in this case more expected than unexpected.

4.17 Piper A Explosion and Fire

4.17.1 Event Sequence

This accident is well known and well documented through the investigation conducted by Lord Cullen [17], and just a brief summary is presented in the following.

A gas leak from a blind flange in the gas compression area occurred just before 22.00 on 6th July 1988, and was ignited within few seconds. The explosion load has been back calculated to be in the range 0.3–0.4 bar overpressure. The initial blast caused a subsequent oil fire due to the escalation. No fire water cooling was, however, available because all the fire pumps were in manual standby mode due to the ongoing diving activities. Thus further escalation could not be stopped. The first gas riser rupture occurred after some 20 min, from which the fire was very dramatically increased, being initially fed by probably a few tons of gas per second. Further riser ruptures occurred subsequently.

The personnel, most of whom appear to have survived the initial explosion, gathered in the accommodation, but were not given further instructions about escape and evacuation other than to wait. Both the onboard and the external communications were severely impaired when the radio room had to be abandoned during the initial stages of the accident. Previous emergency training of personnel had left them with the expectation that evacuation would be by helicopters, which was completely impossible in view of the fire and the smoke around the platform.

In total, 166 of the Piper Alpha crew died in the incident including one who later died in hospital from severe injuries. Most of the fatalities are believed to have occurred due to smoke inhalation inside the accommodation, and were still in the accommodation when it collapsed into the sea approximately 1 h after the initial incident. Most of the 63 survivors jumped from the platform into the sea and were rescued by other vessels, including the fast rescue craft (FRC) launched from rescue vessels. In fact two crew members in one of these FRCs lost their lives during attempts to rescue people from the sea close to the platform.

4.17.2 Barrier Performance

The extent of barrier failures was very considerable in the Piper Alpha accident; below follows a brief summary:

- The leak was caused by repeated attempts to start a compressor that had been taken out of service to undergo maintenance, the crew not knowing that the downstream piping had been isolated. This occurred as the Permit To Work (PTW) system had failed, and the crucial information had not been shared with installation management.
- The barrier function to prevent ignition also failed, it is unknown what caused the ignition of the leak.

- The barrier functions to avoid escalation due to explosion and fire failed to limit the consequences. It should be noted that no fire water whatsoever was available, due to all pumps being in manual standby mode. This was a precaution taken due to divers in the water, but the divers were not diving near to water inlets, and the pumps could have been switched back to automatic standby.
- The barrier function to prevent fatalities through escape and evacuation failed extensively.

Some of the barrier failures occurred due to them being insufficiently robust. On the other hand, when large unprotected pipes are subjected to impinging fire loads for extensive periods, no pipe can withstand this. The failures actually occurred during the design of unprotected pipes, and were repeated, at later stages, when the company management commissioned studies that pinpointed the vulnerability of the pipes, but management failed to take action to rectify the situation. It is quite likely that the complete absence of fire water severely accelerated what some media called the 'spiral to disaster'. There were few barriers that worked in the Piper Alpha accident, but some did:

- The barrier function to isolate the leak and limit the source of fuel worked initially, when the segment was isolated by ESD valves. Later however, this barrier also failed, due to escalation.

Figure 4.2 presents a simple sketch that illustrates some of the problems with the layout of the Piper Alpha platform, compared to a layout that gives much better protection of the accommodation and other safety critical equipment. The layout of the Piper Alpha platform is indicated on the left-hand side of the sketch, with the Central Control Room (CCR) close to gas compression, and accommodation also close to (and above) gas compression.

The layout to the right has a utility module as a buffer between gas compression and safety critical equipment (such as emergency power and fire water supply).

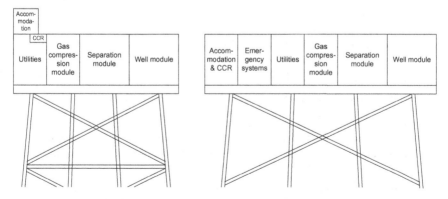

Fig. 4.2 Comparison of installation layouts, with little and extensive protection of accommodation

These systems provide a further buffer for the accommodation, which is also low on the installation.

The following is a summary of the barrier performance:

HL1	Barrier function designed to maintain the integrity of the process system	Failure
HL2	Barrier function designed to prevent ignition	Failure
HL3	Barrier function designed to reduce cloud and spill size	Failure
HL4	Barrier function designed to prevent escalation	Failure
HL5	Barrier function designed to prevent fatalities	Failure

4.17.3   Lessons Learned for Design

The official report [17] presents a very extensive list of findings and recommendations. Some of these are related to design, of which the most important findings are:

- The layout of main areas onboard Piper Alpha was a significant contributor to the accident, in the sense that the central control room, radio room, and accommodation were quite close to the gas compression area and were not protected by fire and blast barriers.
- The failure to protect against escalation of effects of explosion and fire was a severe failure, as already mentioned above.
- The location of the accommodation on top of the installation was a significant cause of smoke accumulation inside the quarters. If the accommodation had been on a low level, it would have avoided filling up with smoke so fast, and would have provided better access possibilities to the lifeboat stations, at least in the initial stages of the accident.

4.17.4   Lessons Learned for Operation

This summary, based on Lord Cullen [17], is intended to be brief, and concentrates only on the findings that are considered important in an analytical context.

- The most obvious finding is the criticality of the oil fire when water for cooling purposes was not available. The failure to keep the fire water pumps on automatic standby for as long a time as possible was a severe failure of the installation management.
- The expectation that many offshore employees still have, that emergency evacuation is realistic by means of helicopters, is a considerable problem. If we disregard the effect of smoke from the accident, and assume (!) that helicopters

could still land after the first gas riser rupture, then by the time the accommodation module collapsed into the sea, there could still have been perhaps 90–100
people waiting for a helicopter seat. This is even more serious when it is realised
that the helicopter would only have needed to ferry people to the mobile
emergency unit just a couple of hundred metres away. In most other cases the
flight time to the nearest installation might easily be 10–15 min or more.
- The Permit to Work system onboard was apparently failing severely. In fact the
  operator's HES management system received severe criticism.
- Nevertheless, quite a substantial number of people were saved by jumping into
  the sea, and being rescued from the water. This experience is similar to that from
  the Alexander Kielland accident (see Sect. 4.25.1) although in the Piper Alpha
  incident the temperature of the water was higher due to the summer season.
  Probably the temperature of the water was not a problem for those who jumped,
  as radiation from the fire was probably a more than sufficient heat source.

4.18   Cidade de São Mateus Explosion

The explosion in the Brazilian offshore sector on the FPSO unit Cidade de São
Mateus (CDSM) on 11th February 2015 caused nine fatalities amongst the 74-man
crew on board. Petrobras is the operator of the field, whereas the FPSO was
operated by the Norwegian company BW Offshore. The death toll was unusually
high and not much less than the 11 fatalities in the Macondo blowout in the US
Gulf of Mexico (GoM) in 2010 [18], but without pollution. The FPSO was after a
long period brought to a yard for repair. It was out of service for several years.

Two investigation reports have been published. Also the operator of the field,
Petrobras, has issued an internal investigation report although this is not in the
public domain. The most comprehensive investigation was performed by the
Brazilian authority National Agency of Petroleum, Natural Gas and Biofuels
(ANP), and was published in Portuguese in December 2015 and in English in early
2016 [19]. This is the main source of the descriptions in this subsection. Earlier, the
Brazilian Coast Guard [20] published a brief report dealing mainly with the role of
the emergency services.

4.18.1   Event Sequence

A condensate leak occurred in the pump room at approximately 11:30 on 11th
February 2015, while the stripping pump was being used to drain liquid waste from
central cargo tank no. 6. The leak occurred in a flange in the piping system inside
the pump room, due to failure of a spade in the flanged connection. The spade had

probably been fabricated on board and it failed due to a pressure overload caused when the pump was operated against a closed valve.

The stripping pump was not designed for shut-off conditions. Such a hazard had been identified in the HAZOP performed during the design stage [19], but the hazard had been resolved by requiring permanent presence of personnel during operation. The FPSO was not operated in compliance with this requirement, because this assumption from the design stage was not reflected in the operating procedures.

Three fixed gas detectors installed at the bottom of the pump room detected the gas immediately, but the production plant was not stopped automatically. The production continued for another ten minutes until a management meeting decided to stop production and thereafter send the first team into the pump room. The crew had heard a bang before the pump was manually stopped. The ventilation system was stopped due to the gas detection. This implied that no dilution of the explosive atmosphere was attempted, which would increase risks for personnel sent to the pump room, as described in the following.

But despite the availability of information that the atmosphere in the pump room was explosive, three different teams were sent to the leakage location at three different times. The first team was sent to the pump room to investigate the reported gas detection. This team identified the source of the leakage as a flange, and the team leader informed the unit's emergency command about the leak. The second team was intended to assess the work required to repair the leak and restore normal operation of the pump room. Whilst this team was in the pump room the portable detector worn by one member recorded 100% of the Lower Explosive Limit (LEL).

The first team reported that the pool of condensate below the leaking flange covered an area of about 2 m^2. This is the only indication we have of the amount of condensate that leaked out. The size of the spill may have increased further before the explosion, as this observation was made less than 30 min after the leak started, and about 45 min before the explosion. The leak is described as 'dripping', which suggests that when this observation was made it was not severe.

If the average depth of the pool of condensate were 5 mm, then the amount of condensate in a 2 m^2 pool would be about 15 kg; if the average depth were 10 mm then it would be about 30 kg. This latter amount is of the order of magnitude required to fill the pump room (but not the access shaft) with a stoichiometric LEL concentration. Although this information is rather imprecise, it suggests that the amount of condensate involved in the explosion was quite small.

Once the second team returned from the pump room the emergency management decided that the situation was under control, and the muster status onboard was stood down, allowing people to return to normal operations and prepare for lunch.

A third team entered the pump room around 12:30, equipped with absorbent blankets, fire hose, ladder and tools, in order to clear the pool of liquid and tighten the connection screws which appeared to be the source of the leakage. This team went to the site of the leakage, four decks below main deck level. Another team (consisting of members from other emergency teams) was assembled on the main deck, near the entrance to the pump room, to support the third team.

The third team attempted initially, unsuccessfully, to mop up the leaked condensate using absorbent blankets. They also tightened the bolts of the flange connection. Then they started to use the fire hose to dilute the condensate and remove it to the drain system. A strong explosion occurred at around 12:38, whilst they were doing this. The explosion ruptured the bulkhead between the pump room and the engine room, causing substantial damage to the pump room, access shaft, engine room, the area near the entrance to the access shaft and some accommodation rooms. The report suggests that the likely source of ignition was static electricity from the fire water jet. Ignition occurred when use of the water jet was initiated. Static electricity from fire water is also known from the explosion on Brent A in 1988 (see Sect. 4.16).

The timeline of the accident is shown in Fig. 4.3, which covers the period from the occurrence of the condensate leak until the explosion.

The pressure overload caused the destruction outside of the single access hatch on the top of the pump room, causing the destruction on the main deck level and the immediate death of four members of the emergency teams gathered near the entrance to the pump room. Most of the emergency management personnel were lost in the accident. Therefore, there was little help available to survivors and only limited efforts to search for missing personnel until around 23:30, when external firefighting personnel arrived. It took several days to find all the missing personnel. The final death toll was nine and there were seven cases of serious injury and 17 with light injuries.

Fig. 4.3 Timeline for the CDSM accident

Four of the members of the third team sent into the pump room died (one survivor only), four persons on the main deck near the entrance to the access shaft for the pump room were fatally injured, and one fatality in the engine room muster area [19]. Another twenty-four people were wounded, some of them inside the accommodation. The evacuation of injured persons to Victoria onshore was delayed significantly, due to the lack of backup for an unavailable service. Further discussion of causes may be found in [19, 21, 22].

4.18.2 Barrier Performance

The flange where the loss of containment occurred had an onboard made spade which failed after the pump had been operated with its discharge blocked. The cargo system lines were from the vessel's construction made using Japanese Identification Standards (JIS) joints, which was different from the American Society of Mechanical Engineers standards (ASME), that were widely used in the process plant. There was no strict control of spare parts, as spades and blinds. In addition, in many cases there was not enough space between joints to install spades and the crew used to manufacture low profile parts aiming to reduce the required maintenance work. Organizational barrier failures are discussed in Sect. 4.18.4 below.

The following is a summary of the barrier performance:

HL1	Barrier function designed to maintain the integrity of the process system	Failure
HL2	Barrier function designed to prevent ignition	Failure
HL3	Barrier function designed to reduce cloud and spill size	Failure
HL4	Barrier function designed to prevent escalation	Success
HL5	Barrier function designed to prevent fatalities	Failure

4.18.3 Lessons Learned for Design

The accident had two very crucial lessons for input to design, which were not addressed by the investigation [19], but are discussed by Vinnem [22]:

- Reliance on a pump room rather than individual deep well pumps in each storage tank
- Having accommodation located in the vessel's stern rather than her bow.

Pump room explosion has for a long time been one of the most important hazards on commercial tankers [23]. Most Norwegian FPSO vessels have no pump room, instead having deep well (submerged) pumps in each liquid storage tank. Using submerged deep well pumps for each storage tank completely eliminates the pump room explosion hazard. Individual submerged pumps may be considered an inherently safe design because there is no possibility of explosion and fire when the pump is submerged in hydrocarbons. The penalties include significant investment cost and more complicated maintenance operations. This underlines the fact that safety improvement sometimes entails a substantial cost penalty.

There are no known previous cases of pump room explosions or fires on FPSO vessels [23], but there have been several similar incidents on commercial tankers. In the risk matrix presented by Vinnem and Kirwan [23], the pump room explosion hazard is placed in the highest category, as one out of six scenarios in that category.

The criticality of pump room explosion may be illustrated as follows: If a pump room explosion affects the engine room, as it did in the CDSM accident then it may cause extensive material damage and result in long repair times (as in the CDSM case). This is avoided entirely with deep well pumps.

It is standard design practice on commercial tankers for excess pressure in the pump room to vent itself on the main deck, as in the CDSM accident, in order to minimize the chance rupture of the hull, which could have catastrophic consequences for the vessel's integrity.

The other lesson learned from the CDSM accident relates to the location of the accommodation, and is associated with the difference between marine and offshore design principles. All Norwegian FPSOs, whether purpose-built or converted, have accommodation and the helideck in the bow of the vessel [23]. This may be considered an inherently safe design, because in a weather-vaning vessel the accommodation will always be upwind of any source of fire or explosion. This design embodies the offshore principle of siting fire and explosion hazards as far away from the accommodation as possible but although it has been the basis of Norwegian offshore design for more than 30 years, it has not been adopted in many other countries, except the UK.

Both individual submerged pumps and bow-end accommodation are somewhat specific to the Norwegian sector [23]. Some of the FPSO vessels in the UK sector have accommodation in the bow, but most have accommodation towards the stern. It has been argued that siting accommodation in the bow means that the helideck on top of the living quarters will be in the part of the vessel subject to the most extensive motion in severe weather conditions, and hence the usability of the helicopter and the safety of helicopter operations are more likely to be affected [23]. Having the helideck in the bow makes landing in severe weather conditions more difficult, but most floating offshore installations in Norway make extensive use of sensors that automatically transmit data on helideck movement to helicopters so that the pilots have a better understanding of conditions before they land. There are no known incidents relating to landing on helidecks in exposed locations in bad weather.

Having the pump room next to the engine room and in close proximity to the superstructure is unfortunate, as was clearly demonstrated in the CDSM accident. The majority of fatalities and injuries occurred on the main deck and in the living quarters. With living quarters in the stern of the vessel, the superstructure will inevitably be close to the pump room and engine room(s), as well as being downwind of any source of fire or explosion in the hydrocarbon process and storage areas. Thus if an explosion occurs in the pump room, such a layout ensures that excess pressure is vented next to the quarters.

Both the existence of a pump room and the siting of accommodation are fundamental design factors that contributed to the high death toll of the FPSO explosion (9 fatalities). Organizational errors also contributed to the high number of fatalities (see below) on the main deck and in the living quarters, but under a different design they would have had less impact. Had the pump room not been next to the engine room, there might not have been an explosion; had the living quarters not been in the stern the number of fatalities would probably have been lower.

4.18.4 Lessons Learned for Operation

Most of the lessons from the CDSM accident are of an operational nature, in addition to the two important lessons for design discussed above. Overall there were several severe breaches of safe operating practices onboard, mostly probably due to organizational faults. The main operational lessons are:

- Tanks not used in accordance with design provisions, no management-of-change process conducted. The cargo tanks were designed to store oil or an oil-condensate mixture. However, since its initial production the FPSO produced only gas, stored condensate and exported it by offloading. This changed all considerations regarding the major hazard risk identification and assessment, as all the consequence studies considered oil production. The decision to store condensate was made just before the commissioning phase and it was intended to be just for a short period of time. But the condensate storage was in fact carried out for more than five years without any hazard or risk review regarding the effects of the changes to operational mode.
- Degradation of the cargo tank system. The stored condensate started to react with the sealing of cargo tank valves. This led to the valves not retaining liquid even in the closed position and the crew was not able to sustain different tank levels. Due to the number of valves inside the cargo tanks, the repairs demanded line isolation and tank entrance by the repair team. Thus, as repairs were going on, many restrictions to operate the system were included. Even though the repair scope was huge and too many changes had been put in place without proper management, it was decided to maintain platform production. In addition, there was a lack of upgrade information about the actual system status in diagrams and handover reports.
- Lack of competence. At the time of the accident, the marine positions in the platform's organogram were not filled with competent personnel. Since the FPSO had started its production, there were no professionals in supervisory role, even with job activities designed for this position. In addition, the marine superintendent on duty on the day of the accident was not properly prepared, as

he had been on board just for ten days before the accident. He was left without any specific guidance, as all previous superintendents had left the installation staff months before.

- Operating a pump with a closed discharge valve. It was not unusual for operators to consider the pump stopped even with some gas supply, as similar operations had occurred sometimes before the accident. Moreover, to operate the pump from the marine system control console in the control room, the marine operator had just the opening percentage reading from the pump gas supply valve and no pump strokes or pressure information. Remotely operated valves were planned, but not installed. The design risk assessment had assumed the presence of a 'pumpman', but this was not reflected in the procedures.
- Personnel risk exposure. Once the second team returned from the pump room, the emergency management decided that the muster status onboard could be stood down, allowing people to return to normal operations and prepare for lunch. This caused a dramatic increase in personnel risk exposure and contributed to the high death toll.
- Ignition of flammable atmosphere through emergency response actions. As the flammable atmosphere were inside the pump room and no planned actions to control this scenario were stablished in the emergency plan, the platform incident command decided to send people into the pump room to clean up and fix the leaking joint. However, no consideration was given to possible sources of ignition created by the response actions inside the pump room.

4.19 Explosion and Fire, Abkatun A (2015) Production Platform

Early in the morning of 1st April 2015 the dehydration and pumping area caught fire in the Pemex operated Abkatun A platform in the Bay of Campeche, Mexico. Pemex's Emergency Response Plan was put in action and approximately 300 workers were evacuated and transferred to other platforms in the area.

Seven fatalities resulted and at least 45 workers from Pemex and other companies were injured. The rvacuation of 301 workers was performed by 10 ships and 28(!) helicopters, some personnel jumping into the sea. Eight firefighting vessels were reported to be battling and controlling the emergency. The fire was extinguished after about four hours.

Agencia Nacional de Seguridad Industrial y de Protección al Medio Ambiente del Sector Hidrocarburos (Agency of Safety, Energy and Environmental Enforcement, ASEA, established in March 2015) has investigated the accident, but the investigation report is unavailable, as is a root cause analysis performed by a third party on behalf of Pemex, based on a request from ASEA. Some information has been received from ASEA.

The accident occurred in the 'dehydration and pumping area' (gas plant). The explosion was caused by a leak from a corroded fuel gas line, due to an unusual kind of accelerated corrosion caused by the presence of micro-organisms and sulphuric acid within the gas, the ignition apparently being caused by a hot surface.

The presence of such micro-organisms was apparently an unknown source of corrosion, causing further studies to be conducted, and inspection of large amounts of piping on the installation.

4.20   Fire, Abkatun A (2016) Production Platform

Three workers were killed after a fire on the compression unit of the Abkatun A platform in the Bay of Campeche on 6th February, 2016.

The fire was controlled and the rest of the platform was not evacuated. Several other workers, for both Pemex and contractors, were injured.

The investigation report is unavailable, as is a root cause analysis performed by a third party on behalf of Pemex, based on a request from ASEA. Some information has been received from ASEA.

4.21   Ekofisk A Riser Rupture

4.21.1   Event Sequence

The test riser on the steel jacket wellhead platform Ekofisk Alpha ruptured due to fatigue failure on 1st November 1975, a few months after operations started. The failure occurred, it was subsequently discovered, due to insufficient attention to corrosion protection in the splash zone, thereby allowing rapid corrosion to occur.

The leak occurred immediately below the platform living quarter, causing an explosion followed by fire. The intense fire had a relatively short duration (25 min), as the flow of gas from the riser was immediately shut down on the Ekofisk Centre. The fire was completely extinguished after 2 h.

One life saving capsule with only six men in it, was inadvertently released from the hook at full height, probably due to panic. Three of those inside perished, the others were injured. The evacuation of the rest of the crew (65 persons) was accomplished in a safe and orderly manner, in virtually calm conditions and no waves.

There was only minor fire damage to the platform, although a section of the riser obviously needed to be replaced. The most extensive damage to the platform was caused by the fire water impact, due to the high pressure monitors on the standby vessel.

4.21.2 Barrier Performance

The barrier failures that contributed to this accident are the following:

- The rapid loss of integrity of the riser due to severe corrosion. During installation work, the riser had been damaged by a vessel and dented. A section of the riser in the splash zone was therefore removed and replaced with new pipe. The corrosion protection provided on this part of the riser was very superficial, and the corrosion speed was quite severe when production started, also due to the increased temperature of the hydrocarbons. The reliability of the corrosion protection was obviously far from sufficient.
- The barrier function to prevent ignition also failed, although the ignition source was unknown. It could be noted that the gas cloud was around the platform, and it is not unlikely that the ignition was in unclassified areas, such as by the flare. As such, there may not have been any failure involved in the occurrence.
- The barrier function to protect personnel had a partial failure, when the capsule severely injured personnel due to a free fall of more than 20 m to sea level. The robustness of the evacuation system was not sufficient.

The following barriers performed successfully in the riser rupture and subsequent gas explosion and fire on Ekofisk A:

- The barrier function to limit the amount of hydrocarbons released through isolation of streams performed successfully. This included shutting off the possible backflow from the Ekofisk complex, through isolation of the particular pipeline. The distance and the pipeline length are short, around 1 km, which also limited the hydrocarbon reservoir.
- The barrier function to limit escalation of explosion and fire performed successfully. It is noted that external fire fighting by means of a standby vessel also performed well.

The following is a summary of the barrier performance:

HL1	Barrier function designed to maintain the integrity of the process system	Failure
HL2	Barrier function designed to prevent ignition	Failure
HL3	Barrier function designed to reduce cloud and spill size	Success
HL4	Barrier function designed to prevent escalation	Success
HL5	Barrier function designed to prevent fatalities	Partial failure

4.21.3 Lessons Learned for Design

There was no official investigation of this accident, and the findings were only reported in internal documents. This was the first major accident in the Norwegian offshore operations, and routines were probably not well established. The following lessons may be concluded:

- The location of the riser immediately below the accommodation was demonstrated as a critical issue. The fire exposed the living quarters immediately, requiring rapid emergency evacuation. This may have contributed to the panic, and the failure to lower the capsule in an orderly and safe manner.

4.21.4 Lessons Learned for Operation

In respect to the analysis of major hazards, the following lessons may be learned:

- Panic was present among the crew members during the emergency evacuation. It should, however, be realised that evacuation training and exercises have improved considerably since the time of this accident.
- The evacuation of the remaining 65 persons was successful with one capsule and one work boat in good weather conditions.
- Neither injuries nor fatalities were caused due to the explosion and fire.
- The platform only suffered limited fire damage due to the short duration of the intensive fire loads.

4.22 Jotun Pipeline Rupture

The description of this accident is based on a summary of the investigation report published by PSA [24].

4.22.1 Event Sequence

At 11:20 h on Friday, 20th August 2004, a pressure drop in the gas export pipeline from the FPSO Jotun A was noticed as a consequence of a rupture in the pipeline about 10 km from Jotun A. This is close to where the Jotun gas export line had been 'hot tapped' into the Statpipe gas line. The gas flowed into the sea and reached the surface.

Actually, it was later discovered that two ruptures had occurred, the second rupture as a consequence of sweeping motions of the pipe end resulting from the first rupture. Both pipe ends were deformed during the incident, and as such severely restricted the flow from the pipeline. It was later calculated that the flow had been 0.6 MSm^3/d, whereas the theoretical maximum with 6 inch full opening would be about 50 MSm^3/d.

25 min after the first pressure drop, automatic shutdown of valves on the Jotun leg of the pipeline occurred, due to extensive pressure drop in the line. Gas bubbles on the surface around the 'hot tap' area were observed by a vessel around 14:00 h. Production from Jotun was stopped in a controlled manner around 20 min later. The continued gas leak was fed from the Statpipe line, since gas export from Jotun was stopped 2.5 h earlier.

The resources required to close the leak by shutting valves using a Remote Operated Vehicle (ROV) were made available some 20 h after the occurrence of the leak. Calculations had to be made in relation to gas concentration in the gas plume in the sea and the atmosphere immediately above, in order to ensure that no ignition hazard was present. Almost exactly 24 h after the initial leak, an ROV was launched. The attempt to close the valves failed due to lack of the required resources.

A new vessel arrived the next morning with all required tools and resources, and the two valves were closed at 10:00 and 10:30 h on the morning of 22nd August. The actual safety and environmental consequences of the incident were limited, although the incident had the potential for significant safety consequences.

The consequence of the event was a gas volume released to atmosphere, equal to 1.3 million Sm^3. The direct repair cost was about 150 million NOK, in addition to which was the lost revenue from shut down of gas transport until the repair was completed. The environmental effects were limited, including the flaring of gas and CO_2 production for some hours.

No vessel was present in the area when the rupture occurred. If a vessel had been right on top of the plume, an ignition could have occurred. Loss of buoyancy would have been unlikely.

4.22.2   Barrier Performance

The only barrier failure in the present case was the failure of containment i.e., the pipeline ruptures. The rupture was probably caused by repeated (at least two) impacts by fishing trawl on a particular flange, during the three years prior to the event. The flange had not been covered by rock dumping during installation, but this was not recorded by the operator. This was not recorded as a result of the visual inspection in 2001. Annual inspections had been suspended after the 2001 inspections, and the next inspection was scheduled for 2006.

There are no restrictions on fishing activity around the 'hot tap' area, all installations should be compatible with fishing activity, including trawling. For a similar flange near to the affected flange, a risk assessment during installation had concluded that it was not a problem that the flange was unprotected against fishing gear.

The barriers to isolate the leak and shut in the wells (Jotun B) all performed successfully. No other barriers were involved, due to the location of the leak far away from the installation. The following is a summary of the barrier performance:

HL1	Barrier function designed to maintain the integrity of the process system	Failure
HL2	Barrier function designed to prevent ignition	Not applicable
HL3	Barrier function designed to reduce cloud and spill size	Not applicable
HL4	Barrier function designed to prevent escalation	Not applicable
HL5	Barrier function designed to prevent fatalities	Not applicable

4.22.3   Lessons Learned for Design

The lesson learned for design is the need to protect subsea installations against impact from fishing gear. It was noted in the present case that a similar flange on the pipeline had intentionally been designed without protection from rock filling or similar, and a risk assessment had concluded that protection was not required. This accident probably indicates that the conclusion from the assessment was incorrect.

4.22.4   Lessons Learned for Operation

There are no particular lessons for operation of installations from this event. It could be noted though that there was considerable uncertainty in the emergency management team whether the ROV vessel could safely manoeuvre freely within the gas plume and the gas cloud without any ignition hazard.

4.23 Mumbai High North Riser Rupture

4.23.1 Event Sequence

The production complex Mumbai High North (previously Bombay High North) located 160 km west of the Mumbai coast, consists of four bridge linked installations:

- Wellhead platform (1976)
- Processing platform (1981)
- Additional processing platform (recent)
- Accommodation platform (1978).

Wellstreams are received from several normally unmanned wellhead platforms, and oil is exported to shore by pipeline, after processing.

The accident occurred on 27th July 2005 in the monsoon season. A jack-up was positioned over the wellhead platform for drilling purposes. A cook sliced off the tips of two fingers onboard a multi service vessel (MSV) working elsewhere in the field. Helicopters were unavailable due to weather conditions, and the injured person was due to be transferred to the complex by crane.

While approaching the processing platform on the windward side (leeward side crane not working), the MSV experienced problems with its computer-assisted azimuth thrusters. The MSV was brought in stern-first under manual control and the injured person was transferred off the MSV.

Strong swells pushed the MSV towards the platform (around 16:05), causing the helideck at the rear of the vessel to strike and sever one or more gas export risers on the jacket, causing a leak that ignited after a short time. The fire escalated to more risers due to the close distance and lack of fire protection.

The subsequent fire engulfed the processing and accommodation platforms, causing the total destruction of the latter. Further, the MSV was engulfed by fire, and the wellhead and jack-up platforms received high heat loads.

Altogether, 22 people were lost (11 bodies recovered, 11 missing) and 362 were rescued over a 15 h period. The fire significantly affected rescue efforts, with only two out of the eight complex lifeboats able to be launched, and only one out of 10 liferafts. Similarly, only half of the jack-up's rescue craft could be launched. As said, rescue helicopters were unavailable due to weather conditions. Six divers in saturation chamber on the vessel were rescued 36 h later. The MSV sunk four days afterwards.

As an additional comment about this accident, one is reminded of the proverbial rhyme "For Want of a Nail":

For want of a nail the shoe was lost.
For want of a shoe the horse was lost.
For want of a horse the rider was lost.
For want of a rider the message was lost.
For want of a message the battle was lost.

For want of a battle the kingdom was lost.
And all for the want of a horseshoe nail.

This should not be interpreted such that the fingers of a cook are not important, but rather that the unavailability of the leeward side crane is the actual cause of the accident.

4.23.2   Barrier Performance

The barrier failures that contributed to this accident include the following:

- Protection of risers was inadequate to prevent loss of integrity due to vessel impact.
- Rupture of risers downstream of platform located isolation valves implied that inventory could not be minimised.
- The barrier function to prevent ignition also failed, although the ignition source was unknown. The mechanical impact by the vessel is a likely ignition source, but it remains impossible to verify.
- Separation of risers was such that the barrier function to prevent escalation did not function.
- The barrier function to protect personnel had a partial failure, when the majority of lifeboats, rafts and capsules could not be launched. Still 362 out of 384 were rescued.

The following is a summary of the barrier performance:

HL1	Barrier function designed to maintain the integrity of the process system	Failure
HL2	Barrier function designed to prevent ignition	Failure
HL3	Barrier function designed to reduce cloud and spill size	Failure
HL4	Barrier function designed to prevent escalation	Failure
HL5	Barrier function designed to prevent fatalities	Partial failure

4.23.3   Lessons Learned for Design

No official investigation report is publicly available, and thus information is sought from various sources on the Internet. The lessons for design include the following:

- The location of risers not sufficiently well protected within the jacket structure is an obvious lesson which had catastrophic consequences.

- Even if risers were mounted on the outside of the structure, protective structures and/or fender systems could have prevented the loss of containment.
- Subsea isolation valves are essential if riser rupture occurs.

4.23.4 Lessons Learned for Operation

There are also very important lessons to be learned for operations including vessel management:

- In the case of severe weather conditions, vessel management needs to be proactive in order to avoid contact between vessel and structure
- The nature of the risk assessment process, which allowed a vessel with thruster problems to come in on the windward side during strong winds and heavy swell.

4.24 Deep Sea Driller Capsize

The description of this accident is based on a report on evacuation means [1] as well as a newspaper article in 2004 [25].

4.24.1 Event Sequence

The rig was in transit from the offshore location to a yard on the Norwegian West coast (Fedje, outside Bergen), and was escorted by two vessels; a supply vessel and a converted fishing vessel, which did not have sufficient towing power. When approaching the coast in the middle of the night, the weather increased to storm, gusting at hurricane level, and the engines of the rig were insufficient to prevent the rig drifting against the rocks. One engine was unavailable, and the rig owner had declined the request for a pilot to manoeuvre the rig into the fjords.

Towing lines could not be transferred to the supply vessel, due to missing equipment, several attempts to establish towing connection were unsuccessful, and the rig grounded around 02:00 on 1st March 1976. Actually, the supply vessel got too close to shore itself, and soon had its own problems.

The 50 men crew mustered on the helideck in order to await arrival of Search and Rescue (SAR) helicopters, which never turned up. After waiting for a couple of hours, the crew had to launch one of the enclosed lifeboats. The battery was discharged ('flat'), but the crew managed to start the engine manually. Due to the darkness and the dangerous rocks, the manoeuvring of the boat was extremely difficult, and seven persons had to sit on top of the harness in order to guide the helmsman.

When the lifeboat grounded it overturned and the persons on top were thrown into the sea. Of the seven on top of the boat, only one survived, the others were lost in the sea.

Some 20–25 of crew were airlifted by helicopters that had arrived on the scene. Others managed to jump ashore and climb out of the sea, when the lifeboat drifted into sheltered waters. One man was taken by the waves, but managed later to grip a rope from a helicopter and was lifted ashore.

The rig was salvaged after some two months aground, was declared 'constructive total loss' but was still repaired and has been operating under the name "Byford Dolphin" ever since.

4.24.2 Barrier Performance

This accident occurred very early in the petroleum operations in Norway, where knowledge was substantially weaker than it is at present. Several barrier functions and elements failed:

- Failure to ensure that the installation had sufficient propulsion in order to manoeuvre in transit close to the coast in extreme weather conditions.
- Failure to manoeuvre the installation in difficult coastal waters without a pilot in charge.
- Failure to realise the extreme risk level involved in navigating the rig close to the coast in extreme conditions.
- Failure to ensure that the installation had necessary equipment for emergency tow assistance in critical conditions.
- Failure to prevent a large number of people being exposed to hazards during tow.

For a long time it was considered an error of judgement that some people were allowed to sit on top of the enclosed lifeboat, such that they would be thrown over board when the boat capsized. After a long time, it was acknowledged that this was critical in order to try to manoeuvre the boat clear of the rocks, and that these people performed heroically.

Virtually all of the barriers that failed in the Deep Sea Driller accident were of an organisational nature.

To some extent, it may be claimed that even though the conditions were extremely difficult, the rescue operations of all but those who fell overboard were successful, probably more due to luck, very skilled crew members in SAR helicopters and courageous efforts by volunteers from nearby communities. The following is a summary of the barrier performance:

STR1	Barrier function designed to maintain structural integrity and marine control	Failure
STR2	Barrier function designed to prevent escalation of initiating failure	Failure
STR3	Barrier function designed to prevent total loss	Failure
STR4	Barrier function designed to prevent fatalities	Partial failure

4.24.3   Lessons Learned for Design

The lessons learned from this accident were mainly related to operational aspects. One lesson is that the propulsion capacity needs to be sufficient for manoeuvring the installation in whatever waters it may be subjected to, and with sufficient redundancy, such that it still will be seaworthy with one single failure, such as an engine failing.

The Deep Sea Driller accident was the first instance that the vulnerability of davit launched lifeboats for evacuation purposes was demonstrated. The lifeboat launching was actually successful, but the vulnerability of this barrier element was nevertheless demonstrated clearly.

4.24.4   Lessons Learned for Operation

This accident was one of the first major accidents in Norwegian offshore operations, and a number of lessons were learned from this event. The most fundamental experience was that mobile units in transit have a much higher risk level, compared to on station offshore. Consequently, the manning level should be as low as possible during transit, only the manning level needed for sailing and marine operations.

4.25   Alexander L. Kielland Capsize

This accident was the first instance in the Norwegian offshore operations where an official commission of enquiry was appointed to investigate a severe offshore accident, resulting in the presentation of a very extensive report in 1981 [26].[2] Attention was mainly focused on the cause of the failure, but considerable attention was also paid to the evacuation and rescue operations that had revealed extensive shortcomings.

---

[2]It is known that a French investigation report, published some years later, had different causes for the accident. This is not in the public domain, and is not reflected below.

4.25.1 Event Sequence

The semi-submersible flotel Alexander L. Kielland capsized on 27th March 1980 while bridge connected to the steel jacket Ekofisk Edda platform. The flotel lost one of its five legs (with buoyancy elements at the lower end) in severe gale force winds, but not an extreme storm.

The accident started with one of the bracings breaking off due to fatigue, thereby causing a succession of failures of all bracings attached to this leg. It was discovered during the investigation that the weld of an instrument connection on the bracing had contained cracks, which had probably been in existence since the rig was built. The cracks had developed over time, and the remaining steel was less than 50%.

When the leg came loose, the rig almost immediately developed a severe listing. Within 20 min of the initial failure it capsized completely, floating upside down with just the bottom of the columns visible in the sea.

Both the escape and evacuation operations were far from orderly and had only limited success. Only one lifeboat was in fact launched successfully, one was totally unavailable due to the listing, and others were smashed against the platform during launching in high waves. The final death toll was 123 fatalities and 89 survivors.

Attention in the official enquiry report was mainly focused on the cause of the failure of the leg, but considerable attention was also paid to the evacuation and rescue operations that had revealed extensive shortcomings.

4.25.2 Barrier Performance

There were a number of severe barrier failures during this accident, which together caused the extremely severe consequences:

- The barriers intended to maintain structural integrity of the installation failed rapidly.
- The barrier function to provide stability in damaged condition in order to allow safe evacuation failed. In fact the capacity of the stability barriers was not intended to cover the conditions which prevailed.
- The barrier function to prevent fatalities failed severely. These failures probably occurred due to lack of robustness in the prevailing conditions.

There were no barriers that operated successfully in this accident. This implies that the following summary of the barrier performance may be stated:

STR1	Barrier function designed to maintain structural integrity and marine control	Failure
STR2	Barrier function designed to prevent escalation of initiating failure	Failure
STR3	Barrier function designed to prevent total loss	Failure
STR4	Barrier function designed to prevent fatalities	Failure

4.25.3   Lessons Learned for Design

The following lessons emerged from the accident, and have applications for the design of semi-submersible floating units:

- Cracks introduced during construction must be detected before the unit is launched.
- When fatigue cracks might grow, means to detect such cracks before they grow to a critical size must be implemented.
- If a floating unit develops severe listing, there should be a last barrier (i.e. buoyancy volume or a righting force) in order to allow time for organised and safe evacuation of personnel.
- Conventional lifeboats are not satisfactory in bad weather conditions. The experience from this accident was the driving force behind the development of free fall lifeboats for offshore applications.
- It was clearly demonstrated that the rapid and steep inclination angle makes orderly escape and evacuation very difficult.

4.25.4   Lessons Learned for Operation

The main lessons to be learned, from an analytical point of view, are as follows:

- It was realised that the rescue of survivors from lifeboats by traditional vessels was impossible in bad weather conditions.
- The role and the capabilities of the standby vessel were questioned after the accident, when it was realised that it took the vessel 1 h before it could attend the scene of the accident.

4.26   Ocean Ranger Capsize

The basis for the description of this accident is the Canadian Investigation report [27].

4.26.1   Event Sequence

The semi-submersible mobile drilling unit Ocean Ranger capsized on 15th February 1982 off Newfoundland in Canadian waters. The ballast control room in one of the columns had a window broken by wave impact in a severe storm. Short circuits occurred in the ballast valve control systems, when seawater entered the room, thereby starting spurious operations of the ballast valves.

The crew then had to revert to manual control, but were probably not well trained in this, and did actually leave the valves in the open position for some time, when it had been assumed that they were in the closed position. Correction of this failure did not occur sufficiently soon to avoid an excessive heel angle. Due to this excessive heel angle, the rig could not be brought back to a safe state, because only one ballast pump room was provided in each pontoon, at one end. The heel angle was such that the suction height soon exceeded the maximum of 10 m, and water from the lowest tanks could not be removed.

The onshore based SAR helicopters could not assist due to the severe weather conditions involving strong wind and low visibility. The rig therefore capsized and sank before any assistance could be provided.

The personnel (84 men crew) apparently evacuated, probably to two lifeboats, which at least were seaborn, although the exact state is not known, but only one was sighted. One boat collided with the standby vessel during the transfer attempt from the lifeboat onto the deck of the larger vessel. Within a short time the boat started to drift away, and was never seen again. No survivors or bodies were ever found.

4.26.2 Barrier Failures

The following barriers failed in this accident:

- The ballast control system which is essential for control of the stability of floating structures, failed in several ways:

  - Remote control of the ballast system failed due to water ingress and short circuits. This implies that the ballast control system did not have the required robustness, mainly due to the location of the control room in a column.
  - Manual control of ballast valves failed due to lack of competence and/or knowledge.

- No reserve buoyancy had been provided in the design in order to give a last barrier against capsize.

There were no barriers that performed successfully in this accident. The following is a summary of the barrier performance:

STR1	Barrier function designed to maintain structural integrity and marine control	Failure
STR2	Barrier function designed to prevent escalation of initiating failure	Failure
STR3	Barrier function designed to prevent total loss	Failure
STR4	Barrier function designed to prevent fatalities	Failure

4.26.3   Lessons Learned for Design

There are a number of lessons that may be drawn from this accident which have applicability for design:

- Ballast pumping needs system flexibility in order to enable rectification of serious accidental conditions in unforeseen circumstances.
- Reserve buoyancy or similar functionality should be provided in deck as a last barrier against capsize and sinking.
- Conventional lifeboats (whether one or two was used is unknown) could apparently be launched even in bad weather conditions.

4.26.4   Lessons Learned for Operation

There are some very important lessons to be learned from this accident, not the least regarding operational safety. The main lessons are the following:

- Competence and training are important in order to enable manual control when automatic systems fail.
- Rescue of people from lifeboats by traditional vessels was virtually impossible in bad weather conditions, without special equipment.

4.27   Glomar Java Sea Capsize

The description of this accident is based on a report on evacuation means [1].

4.27.1   Event Sequence

The 5930 GRT drill ship Glomar Java Sea capsized in a storm (tropical typhoon 'Lex') in the South China Sea on 25th October 1983. In fact, conditions reached 75 knots with 12 m wave height. It was later found resting upside down on the sea floor in 96 m of water, 96 km south of Hainan Island. It was in its anchored position over the well.

On 23rd October 1983 the drill ship prepared for the forecasted storm. The riser was disconnected and pulled aboard the ship. The decision was taken not to move the ship or evacuate the crew. Later the weather deteriorated to such an extent that evacuation would be too dangerous to attempt.

At 20:00 on 25th October 1983, the ship reported rolling 20–30°. Some time after 23:00, it heeled 15° to starboard, probably due to shifting of the cargo of drill pipe. An alternative theory for the heeling was that flooding had occurred into damaged wing tanks. The crew could not identify the cause, and may have made the list worse as they tried to correct it.

At about 23:50 the ship capsized to starboard under the influence of severe wind and waves, breaking four of its anchors. The 81 crew mustered and put on life-jackets around 23:00. The lifesaving equipment consisted of:

- Two off 64-person enclosed lifeboats with twin falls and onload release gear, located one on each side of the ship.
- Three inflatable throw-overboard liferafts, with a total capacity for 55 people.
- Crew members did not have survival suits, only life jackets.

The starboard lifeboat appears to have been launched with an unknown number of people aboard (not more than 45). The port lifeboat was not launched, and was ripped from its stowed position, possibly as the ship sank. One of the crew was found trapped at the starboard lifeboat, and 35 were found trapped in the accommodation.

The drill ship had no standby vessel. Its supply vessel was 20 miles away at the time, and was only directed to the drill site during the morning of 26.10, finding only debris.

A distress signal was picked up from the lifeboat on 27th October 1983, and it was observed capsized from a helicopter the following day. It was not recovered. One empty liferaft was recovered, without survivors.

Videotape of the wreckage showed major structural failure amidships and starboard side. Fracture from main deck plating down starboard side to the bottom plating. Probable cause of the accident was the decision to keep the ship with all its nine anchors at the well site, which subjected the vessel to the full force of the storm against its starboard side, and allowed it to capsize to starboard as a result of severe rolling while experiencing a 10° starboard list.

4.27.2 Barrier Failures

The following barrier functions failed in this accident:

- Protection of structural integrity from wave overload failed in typhoon conditions.
- Prevention of fatalities (through evacuation) failed in typhoon conditions.

None of the relevant barriers performed successfully in the present case, at least not any of the known barriers. The following is a summary of the barrier performance:

STR1	Barrier function designed to maintain structural integrity and marine control	Failure
STR2	Barrier function designed to prevent escalation of initiating failure	Failure
STR3	Barrier function designed to prevent total loss	Failure
STR4	Barrier function designed to prevent fatalities	Failure

4.27.3 Lessons Learned for Design

There are no particular lessons to be learned for the design of such vessels from the present case.

4.27.4 Lessons Learned for Operation

The following lessons could be learned from this accident with respect to operation of floating installations:

- Evacuation of installations needs to be taken sufficiently early when extreme conditions are forecasted.
- Safeguarding of the installation also needs to be considered when extreme conditions are forecasted.

4.28 Seacrest Capsize

The description of this accident is based on a report on evacuation means [1].

4.28.1 Event Sequence

The drill ship capsized on 3rd November 1989 during the typhoon 'Gay' in the Gulf of Thailand. Wind was 90 knots, causing waves estimated at 11–14 m height. The eye of the typhoon apparently passed directly over Seacrest, causing what was described as 'confused' waves. The peak of the typhoon lasted for 15 min. The capsize occurred after the eye had passed over the vessel.

Out of a total crew of 97 persons, 91 were lost in the accident, 6 survived. The unit sustained damage to the quarter and one crane, followed by an overturn of the unit. Investigation revealed an inadequate warning of the impending storm. No further information about the accident is available.

4.29   West Gamma Capsize

The description of this accident is based on a report on evacuation means [1].

4.29.1   Event Sequence

The Trident II class 518-bed jack-up accommodation platform West Gamma capsized in the North Sea on 21st August 1990.

The platform was towed by a single anchor handling vessel, the 'Normand Drott', which has a bollard pull of 150 tonnes. Waves were reported to be up to 16 m high, with the rig rolling and pitching 30° and the main deck constantly awash. At 12:30 a large wave broke off most of the helideck. Then the towline parted.

Personnel mustered with survival suits. Helicopter evacuation was not possible, due to the collapsed helideck and the rig motions with the legs elevated 120 m above the main deck. Lifeboat launching was also too hazardous, with breaking waves all around. At first, the intention was to remain on board until the weather improved.

One of the lifeboats stowed on deck broke loose and damaged vent pipes and access hatches, allowing down-flooding into the machinery space. By 01:00 on 21st August, the rig had listed 10° with roll of 5–8% and was drifting at 4.5 knots towards the shore. Significant wave height had increased to over 10 m, making the main deck inaccessible.

At 02:00 it was decided to evacuate. All 49 personnel jumped overboard in groups of five or six tied together from the first level of the accommodation. Four rescue vessels attended and launched three fast rescue craft (FRC), which positioned themselves downwind in a horseshoe so that personnel drifted towards them. Helicopters directed searchlights onto the sea. 4 people climbed a scramble net onto a support vessel; the others were picked up by FRCs. One FRC capsized during retrieval due to a rope from the vessel hooking onto the waterjet. All 49 crew were recovered.

At 03:07 the rig capsized in 30 m of water, 70 km west of Sylt Island off the German coast. In 1994, the company 'TV Bergingswerken' was appointed to salve parts of the jackup and this work started in May 1994. Two of the platform's legs were detached and laid on the seabed beside the platform. The rest of the platform has been moved away to water of 25 m depth.

4.29.2   Barrier Performance

The following barriers failed in this accident:

- Failure to limit the number of personnel to be exposed to hazards during tow of jack-up platform in severe weather conditions.
- Failure to take necessary precautions to safeguard the installation during tow of jack-up platform in severe weather conditions.
- Failure to prevent escalation and water ingress into hull.
- Failure to provide organised evacuation of personnel in time.

The following barriers performed successfully in this accident:

- Successful rescue of all personnel who had to evacuate in an uncontrolled manner at sea. It could be noted that the towing vessel which also was certified as a standby vessel, failed in picking up all personnel from the sea, and had to be assisted by Danish sea rescue service Esvagt in order to complete the rescue actions.

The following is a summary of the barrier performance:

STR1	Barrier function designed to maintain structural integrity and marine control	Failure
STR2	Barrier function designed to prevent escalation of initiating failure	Failure
STR3	Barrier function designed to prevent total loss	Failure
STR4	Barrier function designed to prevent fatalities	Planned system failed. Success with improvised actions

4.29.3   Lessons Learned for Design

The lessons for design are focused on location of the helideck on such an installation. When the jack-up is in transit, the legs have been retracted, and may pose a very large obstruction for helicopter approach, if the helideck has not been located with this in mind.

Similar restrictions may also occur when installed close to an existing installation, if structures on the other installation are causing similar obstructions.

4.29.4 Lessons Learned for Operation

The Norwegian vessel with standby capabilities had difficulties with its use of Fast Rescue Crafts (FRC), as noted above. The most critical aspect was a sufficiently high speed for the lowering and retrieval of the FRC.

The crane used for deployment and retrieval of the FRC should also be located as close to midships as possible, where movements are least.

A need was also demonstrated to be able to retrieve the FRC with more than nine persons onboard.

4.30 Norne Shuttle Tanker Collision

The description of this accident is based on a report on an NTNU report [28].

4.30.1 Event Sequence

The incident occurred on 5th March 2000, in the Norwegian Sea, when a shuttle tanker impacted the stern of the FPSO during a tandem off-loading operation. The DP2 shuttle tanker was on her first shuttle tanker mission and had almost completed off-loading. Normal distance apart was 80 m, and at the time of the incident was about 77 m. The relative headings were:

- FPSO: 250°
- Shuttle tanker: 226°.

Bad weather resulted in disruption of loading and the start of nitrogen flushing. The shuttle tanker had to be in line with the FPSO when the hose was sent back from the shuttle tanker. The DP system was transferred from 'Weather vane' mode to 'Auto position' mode, in order to achieve alignment, and a number of navigation commands were given.

Three crew members were on the bridge when the incident occurred. The captain was operating the DP system, and the advising captain was observing the operation. A junior officer was operating the loading system. The advising captain had long experience with tandem off-loading. The residual crew had no experience and little training in operating DP systems.

The captains discovered that one DP-monitor showed increased current from 1.3–2.9 knots. They discussed this and did not pay attention to the monitor that showed the thrust from each propeller. About 50 s after Auto position was activated, the captains noticed that the thrust on both main engines showed 'red forward', and the shuttle tanker had gained a significant forward speed. This observation was then confirmed with analogue indicators showing the propeller torsion. The DP control was then changed to 'Manual' mode, 'high gain' and the joystick put full astern. The speed decreased, but the shuttle tanker still hit the FPSO.

The impact speed was 0.6 m/s (1.2 knots), and the impact energy was 31 MJ, which at the time was the most powerful collision ever in North-west European waters. Only minimal structural damages were caused on FPSO and shuttle tanker.

4.30.2 Barrier Performance

The movement attempted was a complex operation, which could have been easier to implement if taken in smaller steps. The software had an error, causing the erroneous movement to be initiated.

The only barrier against collision in the case of abnormal excursions or drive-off cases is rapid intervention by the DP operator. There are warning and alarm limits, but these also require DP operator invention. The barrier in this case failed, maybe due to inattention or inexperience.

The discovery of the forward movement was made too late. It was also noted that no side thrusters were activated in order to turn the bow of the shuttle tanker away from the stern of the FPSO.

No barrier successes are relevant for this case. The following is a summary of the barrier performance:

STR1	Barrier function designed to maintain structural integrity and marine control	Failure
STR2	Barrier function designed to prevent escalation of initiating failure	Success
STR3	Barrier function designed to prevent total loss	Success
STR4	Barrier function designed to prevent fatalities	Success

4.30.3 Lessons Learned for Design

It is noted that the forward movement was caused by the DP software, which had a fault that had to be corrected. Other necessary improvements of the DP software were also identified through this incident.

4.30.4   Lessons Learned for Operation

It should be noted that this incident occurred when there was limited experience with tandem off-loading operations from FPSO vessels. A number of lessons were learned relating to:

- Improved operational instructions for DP operations
- Revision of alarms and set-points
- DP calibration needed
- Signal disturbances need to be identified and corrected
- Improvement of crew competence in off-loading operations
- Experience transfer for crew members
- Improvement of operation manuals.

4.31   P–36 Capsize

The description of this accident is based on a report on an investigation report published by the Brazilian National Petroleum Agency [29].

4.31.1   Event Sequence

The P–36 accident in the Roncador field, Campos Basin, Brazil on 15th March 2001 was a catastrophic loss, which led to an official investigation of the sequence of events and causes. Analysis of the most likely causes for the accident allowed the authorities to identify the critical event as the water depletion operation for the port stern emergency drain tank, which started on the evening of 14th March 2001. Water contaminated with waste oil present in the tank was planned to be pumped to the platform production header receiving the oil and natural gas flow from production wells. From there, together with the hydrocarbon production, it would flow to the processing plant. However, operational difficulties in getting the depletion pump of the tank started allowed the reverse flow of oil and gas through the tank flow lines and into the other tank (starboard stern) through a presumably damaged or partially open valve. The pump started after 54 min and reduced the reverse flow of hydrocarbons, and pumped water entered the starboard stern tank. Continuous pressurization of this tank led to its mechanical rupture around two hours after the beginning of the depletion operation for the other tank, thus characterizing the reported event as the 'first explosion', which occurred at 0:22 h on 15th March 2001.

Fluids from the ruptured tank and from similarly damaged lines and other equipment accumulated in the column's fourth tier compartment. Gas escaped to the upper decks through openings in the compartment and ruptured exhaust and blower lines. Around 20 min after the tank suffered rupture, the gas that reached the tank top deck area and the second deck close to the column exploded. This characterised the occurrence of the 'second explosion', which killed eleven people from the platform's fire brigade.

The mechanical collapse of the starboard emergency drain tank, immediately followed by rupture of the sea water outlet pipe passing through the fourth tier, initiated the flooding of the column. Water migration towards the lower part of the column occurred as the water in the fourth tier compartment reached the exhaust and blower system dampers that should have closed automatically; however, due to a failure in their activators the passage of fluids was allowed.

The amount of liquids inside the column and partly in the pontoon caused the platform to list, which intensified as the water progressed towards the ballast tank in the starboard aft column and the contiguous stability box. These spaces were flooded because the access ellipses (man holes) had been open since the previous day for inspection of the repair performed on a crack found in the stability box.

To make up for tilt caused by water entering the damaged column, water was allowed into the diametrically opposite ballast tank. This measure accelerated the undesirable increase in the platform's draft.

Continuous submersion was intensified by the flooding of the damaged column, the starboard stern ballast tank, the contiguous stability box, and the deliberate intake of ballast water into port bow tanks.

Evacuation of 138 people not regarded as essential to the emergency operations began at 01:44 h on 15th March and lasted around 2 h 30 min, a crane and a personnel basket being used for the purpose. At 06:03 h on the same day, after running out of alternatives for levelling the platform, the remaining emergency operations crew aboard abandoned the unit.

After the platform was abandoned, several attempts were made to save the unit, particularly by means of injection of nitrogen and compressed air into the flooded compartments for water depletion. However, these measures were not successful in keeping the unit stabilised and its slow and progressive submersion went on until 11:40 h on 20th March, when it finally sank. No attempts have been made to salvage the installation due to the water depth of 1,360 m.

4.31.2 Barrier Performance

The analysis performed by the enquiry commission allowed for the identification of several non-conformities pertaining to standard operation, maintenance, and project procedures, particularly those referring to frequent flow of water inside emergency drain tanks, to the port stern tank depletion operation, and to the hazard area classification around these tanks.

The main non-conformity identified pertains to storage of a large quantity of contaminated water inside emergency drain tanks during a considerable part of the production period of the platform, which is not in compliance with the Process Operating Manual. According to this manual, during normal operation the tanks must remain isolated and should only be used for the emergency depletion of large volumes of oil from process vessels or, in an emergency situation, the storage of large quantities of produced water.

The barrier to prevent bringing large volumes of water inside the watertight compartments, which was an essential measure to avoid stability or buoyancy problems, was thus breached. In addition, the following non-conformities were identified pertaining to standard operation and maintenance procedures:

- Systematic errors in manual volumetric survey and inoperative level indicators in emergency drain tanks.
- Clogging of open drainage vessel, which holds the water flowing through platform equipment trays.

Analysis of the crucial events pertaining to the flooding of the platform led to the identification of several non-conformities regarding standard operation and maintenance procedures, mainly non-compliance with watertight compartment procedures in areas that were critical for maintaining unit stability. In addition, it is worth highlighting the inefficacy of the measures taken in order to contain flooding or empty tanks before the platform was completely abandoned.

The inquiry commission found deficiencies in the operational management system for Petrobras' offshore oil and natural gas activities. In addition, contingency plans for major accidents and high risk emergency response schemes needed immediate improvement, as well as review of the criteria for engineering projects on floating production units for greater intrinsic protection.

The barrier function to protect personnel who were not affected by the initial events ('second explosion') performed successfully, in the sense that all non-essential personnel were evacuated safely. Also the personnel left in order to attempt to salvage the installation were safely removed before the installation capsized and sank.

The following is a summary of the barrier performance:

STR1	Barrier function designed to maintain structural integrity and marine control	Failure
STR2	Barrier function designed to prevent escalation of initiating failure	Failure
STR3	Barrier function designed to prevent total loss	Failure
STR4	Barrier function designed to prevent fatalities	Partial failure

4.31.3   Lessons Learned for Design

It is noted that manholes into watertight compartments which were intended to function as buoyancy compartments were open and allowed flooding of these compartments. It should be considered how, through design, such openings may be prevented from being left open for significant periods.

4.31.4   Lessons Learned for Operation

There are several important lessons to be learned for the operation of floating production installations from this accident:

- Allowing frequent flow of water inside emergency drain tanks should not be performed.
- Storage of large quantities of contaminated water inside emergency drain tanks during considerable part of the production period of the platform should not be allowed.
- Barriers to prevent ingress of large volumes of water inside the watertight compartments should be kept operational at all times.

4.32   P–34 Listing

The description of this accident is based on a weekly magazine news article [30].

4.32.1   Event Sequence

The listing of Petrobras operated FPSO P–34 on 13th October 2002 was so severe that the unit almost capsized before deballasting operations brought the vessel under control.

Failure of electric power caused ballast valves to open, thus causing the vessel to list to a maximum of 32°. 25 crew members of a total of 67 persons, jumped over board, and were subsequently picked up from the water by vessels assisting the FPSO. All personnel survived the incident.

The vessel was gradually brought back to upright position through deballasting. No further information is available from the incident.

4.33   Ocean Vanguard Anchor Line Failure

The description of this incident is based on the investigation report published by PSA [31].

4.33.1   Event Sequence

Two anchor lines failed on the semi-submersible mobile drilling unit 'Ocean Vanguard' formerly known as 'West Vanguard' (see Sect. 4.5) at 22:40 h on the 14th December 2004. Two anchor winches (No. 1 and 2) failed to hold the anchor lines, which ran out in an uncontrolled manner, allowing the rig to drift 160 m off location in 3–5 min. The unit was also listing 7–10°.

The brakes on two anchor lines were failing almost simultaneously in a sea state of about 10 m significant wave height. The band brakes were connected, but malfunctioned. Because of the damage it was impossible after the incident to conclude if the brakes had been correctly adjusted. The supplier had previously recommended changing the brake band, but it had not been done. The pawl stopping mechanism did not function, because it had not been installed according to the procedures.

The movements of the platform caused the drilling riser to fail, the tension system to rupture, the BOP at the sea floor got a six degree permanent inclination, the anchor winch system was damaged and the well was lost. This event has been through a detailed investigation by PSA [32].

There was a 60 knots west south-westerly wind and 15 m waves at 22:00 h on the 14th December. The delineation well in the Norwegian Sea had been drilled virtually to plan. The drilling of the exploration well had almost reached the target depth, but drilling operations had been suspended and the crew was waiting on weather at the time of the accident.

The riser was connected to the subsea wellhead at the time of the accident, and could not be disconnected sufficiently rapid, to the extent that damages were sustained to riser and wellhead. Also the heave compensation system for the riser was destroyed.

Thrusters were activated in order to compensate for the lost anchor lines. Actions were also taken to correct the listing of the unit. During the following night, 23 non-essential personnel were transported by helicopter to a near-by production installation, as a precaution.

There were no further consequences to installations and equipment, and no consequences for personnel. But the incident had quite severe potentials, as far as consequences were concerned:

- If the rig had been drilling in the 'pay zone' when the anchor line failure occurred, a blowout could have occurred.

- If a similar event sequence had occurred on a flotel which was bridge connected to a production installation, the drifting of the unit could have produced a collision between the flotel and the platform with severe structural damage.

The mobile unit had to be taken to shore for repairs, and it was several months before the unit was back in operation.

4.33.2 Barrier Performance

The barrier failure in this incident was the anchor line holding function by anchor winches 1 and 2, and the subsequent failure of quick disconnect functions of the marine riser.

The installation was more than 20 years old, and the anchor winches were possibly not in the best condition. The unit actually experienced a shallow gas blowout in October 1985, when it was operating under the name 'West Vanguard' (see Sect. 4.5).

The quick disconnect functions also failed. Obviously these were also old, but one would not expect this equipment to be worn out. As such, these failures are equally severe.

The PSA investigation identified 11 deviations, most of which are related to the management systems followed by the operator and the rig owner. Essentially, these are also barrier failures (see Sect. 4.33.4).

There were luckily a number of barriers that succeeded in their performance:

- The anchor system with six remaining anchors and thrusters managed to hold the unit in position.
- Well control barriers functioned, even though the blowout preventer was damaged, but since the reservoir had not been penetrated, this was not particularly critical.
- All actions to safeguard personnel functioned as intended.

The following is a summary of the barrier performance:

STR1	Barrier function designed to maintain structural integrity and marine control	Failure
STR2	Barrier function designed to prevent escalation of initiating failure	Success
STR3	Barrier function designed to prevent total loss	Success
STR4	Barrier function designed to prevent fatalities	Success

4.33.3 Lessons Learned for Design

No particular lessons with respect to design could be drawn from this incident.

4.33.4 Lessons Learned for Operation

The following were the direct and indirect causes according to the PSA investigation report:

- Direct causes:
  - The band brake had holding power less than specified
  - The brakes were not operated correctly.

- Indirect causes:
  - Lack of insight into how to operate second brake
  - Not following procedures for operation of anchor winches
  - Lack of insight into how to use winches and when to disconnect
  - Insufficient maintenance of anchor winches
  - Not following criteria and procedures for disconnection of drilling riser
  - Insufficient knowledge about disconnection from the operator representative
  - Insufficient personnel qualifications
  - Insufficient accuracy on the measurement of anchor line tension
  - Insufficient measurement and reporting of weather data
  - Insufficient implementation of the operator's duty to ensure that operations are carried out safely.

All of these causes are 'MO' factors, with the majority of causes in the management system. Based on the findings, both the operator and the rig owner were given notifications of injunctions, the rig owner received the following notification which later was confirmed as an injunction:

- 'Diamond Offshore is ordered to review its own management system and its own organization to identify causes of potential non-conformities, as well as to develop a binding plan with specific times for activities intended to prevent recurrence of this type of non-conformity.'

4.34 Gryphon Alpha FPSO Multiple Anchor Line Failure

The text in this subsection is mainly based on the text from Step Change in Safety [33].

4.34.1   Event Sequence

The Gryphon Alpha FPSO is located 280 km northeast of Aberdeen, in a water depth of 112 m. The vessel is a purpose built FPSO of with a length of 260 m and 41 m beam. It is capable of storing 86,000 m³ of oil.

The vessel has a turret just forward of midships equipped with a 10 point mooring system with 84 mm diameter, K4 grade chains with a design break load of about 730 tons. The turret is maintained in a fixed orientation, the vessel rotates around it. To maintain station keeping and to minimise environmental forces, the heading of the vessel is changed by five thrusters (3 aft and 2 forward) to align the bow into the prevailing seas. The control of the thruster system for heading control is maintained by a position mooring (PM) system.

An incident occurred at 07:05 h on Friday the 4th of February 2011 whilst the Gryphon Alpha FPSO was engaged in production operations. The vessel lost heading and position during stormy conditions; about 60 knots maximum wind speed with a significant wave height of between 10 m to 15 m. The initiating event was the low tension failure of windward mooring line 7. The PM system then drove the vessel beam on to the prevailing weather. This resulted in three further windward mooring lines (in order, 6, 5 and 4) failing progressively as the vessel heading turned beam on to the environment. The mooring lines failed due to the high environmental forces they were subjected to, which exceeded the design criteria. The subsequent loss of position (180 m) resulted in significant damage to the subsea infrastructure. The vessel was also claimed to have done several 21° rolls.

A gas cloud formed, but was quickly dispersed due to the wind.

The general mustering alarm was manually initiated and production was shutdown and blown-down. Subsequently 74 non-essential personnel were evacuated by helicopters to nearby installations.

Two tugboats were used to reconnect the vessel to secure it while work was underway to reattach chains to the FPSO anchors. Risers, flowlines and anchor lines were left in a total chaos on the seabed.

The FPSO is aiming for restart at the end of 2012, after comprehensive refurbishment.

The position mooring system drove the vessel beam on to the prevailing weather due to a chain of coincident events producing inaccurate inputs to the PM system models. This caused erroneous calculation of the forces and moments acting on the vessel. These events included:

- Underestimate of wind and wave forces as a result of inaccurate anemometer readings
- Loss of Differential Global Positioning Systems (DGPS) position reference, leading to the mooring model using wrong position to calculate mooring forces
- Manual heading change
- Failed detection of mooring line break
- Mooring line failure resulting in change in position

- Automatic PM system mooring model refresh failed
- Turret auto rotation.

Due to hard drive memory settings, only limited data was recoverable from the PM alarm historian, hindering the investigation into the root causes.

4.34.2 Barrier Performance

Failure of one anchor line is a dimensioning incident, which should not lead to subsequent line failures. There were several system failures in the present incident, which implied that unwanted escalation of the incident occurred. The following is a summary of the barrier performance:

STR1	Barrier function designed to maintain structural integrity and marine control	Failure
STR2	Barrier function designed to prevent escalation of initiating failure	Success
STR3	Barrier function designed to prevent total loss	Success
STR4	Barrier function designed to prevent fatalities	Success

4.34.3 Lessons Learned for Design

One aspect of turret design on FPSO vessels is related to the location of the turret. It this location is close to the bow of the vessel, the vessel will be capable of maintaining heading into the weather without active assistance from thrusters. In the present case, the location of the turret is just forward of midships, implying that the vessel is completely dependent on active use of thrusters to maintain heading. Power failure in adverse weather conditions may have dramatic effects, as fully demonstrated by this incident.

The maintenance and functionality checks for a moored PM vessel are less stringent and comprehensive than those implemented by full dynamically positioned vessels such as drilling support vessels. It may be prudent to consider employing a similar philosophy for moored vessels.

The Gryphon Alpha FPSO has a dragchain transfer system between the geostationary turret and the rotating vessel. Most of the purpose built FPSOs in the North Sea and Norwegian Sea have high pressure, multi-path swivel for the transfer of hydrocarbons between the turret and the vessel. This was new technology some 15 years ago, and possibly not as common as today. The swivel is considered to be a more flexible and robust solution, less prone to operational errors.

4.34.4 Lessons Learned for Operation

The incident highlights the need for duty holders to be aware of the potential for loss of heading and position on vessels fitted with Position Mooring systems.

During this incident a number of inputs to the PM model were inaccurate, contributing to error forces building up within the model and resultant incorrect forces being applied by the PM system. It may be useful for operators of FPSO vessels to ensure that:

- The data recording system is sufficient; the model is reset or refreshed with a suitable frequency; the maintenance and functionality checks are suitable and sufficient;
- Control room (bridge) operators are drilled in the actions necessary in all foreseeable emergency scenarios; and inputs to the PM system are accurate, reliable and have sufficient redundancy.

The basis of design of a PM vessel mooring system assumes that the vessel will not lose heading control, so will remain within approximately 10 degrees of head into the environment.

Analyses of the mooring system showed that the forces placed on the anchor lines would have approached or exceeded the break load of the chain at angles seen during the incident.

Operators of FPSO vessels should review their mooring system's failure mode analysis in order to ensure that the forces applied to the vessel, at reasonably foreseeable angles and environmental conditions are understood and appropriate procedures and processes are in place to manage the risks from loss of heading and those forces.

During this incident anchor chain number 7 failed at the flash butt weld of one of its links. This failure mode is unlikely to be picked up by visual inspection. Inspection and discard criteria for anchor chain should be reviewed to ensure that they remain appropriate in the light of this failure type.

During the incident, the power management set up was not optimal for the prevailing weather conditions. The power management procedure and processes may need to be reviewed to ensure that they are appropriate for all reasonably foreseeable operating and weather conditions.

4.35 Exxon Valdez Oil Spill

The description of this incident is mainly based on information published by US Environmental Protection Agency [34]. This event is different from the proceeding incidents discussed, as the main hazard was environmental damage, and some threat to the crew onboard.

A detailed discussion of oil spill hazards is outside the scope of the book, and the event is therefore not discussed as thoroughly as the proceeding events, when it comes to barriers and lessons learned.

4.35.1 Event Sequence

At 21:12 on 23rd March 1989, Exxon Valdez was under way from Valdez, Alaska, under Captain Joseph Hazelwood with a cargo of 180,000 tons of crude oil. After dropping her pilot, she left the outbound shipping lane to avoid ice. Owing to poor navigation, at 00:04 on 24th March, the tanker ran aground on Bligh Reef in Prince William Sound, just 40 km from Valdez. The grounding punctured eight of the eleven cargo tanks, and within four hours 19,000 tons had been lost. By the time the tanker was refloated on 5th April 1989, about 37,000 tons had been lost and 6,600 km^2 of the country's greatest fishing grounds and the surrounding virgin shoreline were sheathed in oil. Captain Hazelwood, who had a record of drunk driving arrests, was charged with criminal mischief, driving a vessel while intoxicated, reckless endangerment, and negligent discharge of oil. He was found guilty of the last count, fined 51,000 USD, and sentenced to 1,000 h of community service in lieu of six months in prison.

The size of the spill and its remote location made government and industry efforts difficult. This spill was about 20% of the 180,000 tons of crude oil the vessel was carrying when it struck the reef. The salvage effort that took place immediately after the grounding saved the vessel from sinking, thus preventing a far larger oil spill from occurring.

The spill posed threats to the delicate food chain that supports Prince William Sound's commercial fishing industry. Also in danger were ten million migratory shore birds and waterfowl, hundreds of sea otters, dozens of other species, such as harbour porpoises and sea lions, and several varieties of whales.

Alyeska, the association that represents seven oil companies who operate in Valdez, including Exxon, first assumed responsibility for the cleanup, in accordance with the area's contingency planning. Alyeska opened an emergency communications centre in Valdez shortly after the spill was reported and set up a second operations centre in Anchorage, Alaska. The Coast Guard quickly expanded its presence on the scene, and personnel from other Federal agencies also arrived to help. Three methods were tried in the effort to clean up the spill:

- Burning
- Mechanical Cleanup
- Chemical Dispersants.

A trial burn was conducted during the early stages of the spill. A fire-resistant boom was placed on tow lines, and two ends of the boom were each attached to a ship. The two ships with the boom between them moved slowly through the main portion of the slick until the boom was full of oil. The two ships then towed the boom away from the slick and the oil was ignited. The fire did not endanger the main slick or the Exxon Valdez because of the distance separating them. Due to unfavourable weather, no additional burning was attempted in this cleanup effort.

Shortly after the spill, mechanical cleanup was started using booms and skimmers. However, skimmers were not readily available during the first 24 h following the spill. Thick oil and heavy kelp tended to clog the equipment. Repairs to damaged skimmers were time consuming. Transferring oil from temporary storage vessels into more permanent containers was also difficult because of the oil's weight and thickness. Continued bad weather slowed down the recovery efforts.

In addition, a trial application of dispersants was performed. The use of dispersants proved to be controversial. Alyeska had around 15,000 L of dispersant available in its terminal in Valdez, but no application equipment or aircraft. A private company applied dispersants on 24th March, with a helicopter and dispersant bucket. Because there was not enough wave action to mix the dispersant with the oil in the water, the Coast Guard representatives at the site concluded that the dispersants were not working and so their use was discontinued. The spill ended up polluting more 320 km shore line with considerable amounts of crude oil.

4.35.2 Barrier Failures

The main barrier failure that occurred and caused the extensive spill was that of navigation failure, which led to the grounding of the tanker on a reef in Prince William Sound. The navigation failure was caused by a human error (allegedly, but not proven, due to alcohol use) by the captain. It could be observed that when it comes to navigation as a barrier against accidents involving ships, this is usually strongly influenced by human operator behaviour and performance. Many disasters, accidents and incidents have been caused by human errors on the bridge.

The captain allegedly had a record of alcohol abuse, and it could be argued that the failure to ban him from being in charge of ships and tanker navigation should be considered an organisational failure by the Exxon company. It could also be taken as an indication of a faulty safety culture.

4.36 Penglai 19-3 Oil Spills

In June 2011, two severe oil spill events occurred successively at two separate wellhead platforms (B&C) in the Penglai 19-3 oilfield in the Bohai Sea (the innermost gulf of the Yellow Sea, between Korea and China). The operator is ConocoPhillips China Inc. (COPC). It was reported that about 6200 km^2 of clean coastal water were polluted by these two severe oil spill events. It should be noted that the Chinese Bohai Sea is highly vulnerable to any size of oil spills as it is semi-closed, and its own water exchange is abnormally slow. The description of these two accidents is based on a summary of the investigation report published by the State Oceanic Administration (SOA, [35]) and response to Penglai 19-3 incidents from COPC (see http://www.conocophillips.com.cn/EN/Response/Pages/default.aspx).

4.36.1 Event Sequence

At 19:00 on 4th June 2011, a small amount of oil was observed on the sea surface near the Penglai 19-3 Wellhead Platform B (the B-Event). Then, emergency monitoring measures for oil spill were taken by COPC. Skimmers, absorbent booms and other clean-up equipment were deployed. When the source was determined to be an existing geological fault that had opened slightly due to pressure from water injection into a subsurface reservoir during production activities, COPC began reducing the reservoir pressure to seal the source. A steel subsea capture structure was temporarily developed for future deployment at the 'B' site seep. The water injection well B23 was shut down. On 19th June 2011, the oil spill was controlled and thus limited. On 21st June 2011, the reservoir pressure reduction plan at the location of the B-Event successfully closed the fault thereby isolating the reservoir from the surface and stopping the seepage. On 3rd July 2011, a steel subsea seep containment structure was placed over the seep at the B-Event as an additional precautionary measure.

At 11:00 on 17th June 2011, a large amount of oil spill around the Penglai 19-3 Wellhead Platform C was found by the Chinese maritime surveillance vessel 22. Subsequently, it was confirmed that a well kick accident had occurred on the Penglai 19-3 Wellhead Platform C (the C-Event). An unanticipated high-pressure zone was encountered at a water injection well C20. The C-Event was separate and unrelated to the B-Event. The well was immediately shut-into ensure there was no danger to the personnel or the platform, but the pressure caused reservoir fluids and mineral oil-based drilling mud to flow to the seabed. On 21st June 2011, the oil spill was controlled and thus limited.

Actually, the situation of the oil spill was not under full control. According to the statement of SOA [35], even on 31st August, 2011, COPC failed to meet the requirements of "screening out all potential sources for oil spills and blocking leaks once and for all".

ConocoPhillips estimates that the total amount of material released from the two events was approximately 723 barrels (115 m^3) of oil and 2,620 barrels (416 m^3) of mineral oil-based drilling mud.

On 24th January 2012, ConocoPhillips, together with China National Offshore Oil Corporation (CNOOC), announced an agreement with the Ministry of

Agriculture (MOA) to settle the public and private claims of potentially affected fishermen in relevant Bohai Bay communities. Under this agreement, RMB 1 billion (approximately US $160 million) was paid to the MOA. The MOA is administering the disbursement of these funds. ConocoPhillips also designated RMB 100 million (approximately US $16 million) to help improve fishery resources—this is being administered by the MOA. On 27th April 2012, ConocoPhillips, together with CNOOC, announced an agreement with the State Oceanic Administration. Based on the agreement, the SOA will receive RMB 1.090 billion (USD $173 million) to resolve claims related to the possible impacts on the Bohai Bay marine environment. As part of this agreement, COPC also committed RMB 113 million (approximately US $18 million) to a fund being administered by the SOA, to support initiatives that enhance marine ecological protection and reduce pollutants in the bay.

4.36.2   Barrier Failures

In the B-Event, the SOA investigation [35] found four main deviations that had contributed to the oil spill. These deviations directly or indirectly cover a number of barrier functions that failed. The full list of deviations is as follows:

1. On 2nd June, an abnormal situation was found that the water-injection rate for the Well B23 significantly increased while the water-injection pressure significantly dropped. In this situation, COPC did not take immediate measures, such as stopping the water injection. Instead, the pressure was maintained and water injection was continued. Finally, an existing geological fault opened slightly due to the high pressure from water injection into a subsurface reservoir and extended to the seafloor with oil spills.
2. COPC breached the overall development programme. The water injection for the Well B23 had been conducted in a general way without implementing the separated layer water injection. Hence, there was a risk of high pressure in some oil layers due to the improper water injection.
3. The wellhead pressure monitoring system for the water injection wells was defective in regulations and management. An upper wellhead pressure limit for the safety was not set up.
4. Manometric tests were not carried out to ensure the stability of existing faults. Especially, there were no safety alarms for the geological fault 502 which extends to the seabed and is in contact with several oil layers. Furthermore, the limits of pressure-bearing and fracture of this fault were not numerically analysed and demarcated.

In the C-Event, the SOA investigation [35] found three main deviations that had contributed to the oil spill. The deviations directly or indirectly cover a number of barrier functions that failed. The full list of deviations is as follows:

1. COPC broke the regulations related to the debris reinjection in the overall development programme, which resulted in the ultra-high pressure close to the bottom of the reservoir. A well control incident occurred while drilling into a fault block containing high reservoir pressure. Further, the surface casing did not have sufficient pressure-bearing capability. The result was a release of oil to the surface and mineral oil-based drilling mud to the sea floor in the operating area.
2. The emergency response was unsuitable. When drilling into the fault block containing high pressure, emergency measures were not analysed in a timely and intermediate casings were not set.
3. The drilling design department for the Well C20 did not execute the environmental impact assessment (EIA) report with respect to the design depth of surface casing.

The following is a summary of the barrier performance:

HL1	Barrier function designed to maintain the integrity of the well system	Failure
HL2	Barrier function designed to prevent ignition	Success
HL3	Barrier function designed to reduce cloud and spill size	Success
HL4	Barrier function designed to prevent escalation	Failure
HL5	Barrier function designed to prevent fatalities	Success

4.36.3 Lessons Learned for Design

In the B-Event, the following are the main lessons learned for design:

- The upper wellhead pressure limit for the safety was not defined in the design phase.
- Numerical analysis of the limits on the pressure-bearing and fracture of the fault was not carried out.

In the C-Event, the main lesson learned for design relates to the design depth of the surface casing. The reliability and availability of the setting depth of the surface casing is important for preventing oil spill in the drilling phase.

4.36.4 Lessons Learned for Well Operation

The following are the main observations from the investigation report [35] presented by SOA, as follows:

- The operator's HES management system should work efficiently.
- Two high level aspects are crucial for the safety of well drilling, namely the well planning ahead of commencement of operations, and the management of change (or continuous risk assessment) during the execution of well drilling operations.
- Warning alarms should be made for the existing geological faults.
- Manometric tests should be carried out to ensure the stability of several existing faults.
- The emergency response plans need to address the specific risk to the platform when relevant, such as well kick incident.
- A barrier function designed to prevent escalation should be established and maintained.

4.37 Summary of Barrier Performance

The following Tables 4.2, 4.3, 4.4 and 4.5 contain a summary of the main barrier function performance during the accidents and incidents reviewed in this chapter and the Macondo blowout from Chap. 5.

Table 4.2 Overview of barrier performance for process and riser leaks

Barrier functions		Brent A	Piper A	Ekofisk A	Jotun A	Mumbai High North	CDSM
HL1	Maintain integrity of well system	Failure	Failure	Failure	Failure	Failure	Failure
HL2	Prevent ignition	Failure	Failure	Failure	Not applicable	Failure	Failure
HL3	Reduce cloud and spill size	Success	Failure	Success	Not applicable	Failure	Failure
HL4	Prevent escalation	Success	Failure	Success	Not applicable	Failure	Success
HL5	Prevent fatalities	Success	Failure	Partial failure	Not applicable	Partial failure	Failure

Table 4.3 Overview of barrier performance for marine and structural accidents considered

Barrier functions		Deep Sea Driller	Alex L. Kielland	Ocean Ranger	Glomar Java Sea	West Gamma	P–36
STR1	Barrier function designed to maintain structural integrity and marine control	Failure	Failure	Failure	Failure	Failure	Failure
STR2	Barrier function designed to prevent escalation of initiating failure	Failure	Failure	Failure	Failure	Failure	Failure
STR3	Barrier function designed to prevent total loss	Failure	Failure	Failure	Failure	Failure	Failure
STR4	Barrier function designed to prevent fatalities.	Partial failure	Failure	Failure	Failure	Planned system failed	Partial failure

Table 4.4 Overview of barrier performance for blowouts considered, 1977–1989

Barrier functions		Ekofisk B	IXTOC	Enchova	West Vanguard	Ocean Odessey	2-4-14 Treasure Saga
HL1	Maintain integrity of the well system	Failure	Failure	Failure	Failure	Failure	Failure
HL2	Prevent ignition	Success	Failure	Failure	Failure	Failure	Not applicable
HL3	Reduce cloud and spill size	Not applicable	Not applicable	Failure	Not applicable	Not applicable	Not applicable
HL4	Prevent escalation	Success	Failure	Success	Failure	Success (?)	Not applicable
HL5	Prevent fatalities	Success	Success	Failure	Mainly success Limited failure	Mainly success Limited failure	Not applicable

Table 4.5 Overview of barrier performance for blowouts considered, 2007–2011

Barrier functions		Usumacinta	Snore Alpha	Montara	Macondo	Gullfaks C	Frade
HL1	Maintain integrity of the well system	Failure	Failure	Failure	Failure	Failure	Failure
HL2	Prevent ignition	Success	Success	Success	Failure	Success	Success
HL3	Reduce cloud and spill size	Failure	Success	Not applicable	Failure	Not applicable	Success
HL4	Prevent escalation	Failure	Success	Success	Failure	Success	Success
HL5	Prevent fatalities	Partial failure	Success	Success	Partial failure	Success	Success

References

1. NPD (1998) Evacuation means, strengths, weaknesses and operational constraints. DNV, Høvik, Norway, DNV Report 98–0561, Dec 1998
2. NOU (1986) West Vanguard accident, NOU Report 1986:16. Oslo
3. Ireland RD (1991) Determination by R.D. Ireland, Sheriff Principal of Grampian, Highland and Islands. In: Fatal accident inquiry into the death of Timothy John Williams on board Ocean Odyssey, 8.11.1991
4. Blaauw K (2012) Management of well barriers and challenges with regards to obtaining well integrity. Master thesis, University of Stavanger, 24.5.2012
5. Aadnøy B, Bakøy P (1992) Relief well breakthrough in a north sea problem well. Jof Pet Sci Eng 8:133–152
6. Aadnøy B (2010) Modern well design. CRC Press, Taylor & Francis Group, London
7. Versatel (2012) Adriatic IV. http://home.versatel.nl/the_sims/rig/adriatic4.htm. Accessed 29 Sept 2012
8. PSA (2005) Investigation of the gas blowout on Snorre A, Well 34/7–P31A 28 November 2004, undated, Petroleum Safety Authority
9. Equinor (2005) Cause analysis of the Snorre A incident 28 November 2004 (in Norwegian only), Statoil, undated, 2005
10. Batelle (2008) Root cause analysis of the Usumacinta—KAB-101 Incident 23 October 2007, Columbus, Ohio, 30 May 2008
11. Commission of Inquiry (2010) Report of the Montara commission of inquiry. Commissioner David Borthwick AO PSM, June 2010
12. PSA (2011) The deepwater horizon accident—evaluation and recommendations for Norwegian petroleum activity (in Norwegian only), PSA, 14 June 2011
13. Equinor (2010) Investigation report COA INV, internal accident investigation, Well incident on Gullfaks C (in Norwegian with English summary), Statoil, 4 November 2010. www.statoil.com
14. Austnes-Underhaug R et al (2011) Learning from incidents in Statoil (in Norwegian only), IRIS report 2011/156, 21 September 2011. www.statoil.com
15. ANP (2012) Investigation of the oil leak incident in the frade oil field. Final report, July 2012, Office of the superintendent of operational and environmental safety—SSM, ANP
16. Vinnem JE (1998) Blast load frequency distribution, assessment of historical frequencies in the North Sea. Preventor, Bryne, Norway; 1998 Nov. Report No.: 19816-04
17. Lord Cullen (The Hon) (1990) The public inquiry into the piper alpha disaster. HMSO, London
18. Skogdalen JE, Utne IB, Vinnem JE (2011) Developing safety indicators for preventing offshore oil and gas deepwater drilling blowouts. Saf Sci 49:1187–1199
19. ANP (2016) Investigation report of the explosion incident that occurred on 11/02/2015 in the FPSO Cidade de São Mateus. ANP
20. Brazilian Navy (2015) Maritime accident safety investigation report, FPSO Cidade de São Mateus, explosion followed by flooding with casualties, 11 February 2015. Brazilian Navy, Directorate of Ports and Coasts, Superintendence of Waterway Traffic Safety, Department of Inquiries and Investigations of Navigation Accidents
21. Morais C, Almeida AG, Silva BF, Ferreira N, Pires TS (2016) Explaining the explosion onboard FPSO Cidade de São Mateus form regulatory point of view. ESREL Conference. 2016
22. Vinnem JE (2018) FPSO Cidade de São Matheus Gas explosion—lessons learned. Safety Science. 2018, pp. 295-304
23. Vinnem JE, Kirwan B (1997) Safety of production and storage vessels with emphasis on operational safety. Report MK/R 131, 23 May 1997. NTNU, Department of Marine Technology

24. PSA (2005) Investigation of the gas leak from 6" export pipeline from Jotun A (in Norwegian only) PSA 23 December 2004. Stavanger, Petroleum Safety Authority
25. Borrevik LN, Horve M (2004) When the oil industry lost its 'Innocence' (In Norwegian only). Stavanger Aftenblad. 10 March 2004
26. NOU (1981) Alexander L. Kielland Accident, NOU Report 1998:11. Oslo
27. Royal Commission on the Ocean Ranger Marine Disaster (1984) Report one: the loss of the semi–submersible drill rig ocean ranger and its crew. St. John's, Newfoundland), Aug 1984
28. Vinnem JE, Hokstad P, Saele H, Dammen T, Chen H et al (2002) Operational safety of FPSOs, Shuttle Tanker collision risk. Main report. Trondheim; NTNU; 2002 Oct. Report No.: MK/R 152
29. ANP (2001) P–36 Accident analysis. Petrobras; ANP/DPC inquiry commission report; July 2001
30. Upstream weekly magazine (2002) P–34 alarm sounded in May. 2002 October 14
31. PSA (2005) Trends is risk levels. Main report 2004, Phase 5 (In Norwegian only) Report 05–02, Petroleum Safety Authority; 26.4.2005. http://www.ptil.no/NR/rdonlyres/87F129DD–9FA0–4F7E–A1BB–341D4585B84D/8903/Fase5rapportutenrestriksjoner.pdf
32. PSA (2005) Investigation of the anchor line failures on ocean Vanguard 14 December 2004, Well 6406/1–3. (In Norwegian only) PSA 23.5.2005. Stavanger; Petroleum Safety Authority. http://www.ptil.no/NR/rdonlyres/83A74F56–7F2D–470C–9A36–1153AADE50A7/7950/ovgrrappkomprimertny.pdf
33. Step Change in Safety (2011) Loss of heading, mooring system failure and subsequent loss of position. Alert ID 00289
34. EPA (2004) Exxon Valdez, article in webpages, updated 23 December 2004. http://www.epa.gov/oilspill/exxon.htm
35. SOA (2012) Investigation report with respect to Penglai 19-3 Oil Spill Accident, Final Report, July 2012

Chapter 5
Lessons from Macondo Accident

5.1 The Deepwater Horizon and Macondo Well

In March 2008, BP Exploration and Production Inc. (BP) leased the Mississippi Canyon Block 252 for oil and gas exploration and designated it the Macondo Prospect. BP sold interests in the prospect to Anadarko (25%) and MOEX (10%) but remained the operator and majority owner (65%). BP, as the operator, was responsible for all aspects of the design and development of the Macondo well.

BP's application for permit to drill was submitted to the Minerals Management Service (MMS) on 13th May 2009, and was approved on 22nd May 2009. The plan specified using the Transocean Marianas mobile unit to drill a well about 50 miles off the coast of Louisiana, southeast of New Orleans, in just over 1,500 m water depth. The total depth of the well was planned to be 6,100 m.

During the planning phase BP designed the well in accordance with the geological conditions of the prospect. During this process, BP engineers and geologists determined the type and strength of the well casing, cement, well head, and other equipment, in order to ensure well integrity and prevent its failure during drilling. This is crucial for well safety. Thereafter, selected contractors performed specific procedures such as drilling, cementing, well monitoring, vessel support services, and other well-related tasks.

At Macondo, BP began exploration on 6th October 2009, using the Transocean Marianas rig. Hurricane Ida damaged the Marianas on 9 November 2009, and drilling on Macondo was suspended following the installation and cementing of the well casing. The Marianas was demobilised to a shipyard for repairs, and BP applied to the MMS for permission to use the Transocean Deepwater Horizon rig to continue drilling. MMS approved this change on 14th January 2010.

The Deepwater Horizon, a fifth-generation, dynamically positioned, semi-submersible MODU, was capable of working in water up to 3,000 m. In 2009, the Deepwater Horizon crew drilled the deepest oil and gas well in the world, which had a vertical depth of more than 10.5 km.

© Springer-Verlag London Ltd., part of Springer Nature 2020
J.-E. Vinnem and W. Røed, *Offshore Risk Assessment Vol. 1*,
Springer Series in Reliability Engineering,
https://doi.org/10.1007/978-1-4471-7444-8_5

The Deepwater Horizon entered service in April 2001 and went to work for BP in the Gulf of Mexico in September. With the exception of one well drilled for BHP Billiton in 2005, the rig worked exclusively for BP. The Deepwater Horizon crew drilled more than 30 wells on the US OCS during the course of the rig's career, in water depths between 700 and 2,900 m, and maintained an excellent performance and safety record. BP extended its drilling contract on the Deepwater Horizon to September 2013 in September 2009.

The Deepwater Horizon arrived at Macondo on 31st January 2010. The crew performed maintenance work on the BOP stack, including function and pressure testing, before lowering it onto the wellhead on 8th February 2010. The crew then performed another successful pressure test of the BOP stack after it was attached to the wellhead. Drilling operations resumed on 11th February 2010.

The drilling of the Macondo well was performed as an overbalanced operation, which uses a mud column to prevent the influx of formation fluids into the well. The pressure of the mud column has to exceed the formation pore pressures encountered in the well. In overbalanced operations, the mud column is the primary well barrier, which has to exert a pressure greater than the pore pressure, but lower than the fracture gradient. The secondary well barrier in overbalanced operations is the well containment envelope consisting of selected components of the BOP. The Macondo was planned to be abandoned and left underbalanced by replacing drilling mud with seawater, and with two cement barriers in place.

Many deepwater reservoirs have such narrow drilling windows between the pore pressure and fraction gradient, that resolving one problem often creates another, and the resolution of that problem creates another, and so on until the cycle is broken with hydraulic balance or the well is abandoned. The operational drilling problems most associated with non-productive time include lost circulation, stuck pipes, wellbore instability and a loss of well control.

BP encountered a number of obstacles while drilling the Macondo well. Two cement repair operations, or squeezes, were required because of weak formations and possible problems with cement. On several occasions, fluid losses into the formation necessitated the use of lost-circulation material to stop the escape of fluids. On 8th March 2010, a 35 bbls influx of hydrocarbons, or "kick," occurred, sticking a section of drill pipe in the well. The drill crew had to plug the affected section of the well with cement and drill a side-track in order to continue. In early April, additional fluid losses to the formation prompted BP engineers to change the total planned depth of the well from 6,100 m to about 5,550 m to maintain its integrity.

After the completion of drilling operations on 9th April 2010, Schlumberger conducted a detailed analysis of the well's geological formations, or well logging, for BP over a period of approximately four-and-a-half days. The logging data from the new depth indicated that the well had reached a sizable reservoir of hydrocarbons. BP began planning for the next phase of the development, in which the Deepwater Horizon would run casing and prepare the well for temporary abandonment. On 16th April 2010, BP submitted its proposed temporary abandonment plan to the MMS and received approval the same day.

Upon reaching final well depth five days were spent logging the well in order to evaluate the reservoir intervals. After logging was complete, a cleanout trip was conducted to condition the wellbore and verify that the open hole section was in good condition. This procedure included circulating 'bottoms up' to verify that no gas was entrained in the mud. Upon achieving bottoms up, no appreciable volumes of gas were recorded, indicating that the well was stable.

On 16th April 2010 the MMS approved the procedure for the temporary abandonment of the well. At the time of the accident the $9\frac{7}{8}'' \times 7''$ production casing had been run and cemented in place at 5,527 m, and pressure testing had been completed. The rig crew was preparing for the final activities associated with temporary well abandonment when the accident occurred.

5.2   Organisations Involved

The well drilling project involves as noted above a number of contractors and subcontractors for specific tasks. The following is an overview of the involved in drilling the Macondo well:

BP	BP personnel in Houston, Texas, managed the development and operation of the Macondo well, and provided management and support to their personnel onboard the Deepwater Horizon. These onshore personnel consisted of three engineers, an engineer team leader, an operations team leader, and a manager. BP offshore personnel consisted of two well site leaders, a well site trainee, and three subsea personnel. Well site leaders exercised BP's authority on the rig, directed and supervised operations, coordinated the activities of contractors, and reported to BP's shore-based team
Transocean	Contracted to provide the Deepwater Horizon drilling rig and the personnel to operate it. The Transocean team included the drilling, marine, and maintenance crews. The senior Transocean personnel involved in day-to-day operations were the offshore installation manager (OIM) and captain. The OIM was the senior Transocean manager onboard who coordinated rig operations with BP's well site leaders and generally managed the Transocean crew. The captain was responsible for all marine operations and was the ultimate command authority during an emergency and when the rig was underway from one location to another. The Transocean drilling crew was led by a senior toolpusher, who supervised two toolpushers responsible for coordinating round-the-clock drilling operations. These toolpushers supervised the drillers and assistant drillers, who operated the drilling machinery and monitored the rig instruments. At the time of the accident, there were 79 Transocean personnel onboard the Deepwater Horizon, of which nine lost their lives
Halliburton	Contracted to provide specialist cementing services and expertise and to support the BP teams both onshore and on the Deepwater Horizon. At the time of the accident, two Halliburton cementing specialists were onboard the Deepwater Horizon

(continued)

(continued)

Sperry Sun	Contracted to install and operate a sophisticated well monitoring system on the Deepwater Horizon. Sperry deployed trained personnel, or mud loggers, to monitor the system, interpret the data it generated, and detect influxes of hydrocarbons, or kicks. At the time of the accident, there were two Sperry Sun mud loggers onboard the Deepwater Horizon
M-I SWACO	Contracted to provide specialised drilling mud and mud engineering services on the Deepwater Horizon, which included mud material, equipment, and personnel. At the time of the accident, there were five M-I SWACO personnel onboard the Deepwater Horizon, including two who lost their lives
Schlumberger	Contracted to provide specialised well and cement logging services on the Deepwater Horizon, which included equipment and personnel. At the time of the accident, no Schlumberger personnel were onboard the Deepwater Horizon
Weatherford	Contracted to provide casing accessories, including centralisers, the float collar, and the shoe track on the Deepwater Horizon. Weatherford also provided specialist personnel to advise BP and the drill crew on the installation and operation of their equipment. At the time of the accident, two Weatherford personnel were onboard the Deepwater Horizon
Tidewater Marine	Contracted to provide the offshore supply vessel the Damon B. Bankston. The Bankston carried supplies (such as drilling equipment, drilling chemicals, food, fuel oil, and water) to and from the Deepwater Horizon. At the time of the incident, the Bankston was alongside the Deepwater Horizon and provided emergency assistance. Other personnel onboard the Deepwater Horizon included 14 catering staff, two BP executives, and 14 BP subcontractors for a total of 126 personnel onboard

5.3 Sequence of Events

Twice, prior to the blowout on 20th April, the Macondo well experienced a "kick". The well kicked at 2,798 m. The rig crew detected the kick and shut in the well. They were able to resolve the situation by raising the mud weight and circulating the kick out of the wellbore. The well kicked again, at 4,018 m. The crew once again detected the kick and shut in the well, but this time, the pipe was stuck in the wellbore. BP severed the pipe and sidetracked the well. In total, BP lost approximately 16,000 barrels of mud while drilling the well, which cost the company more than $13 million in rig time and materials. The kicks, ballooning and lost circulation events at Macondo occurred in part because Macondo was a "well with limited offset well information and the preplanning pressure data [were] different than the expected case" [1].

The crew started on 20th April at around 20:00 replacing mud with seawater. Around 21:00 the drill string pressure started to increase, despite the pump rate being constant. Over the next 40 min there were several signs that they had problems, but nobody reacted to these signals in an appropriate manner. Between 21:40 and 21:43 mud started to spew out on the drill floor, and the driller realised apparently for the first time, that they had a kick.

The crew took immediate action, routing the flow coming from the riser into the mud-gas separator rather than overboard into the sea. Second, they closed one of the annular preventers on the BOP to shut in the well. Their efforts were futile. By the time the rig crew acted, gas was already above the BOP, rocketing up the riser, and expanding rapidly. The flow from the well quickly overwhelmed the mud-gas separator system. Ignition and explosion were all but inevitable. The first explosion occurred at approximately 21:49. On the drilling floor, the Macondo disaster claimed its first victims.

The sudden occurrence and impact of the explosion made it difficult for members on the bridge to assess the situation immediately following the incident. In addition, various alarms were sounding and lights were flashing, making it difficult for the crew to acknowledge what was going on.[1]

Typically as part of the evacuation procedure, once crew members reach the designated muster stations, they register their names so that a proper head count can be conducted and missing members can be accounted for. Based on the testimonies provided, there were efforts to prepare such a headcount, however there were difficulties when trying to accurately account for all members.

While crew members on the bridge were trying to assess the situation, others were already mustering near the lifeboats. Some were urging for the lifeboats to be launched despite them being only partially full. The Deepwater Horizon did have a split command depending on the status of the rig; latched up, underway, or in an emergency situation. The decision to evacuate the rig rested on the captain when the rig was in an emergency situation, but from the testimonies it seems to be unclear who was in charge due to missing procedures of handover and interpretation if the rig was latched up, underway or in an emergency situation [2].

Nevertheless, all persons who were not victims of the explosions and fires in the first few minutes were able to evacuate the installation, mainly by lifeboats, but also by life rafts, and a handful of persons jumping overboard.

The rig sank 36 h later. From then on the focus was on killing the blowout and fighting the pollution. The blowout was stopped by temporary measures, when a capping stack was installed and successfully stopped the flow on 15th July 2010, after 87 days. On 4th August BP reported that a final static condition in the well had been achieved by filling the well with mud. The total amount of oil spilled was evaluated to have been 650,000 tons.

5.4 Investigations

Several investigations have been conducted by various stake holders and official organisations. In 2012 court proceedings started, but this is outside the scope of the present discussion. The best known investigations are the following:

---

[1]A movie from 2016 illustrates quite well the challenges for the crew trying to control the event.

- BP investigation, September 2010 [3]
- National Commission on the BP Deepwater Horizon Oil Spill and Offshore Drilling, Report to the President, January 2011 [4]
- Deepwater Horizon Study Group (DHSG; author member of this group) Final Report on the Investigation of the Macondo Well Blowout [5]
- Transocean investigation, June 2011 [6]
- US Chemical Safety and Hazard Investigation Board, April 2016 [7].

5.4.1 Technical Aspects

According to the Commission, the root technical cause of the blowout was that the cement BP and Halliburton had pumped to the bottom of the well did not seal off the hydrocarbons in the formation. The exact reason why the cement failed may never be known, but several factors increased the risk of cement failure at Macondo. These included [1]:

- Drilling complications forced engineers to plan a 'finesse' cement job that called for, among other things, a low overall volume of cement;
- The cement slurry itself was poorly designed, some of Halliburton's own internal tests showed that the design was unstable, and subsequent testing by the Chief Counsel's team raised further concerns;
- BP's temporary abandonment procedures, only finalised at the last minute, called for rig personnel to severely 'underbalance' the well before installing any additional barriers to back up the cement job.

According to the Commission [8], BP's management process did not adequately identify or address the Risk Influencing Factors (RIFs) created by late changes to the well design and procedures. BP did not have adequate controls in place to ensure that key decisions in the months leading up to the blowout were safe or sound from an engineering perspective. While initial well design decisions undergo a rigorous peer-review process, and changes to well design are subsequently subject to a management of change (MOC) process, changes to drilling procedures in the weeks and days before implementation are typically not subject to any processes. At Macondo, such decisions seem to have been made by the BP Macondo team in an ad hoc fashion without any formal risk analysis or internal expert review [7].

According to the Chief Counsel's Report [1], several of BP's decisions, such as not using drill collars, not using a mechanical plug, setting the plug in seawater and, setting the lockdown sleeve last, may have been made in isolation. However, the decisions also created RIFs, individually and especially in combination with the rest of the temporary abandonment operation. For instance, BP originally planned to install the lockdown sleeve at the beginning of the temporary abandonment. BP's decision to change plans and set the lockdown sleeve last triggered a cascade of other decisions that led it to severely underbalancing the well while leaving the bottom hole cement as the lone physical barrier in place during the displacement of the riser. There is no evidence that BP conducted any formal risk analysis before making these changes, or even after the procedure as a whole [1]. BP's own investigative report agrees that they did not undertake a risk analysis to consider the consequences of its decision. BP's management system did not prevent such ad hoc decision-making. BP required relatively robust risk analysis and mitigation during the planning phase of the well but not during the execution phase [1]. Further, Transocean's crew seems never to have undertaken any risk analysis nor to have established mitigation plans regarding its performance of simultaneous operations after the cement barrier was judged to be safe [1].

5.4.2 Organisational Aspects

The following is a summary of the conclusions on the organisational aspects of the DHSG Macondo report [5] ('Looking back—Organizational factors'):

"The organizational causes of this disaster are deeply rooted in the histories and cultures of the offshore oil and gas industry and the governance provided by the associated public regulatory agencies. While this particular disaster involves a particular group of organizations, the roots of the disaster transcend this group of organizations. This disaster involves an international industry and its governance.

This disaster was preventable if existing progressive guidelines and practices been followed—the Best Available and Safest Technology. BP's organizations and operating teams did not possess a functional Safety Culture. Their system was not propelled toward the goal of maximum safety in all of its manifestations but was rather geared toward a trip-and-fall compliance mentality rather than being focused on the Big-Picture. It has been observed that BP's system "forgot to be afraid." The system was not reflective of one having well-informed, reporting, or just cultures. The system showed little evidence of being a high-reliability organization possessing a rapid learning culture that had the willingness and competence to draw the right conclusions from the system's safety signals. The Macondo well disaster was an organizational accident whose roots were deeply embedded in gross imbalances between the system's provisions for production and those for protection.

The multiple failures (to contain, control, mitigate, plan, and clean-up) that unfolded and ultimately drove this disaster appear to be deeply rooted in a multi-decade history of organizational malfunctions and short-sightedness. There were multiple opportunities to properly assess the likelihoods and consequences of organizational decisions (i.e., Risk Assessment and Management) that were ostensibly driven by BP management's desire to "close the competitive gap" and improve bottom-line performance. Consequently, although there were multiple chances to do the right things in the right ways at the right times, management's perspective failed to recognize and accept its own fallibilities despite a record of recent accidents in the U.S. and a series of promises to change BP's safety culture.

Analysis of the available evidence indicates that when given the opportunity to save time and money—and make money—poor decision making played a key role in accident causation. The tradeoffs that were made were perceived as safe in a normalized framework of business-as-usual. Conscious recognition of possible failure consequences seemingly never surfaced as the needle on the real-time risk-meter continued to climb. There was not any effective industry or regulatory checks and balances in place to counteract the increasingly deteriorating and dangerous situation on Deepwater Horizon. Thus, as a result of a cascade of deeply flawed failure and signal analysis, decision-making, communication, and organizational-managerial processes, safety was compromised to the point that the blowout occurred with catastrophic effects.

In many ways, this disaster closely replicates other major disasters that have been experienced by the offshore oil and gas industry. Eight months before the Macondo well blowout, the blowout of the Montara well offshore Australia in the Timor Sea developed in almost the same way—with very similar downstream effects. The Occidental Petroleum North Sea Piper Alpha platform explosions and fires (1988) and the Petrobras P–36 production platform sinking offshore Brazil (2005) followed roadmaps to disaster that are very similar to that developed during and after the Macondo well blowout. These were major system failures involving a sequence of unanticipated compounding malfunctions and breakdowns—a hallmark of system disasters.

This disaster also has eerie similarities to the BP Texas City refinery disaster. These similarities include: (a) multiple system operator malfunctions during a critical period in operations, (b) not following required or accepted operations guidelines ("casual compliance"), (c) neglected maintenance, (d) instrumentation that either did not work properly or whose data interpretation gave false positives, (e) inappropriate assessment and management of operations risks, (f) multiple operations conducted at critical times with unanticipated interactions, (g) inadequate communications between members of the operations groups, (h) unawareness of risks, (i) diversion of attention at critical times, (j) a culture with incentives that provided increases in productivity without commensurate increases in protection, (k) inappropriate cost and corner cutting, (l) lack of appropriate selection and training of personnel, and (m) improper management of change.

In both cases—the BP Texas City and the BP Macondo well disasters—meetings were held with operations personnel at the same time and place the initial failures were developing. These meetings were intended to congratulate the operating crews and organizations for their excellent records for worker safety. Both of these disasters have served—as many others have served—to clearly show there are important differences between worker safety and system safety. One does not assure the other.

In all of these disasters, risks were not properly assessed in hazardous natural and industrial-governance-management environments. The industrial-governance-management environments unwittingly acted to facilitate progressive degradation and destruction of the barriers provided to prevent the failures. An industrial environment of inappropriate cost and corner cutting was evident in all of these cases as was a lack of appropriate and effective governance—by either the industry or the public governmental agencies. As a result, the system's barriers were degraded and destroyed to the point where the natural environmental elements (e.g., high-pressure, flammable fluids and gases) overcame and destroyed the system. Compounding failures that followed the triggering failures allowed the triggering failures to develop into a major disaster—catastrophe."

5.5 Findings

The following is a summary of the findings of the DHSG Macondo report [5] ('Looking forward' and 'Findings'):

"Short-term measures have been initiated and are being developed by the Department of Interior's Bureau of Offshore Energy Management, Regulation and Enforcement (BOEMRE). The previous Minerals Management Service (MMS) has been reorganized into three organizations, each of which is responsible for different aspects of offshore oil and gas developments (leasing, revenues, and regulation). These measures have addressed both technical and organizational aspects. In some cases, the BOEMRE has proposed long-term technical and organizational measures associated with drilling and production operations in ultra-deepwater (5,000 ft or more) depths.

In addition, the National Commission on the BP Deepwater Horizon Oil Spill and Offshore Drilling has addressed both short-term and long-term government regulatory technical and organizational reforms associated with drilling and production operations in high hazard environments—including those of ultra-deep water and in the arctic. U.S. industrial companies and trade organizations (e.g., American Petroleum Institute, International Association of Drilling Contractors) also have responded with suggestions for a wide variety of technical and organizational reforms that will be considered for implementation in its future operations. Many international governmental regulatory agencies have and are responding in a similar fashion. There is no shortage of suggested technical and organizational reforms.

Finding 1—The oil and gas industry has embarked on an important next generation series of high hazard exploration and production operations in the ultra-deep waters of the northern Gulf of Mexico. These operations pose risks (likelihoods [sic] and consequences of major system failures) much greater than generally recognized. The significant increases in risks are due to: (1) complexities of hardware and human systems and emergent technologies used in these operations, (2) increased hazards posed by the ultra-deep water marine environment (geologic, oceanographic), (3) increased hazards posed by the hydrocarbon reservoirs (very high potential productivities, pressures, temperatures, gas-oil ratios, and low strength formations), and (4) the sensitivity of the marine environment to introduction of very large quantities of hydrocarbons.

Finding 2—The Macondo well project failures demonstrated that the consequences of major offshore oil and gas system failures can be several orders of magnitude greater than associated with previous generations of these activities. If the risks of major system failures are to be ALARP, the likelihoods of major failures (e.g., uncontrolled blowouts, production operations explosions and fires) must be orders of magnitude lower than in the BP Macondo well project and that may prevail in other similar projects planned or underway. In addition, major developments are needed to address the consequences of major failures; reliable systems are needed to enable effective and reliable containment and recovery of large releases of hydrocarbons in the marine environment.

Finding 3—The Macondo well project failures provide important opportunities to re-examine the strategies and timing for development of important non-renewable product and energy resource. This final frontier in the ultra-deep waters of the northern Gulf of Mexico and other similar areas provides access to an important public resource that has significant implications for the future generations and energy security of the United States. These social, economic, and national security Deepwater Horizon Study Group Investigation of the Macondo Well Blowout Disaster interests, as well as safety and environmental considerations, dictate a measured pace of development consistent with sustainable supplies and development and application of the Best Available and Safest Technology (BAST).

Finding 4—Major step change improvements that consistently utilize the BAST are required by industry and government to enable high hazard offshore exploration and production operations to develop acceptable risks and benefits. Future development of these important public resources require an advanced high-competency, collaborative, industrial-governmental-institutional enterprise based on use of high reliability technical, organization, management, governance, and institutional systems."

5.6 Lessons Learned

5.6.1 Lessons Learned for Risk Management in Association with Well Drilling

Section 5.4.2 discussed the organisational factors that failed in the case of Macondo. These aspects provide important lessons for the improvement of the safety of well drilling, especially the drilling of ultra deep wells.

Other studies have shown that two high level aspects are crucial for the safety of well drilling, namely the well planning ahead of commencement of operations, and the management of change (or continuous risk assessment) during the execution of well drilling operations. These findings tie in well with the findings cited above.

5.6.2 Lessons Learned for Emergency Management

This section addresses the lessons that may be concluded from Macondo with respect to escape, evacuation and rescue (EER) as well as technical and organisational aspects. The text is mainly based on Ref. [9].

Explosions, fire and smoke were life-threatening hazards during the EER from the Deepwater Horizon. On offshore installations, the crew is familiar with the facility and escape routes. They also participate in regular muster and lifeboat drills. To determine which measures would reduce the time to make decisions, and which steps would lead to people choosing the right egress routes, information is needed regarding human and organisational factors. Of special interest are the perceptions, intentions and motives of the personnel when faced with such situations.

However, is noteworthy that it seems that all of those who were not seriously injured during the initial explosion, and even a few of those who were seriously injured during the first stages, were able to escape to lifeboats and were rescued.

One of the important roles of the master of the vessel is to take charge during a crisis, and to give the order to abandon ship if necessary. The master should assess the severity of the situation properly, and if the decision to abandon is made, the master would then give the order to launch the lifeboats and evacuate the installation. Lowering the lifeboats at the right time is critically important for an effective evacuation, because there are a limited number of lifeboats on an installation. If proper communication is not achieved lifeboats can be launched only partially filled up, resulting in personnel being left behind. By contrast, if members wait too long to launch the lifeboats, they risk being harmed by explosions, fire, smoke, and falling objects.

The Deepwater Horizon had a split chain of command between the OIM and the captain, which seemed to have caused confusion as the lines of authority and the shift of responsibility in the event of an emergency apparently were unclear to some crew members. A critical question to be considered is how can it be relied upon that the captain of the vessel (or installation) will be in a sufficiently healthy condition to perform critical tasks, such as properly assessing the situation, activating the alarm, and giving the order to muster and abandon ship? Or even to discover that the captain is unable to perform his/her function. This was the problem the crew faced following the explosion on board the Piper Alpha, with fatal consequences. During a crisis, it is possible that situations will occur where bypassing the chain of command is unavoidable and necessary; however, the situation must be properly assessed by individuals such that the result is not detrimental to the safety and success of the operations. This can be accomplished through proper training on the worst case scenarios.

It is expected that in some cases, not all members will be able to evacuate using the primary means of evacuation and therefore rescue means are necessary to ensure the safe evacuation of the personnel left behind or not able to make it to safe refuge. As seen in the case of the Deepwater Horizon, there was a need for secondary means of evacuation. In addition to life rafts, these can be escape chutes and ladders. Personal survival suits with splash protection extend the available rescue time due to increased protection from waves and hydrocarbons in the sea. They also extend time before hypothermia.

Several hazards faced those on the Deepwater Horizon who prepared to jump to the sea. Among those hazards were the height from the platform deck to the surface of the water from which they have to jump, the possible fires on the sea level and smoke inhalation. Ideally, the crew would have to get as close as possible to the water surface before jumping or entering the sea. Under some circumstances jumping into the sea is necessary, and offshore personnel should be trained to do this as safely as possible.

The supply ship Damon Bankston played a vital role in rescuing survivors from the Deepwater Horizon. Given the remote location of deepwater operations, nearby vessels play a critical role in rescuing personnel from offshore installations following major accidents. A fast response is especially important with a high number of personnel in the sea and/or in the case of bad weather. Custom designed third generation rapid response rescue vessels are available. They are specially designed to launch and recover a fast rescue craft or daughter craft from a slipway in the stern. The slipway can also be used to recover a lifeboat from the sea. The sea trials of these vessels are promising and it is generally considered to be possible to operate in sea conditions with significant wave heights of up to Hs = <9 m [10].

The distance from shore to the Deepwater Horizon (66 km) meant that it took several hours for rescue boats from shore to arrive. The US Coast Guard scrambled HH-65C Dolphin helicopters when they received the mayday call from the Deepwater Horizon. These helicopters have a limit of rescuing three to four persons. Helicopters did not contribute to the rescue of personnel in the Macondo accident.

The Norwegian system for area based emergency preparedness arrangements [11, 12] may be a relevant model to consider also for the Gulf of Mexico, in order to quickly and efficiently rescue personnel in emergencies. This system includes use of offshore based Search and Rescue (SAR) helicopters as well as fast rescue crafts, in order to provide rescue capabilities for the relevant number of personnel within 120 min from an emergency.

Training, knowledge, experience, and competence are important throughout all steps of EER operations, and for some steps, it is purely human actions that can ensure the success of the operation. Emergency drills and training have limitations on preparing the crew to deal with real-life emergencies and unanticipated events. However, proper training and knowledge can provide the basic ability to cope with evacuation scenarios.

5.7 Similarity Between Offshore and Nuclear Accidents

One lesson learned from the Macondo blowout which is somewhat special, is the similarity between this accident and nuclear accidents. If we think about nuclear accidents such as Three Mile Island, Chernobyl, Fukushima they have worldwide effects, it does not matter which country it occurred, there will be worldwide repercussions.

For offshore petroleum it has often been claimed that unless it takes place very close to shore, there is normally no 3rd party personnel risk to consider. Before the Macondo blowout, this could have been said without the "personnel" word in. However, Macondo has demonstrated that the accident has worldwide repercussions, and is "everybody's accident". This is a similarity between nuclear and offshore petroleum, when it comes to accidental pollution. Then one country's accident is everybody's accident.

References

1. Bartlit JFH, Sankar SN, Grimsley SC (2011) Macondo—the Gulf oil disaster—chief counsel's report. National commission on the BP deepwater horizon oil spill and offshore drilling
2. DHJIT (2010) USCG/BOEM marine board of investigation into the marine casualty, explosion, fire, pollution, and sinking of mobile offshore drilling unit deepwater horizon, with the loss of life in the Gulf of Mexico April 21–27, 2010. Deepwater Horizon Incident Joint Investigation Team, The U.S. Coast Guard (USCG)/Bureau of Ocean Energy Management, Regulation and Enforcement (BOEMRE) Joint Investigation Team (JIT)
3. BP (2010) Deepwater horizon accident investigation report, 2010
4. Commission (2011) National commission on the BP deepwater horizon oil spill and offshore drilling. Report to the President, January 2011
5. Deepwater Horizon Study Group (2011) Final report on the investigation of the Macondo Well Blowout. UC Berkeley center for catastrophic risk management, March 2011
6. Transocean (2011) Macondo well incident, transocean investigation report, June 2011
7. CSB US (2016) Chemical safety and hazard investigation board Macondo investigation report, April 2016
8. Graham B, Reilly WK, Beinecke F, Boesch DF, Garcia TD, Murray CA et al (2011) Deep water. The Gulf oil disaster and the future of offshore drilling. Report to the President. The national commission on the BP deepwater horizon oil spill and offshore drilling, Washington DC, USA
9. Skogdalen JE, Khorsandi J, Vinnem JE (2012) Evacuation, escape and rescue experiences from offshore accidents including the deepwater horizon. J Loss Prev Process Ind 25(1):148–158
10. Jacobsen RS (2010) Evacuation from petroleum facilities operating in the Barents Sea, 47
11. Norwegian Oil and Gas Association (2012) Recommended guidelines for establishing area based emergency response (in Norwegian only). Norwegian Oil and Gas Association, 10 Sept 2012
12. Skogdalen JE, Vinnem JE (2012) Quantitative risk analysis of oil and gas drilling, using deepwater horizon as case study. Reliab Eng Syst Saf 100:58–66

Part II
Analysis of Main Offshore Hazards

Chapter 6
The Occurrence of Hydrocarbon Leaks—Process Systems

6.1 Statistical Sources

Health & Safety Executive (HSE) in the UK has collected data on hydrocarbon (HC) releases since October 1992 as part of the follow-up of recommendations made by Lord Cullen in the investigation into the Piper Alpha accident in 1988 [1].

The Petroleum Safety Authority started the collection of data on HC leaks (corresponding to the term 'release' in the UK) in 2000, the first year covered the period 1996–2000. There are important differences between the collection schemes of the UK and Norway with respect to the systems covered and the classification of incidents. This is discussed for the two sectors separately in Sects. 6.2 and 6.3.

The International Regulators' Forum has collected key statistics for health and safety for some few years, published on www.irfoffshoresafety.com. A comparison between what is reported from the relevant offshore sectors is presented based on the statistics from the IRF.

6.2 Statistics from the UK Sector

6.2.1 Classification of Releases

HSE collects data on HC releases based on reporting according to its RIDDOR (Reporting of Injuries, Diseases and Dangerous Occurrences Regulations) reporting system [2]. All potential sources are covered, including wells, process systems and all utility systems. All types of hydrocarbons are covered, including gas, two-phase, unstabilised and stabilised liquid petroleum, in addition to refined hydrocarbon products such as lube oil, hydraulic oil, seal oil and diesel.

© Springer-Verlag London Ltd., part of Springer Nature 2020
J.-E. Vinnem and W. Røed, *Offshore Risk Assessment Vol. 1*,
Springer Series in Reliability Engineering,
https://doi.org/10.1007/978-1-4471-7444-8_6

Table 6.1 Overview of HSE classification criteria for hydrocarbon releases

Parameter	Release category limits		
	Minor release	Significant release	Major release
Gas/two-phase releases			
Quantity or combination of • release rate and • duration	<1 kg <0.1 kg/s <2 min	Between major and minor	>300 kg >1 kg/s >5 min
Liquid releases (Oil/condensate/Non-process)			
Quantity or combination of • release rate and • duration	<60 kg <0.2 kg/s <5 min	Between major and minor	>9,000 kg >10 kg/s >15 min

The releases are categorised into three broad categories, which reflect hazard potential. The classification criteria reflect amount, rate and duration, and distinguish between gas/two-phase and liquid releases (Table 6.1).

The most noteworthy aspect is that there is no lower limit. In our view, this is a severe weakness, which causes unnecessary uncertainty and difficulty for the interpretation of trends.

There will obviously be different interpretations made by different organisations (and individuals) about what is too small to report. The number of leaks in the lowest (minor releases) category will therefore never be complete.

It is also noted by HSE that minor leaks are usually too small to cause escalation of accidental effects. Therefore, interest in minor leaks should be less. Norwegian data collection has decided to omit the smallest leaks (see the discussion in Sect. 6.3).

6.2.2 Statistical Overview

Figure 6.1 presents the HC releases recorded by HSE in the period 1996–2015. It should be noted that reporting periods in the RIDDOR system go from 1st April until 31st March the following year. For simplicity, the years are denoted '1996'; '1997', and so on, although the extensive reference would be 1.4.1996–31.3.1997. Figure 6.2 presents the trend over the same period with 'minor' releases omitted.

Figure 6.2 shows that significant leaks have decreased by a factor of three over the 20–year period. Major leaks have decreased more in fact. The three-year average was more than 17 in 1998, which dropped to 4 per year in 2015. UK trends are further discussed compared with Norwegian statistics in Sect. 6.4.

It is important to stress that the UK classification criteria presented in Table 6.1 are quite different from the Norwegian classification criteria shown in Table 6.2. This is demonstrated in Fig. 6.3, which compares UK and Norwegian classifications for the period 2006–2010 (actually 1.4.2006–31.3.2011), as this is a period where comparable data have been available. Figure 6.3 shows that there are more

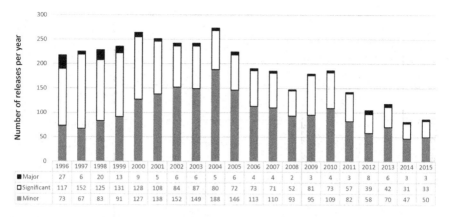

Fig. 6.1 Overview of UK sector HC releases, 1996–15

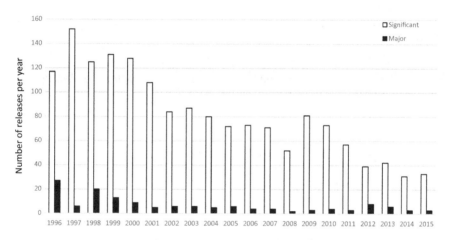

Fig. 6.2 Overview of UK sector HC releases 1996–15, minor releases omitted

Table 6.2 Overview of PSA classification criteria for HC releases

Parameter	Release category limits		
	0.1–1 kg/s	1–10 kg/s	>10 kg/s
Gas/two-phase releases			
Initial release rate	0.1–1 kg/s	1–10 kg/s	>10 kg/s
Liquid releases (Oil/condensate)			
Initial release rate	0.1–1 kg/s	1–10 kg/s	>10 kg/s

releases >1 kg/s compared with 'major' releases, for gas, two-phase and oil releases. When 'significant' releases are compared with releases with a rate in the range 0.1–1 kg/s, there are far more 'significant' releases than 0.1–1 kg/s releases, especially for oil releases. These differences imply that the criteria associated with

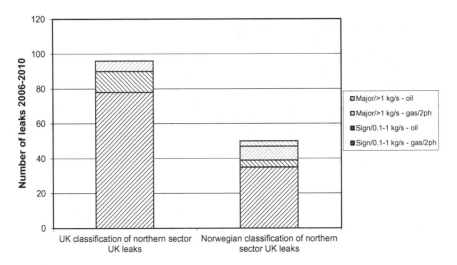

Fig. 6.3 UK and Norwegian classification of significant and major releases in UK northern sector, 2006–10

volume alone (see Table 6.1) must play a strong role in the classification of releases, especially for 'significant' releases.

6.3 Statistics from the Norwegian Sector

6.3.1 Classification of Releases

NPD (later the PSA) started to collect data on HC leaks (corresponding to the UK term 'releases') as part of its Risk level project (RNNP), which started data collection in 2000. Leak data are categorised as 'DFU1' and the data are collected using a special reporting format for RNNP.

All potential sources are covered, including wells, process systems and all fuel gas systems as the only utility system (see also Sect. 6). The only types of HCs covered are gas, two-phase, unstabilised and stabilised liquid petroleum, but none of the refined products used on board.

Leaks were originally categorised into three broad categories, but this has been supplemented by the estimated leak rate values. The classification criteria are limited to initial leak rate, because this is considered to be sufficient in order to characterise hazard potential (see Table 6.2).

For reporting to the PSA, there is a lower cut off limit, 0.1 kg/s. This is done in order to avoid underreporting and uncertainty as far as possible as well as to reflect the inability of leaks below 0.1 kg/s to cause escalation.

6.3.2   Statistical Overview

Figure 6.4 shows the trend for HC leaks with a flowrate above 0.1 kg/s. It can be seen that the highest values occurred in 2000 and 2002, with more than 40 leaks per year. This reduced to ten in 2007, followed by an increase to about 15 in the period 2008–10. The lowest number ever was seen in 2012, with six leaks. In 2015–2017, the number of leaks per year with initial leak rate above 0.1 kg/s has been around 10.

Figure 6.5 presents the total number of leaks normalised against installation years, for fixed and floating production installations, from the PSA based on the

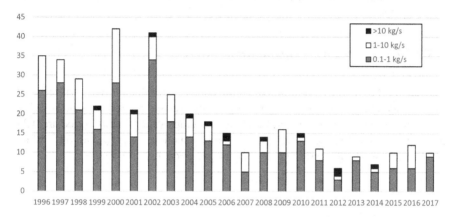

Fig. 6.4   Overview of Norwegian sector HC leaks, 1996–2017

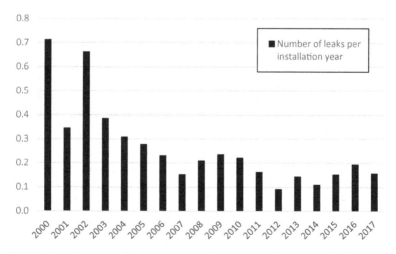

Fig. 6.5   Trend, leaks, normalised against installation year, fixed and floating production installations

RNNP annual report for 2017 [3]. The number of installation years has not changed dramatically during the period, and therefore the shape of the curve is similar to Fig. 6.4. The relative number of leaks per installation year has decreased from over 0.7 per year in 2000 to less than 0.1 per year in 2012. In all years after 2012, there have been between 0.1 and 0.2 leaks per year.

The diagram illustrates a significantly falling trend in the number of HC releases above 0.1 kg/s flowrate since the high levels in 2000–2002. The industry defined an ambitious goal for leak frequency reduction at the beginning of 2003 in order to reduce the major hazard risk. The considerable drop in the frequency of HC leaks is one of the noteworthy results of the RNNP work and efforts made by the industry. When considering the last 5–year period in the diagram from 2013–2017, the trend has been stable or slightly increasing.

6.3.3 Comparison of Installation Types

The data in RNNP allow for different comparisons of installation types with respect to leak frequency. One example of such a comparison is presented in Fig. 6.6, which is based on the period 1996–2017. The average values in this period are 0.28 leaks/year and 0.73 leaks/year for floating production, storage and off-loading (FPSO) and semi-submersible/TLP concepts, respectively.

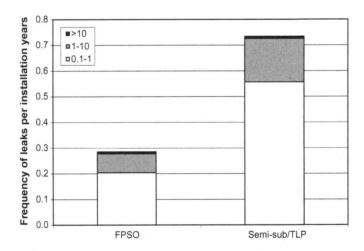

Fig. 6.6 Comparison of leak frequencies for two types of floating production concepts, NCS, 1996–2017

6.3.4 Installations with the Highest Leak Frequency Per Installation Years

Table 6.3 presents the list of installations with highest average frequency (per installation year) on the NCS. This list is in principle the "top 10", but it includes all installations with at least two leaks during the period. The basis for the list is leaks per installation year, corresponding to the listing in RNNP reports.

The listing in Table 6.3 corresponds to the presentation in RNNP and does not differentiate between large and small installations, manned and unmanned, old and new installations, complex and simple installations or installations with extensive or limited amounts of manual interventions. This has been the recognised weakness of the listing in RNNP for years.

In the RNNP 2010 report [4], the corresponding list based on years 2005–10, is AU and AX (joint first), AI, AÆ and AG (joint second). The most significant difference is that installation AG in no longer in the top 10 in Table 6.3.

6.3.5 Installations with the Highest Leak Frequency Per Leak Source

The number of leak sources was collected for the majority of production installations on the NCS during the Risk_OMT research project (available for all installations in Table 6.3 except the last one). This information should to some extent

Table 6.3 Installations on the NCS (anonymous) with at least two leaks in 2008–11, sorted according to falling average leak frequency

#	Installation code	Average number of leaks per year 2008–11
1	AU	1.0
2	AI	0.75
2	AX	0.75
2	BC	0.75
2	BK	0.75
2	BW	0.75
7	AJ	0.50
7	AP	0.50
7	AR	0.50
7	AW	0.50
7	AÆ	0.50
7	BR	0.50
7	AY	0.50
7	D	0.50

Table 6.4 Installations on the NCS (anonymous), top 10 list in 2008–11, sorted according to falling average leak frequency per 1,000 leak points

#	Installation code	Average number of leaks per 1,000 leak points 2008–11
1	AJ	1.12
2	BW	0.81
3	AX	0.64
4	AY	0.59
5	AM	0.46
6	BK	0.45
7	AØ	0.40
8	BC	0.33
9	AU	0.31
10	AN	0.24

reflect the technical complexity of different installations. The number of leak sources is obviously not a perfect representation of complexity, but no better representation can easily be found. This information has not been used so far to normalise the data in Table 6.3. Such normalisation is presented in Table 6.4.

Six of the installations on the top 10 list in Table 6.4 are also present in the top 10 list in Table 6.3. Two of the three top positions are common for the two lists, but there are also significant differences.

6.3.6   Installations with the Highest Leak Frequency Per Number of Operations

Data have been made available with respect to the number of work permits issued for work in the process areas of installations (all the installations in Table 6.3 except one). It has been shown in previous RNNP reports that about 60–70% of leaks are due to manual interventions; therefore the number of work permits should be a reasonable normalisation parameter. The normalisation according to the number of work permits is presented in Table 6.5.

Five of the installations in Table 6.5 are also on the top 10 list in Table 6.3. None of the installations in the top three in Table 6.5 are also in the top three of Table 6.3. Only one the installations in the top three in Table 6.4 is also in the top three of Table 6.3.

Table 6.5 Installations on the NCS (anonymous), top 10 list in 2008–11, sorted according to falling average leak frequency per 1,000 work permits

#	Installation code	Average number of leaks per 1,000 work permits 2008–11
1	AØ	10.2
2	BW	6.0
3	AY	5.0
4	BV	3.8
5	AJ	3.6
6	BC	3.4
7	AX	3.3
8	AU	3.2
9	AK	2.5
10	AV	2.5

6.3.7 Installations with Highest Leak Frequency with Combined Parameters

Table 6.6 is based on leak points as well as work permits, each with an equal (50%) weighting. Six of the top 10 installations in Table 6.6 is also in the top 10 per installation years in Table 6.3. Two of the top three are the same in both tables.

Table 6.6 Installations on the NCS (anonymous), top 10 list in 2008–11, sorted according to falling average leak frequency per 1,000 leak points and 1,000 work permits, each with a 50% weighting

#	Installation code	Average number of leaks per 1,000 leak points and work permits 2008–11
1	AJ	1.7
2	BW	1.4
3	AX	1.1
4	AY	1.1
5	AØ	0.8
6	BK	0.7
7	AM	0.7
8	BC	0.6
9	AU	0.6
10	BA	0.3

6.3.8   Comparison of Different Normalizations

Figure 6.7 summarises the different normalisations, showing that normalisation according to installation years provides an incomplete picture.

Please note that the diagram is somewhat specialised. Installations with the highest leak frequencies are the ones with highest bars, according to Tables 6.3, 6.4, 6.5, 6.6. The values plotted in Fig. 6.7 are the same values as those in Tables 6.3, 6.4, 6.5, 6.6, with the exception that the values in Fig. 6.7 have been divided by 10 in order to fit to a common scale.

The different series in Fig. 6.7 correspond to the data in Tables 6.3, 6.4, 6.5, 6.6. If the rankings were consistently independent of which parameter was used, the rankings would have been similar for all installations in the diagram.

The four installations that altogether had the highest leak frequencies for all parameters combined are BW, AJ, AX, and AY. Interestingly, these installations represent one old (>25 years in operation), two medium aged (10–20 years in operation) and one new (< 5 years in operation) installation at the point of time when the data were collected. Hence, age does not seem to explain the observed differences.

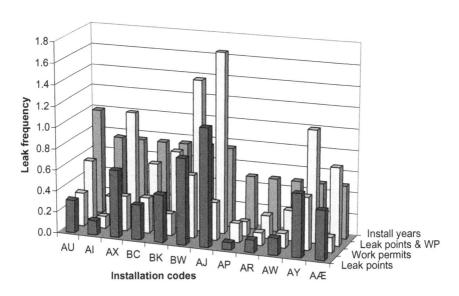

Fig. 6.7 Comparison of ranks according to different normalisations

6.4   Comparison of the UK and Norwegian Sectors

6.4.1   Comparison of Unignited Leaks

The RNNP has carried out several studies comparing Norwegian and UK leak frequencies on a per installation-year basis. The following section is based on such a comparison with data until 2010. The study has been limited to leaks exceeding 1 kg/s and to the northern North Sea, including areas further north in the Norwegian sector. The cut-off limit at 1 kg/s is primarily implemented in order to ensure, as far as possible, that reporting uncertainty is eliminated.

The limitation to the northern North Sea can be explained as follows: there are a large number of small and simple installations in the southern part of the UK Continental Shelf (UKCS). These installations are not comparable to the majority of the installations on the NCS, or even with the installations in the UK northern North Sea. The limitation to the northern North Sea ensures 'like for like' comparison, because the installations are largely of the same types, some large, relatively old installations, and some large, new floating installations. By coincidence, the number of installations is also about the same.

The line separating the northern North Sea from the central North Sea on the UKCS is the 59°N, implying that all fields south of Stavanger are excluded. Normally unmanned installations and mobile drilling units are also excluded.

It is further important to stress that UK leaks from the northern North Sea sector have been reclassified according to the criteria shown in Table 6.2 in order to ensure that classification is consistent. The comparison is presented in Fig. 6.8 (based on PSA 2012).

Figure 6.8 compares the NCS with the UKCS, in which gas/two-phase leaks and oil leaks are both included, and normalised against installation year, for the two respective continental shelves north of 59°N. The figure applies to the periods 2000–10 and 2006–10, respectively. The data included in the figure are limited to process facilities in which oil leaks have occurred. In this period, there was also one leak per year in shafts in connection with storage cells in the northern sector of the UKCS and one leak every third year in connection with tank operations on production ships or storage tankers. No corresponding leaks occurred in this period on Norwegian production installations but in 2008, there was a major oil and gas leak in the shaft on Statfjord A on the NCS. These leaks are not included in the diagram.

The number of leaks on the NCS has been considerably lower in recent years, and thus the period in question is of some significance (this is also illustrated in Fig. 6.8). For example, the following observations can be made from the data on average leak frequency per installation year for all leaks exceeding 1 kg/s:

- For the period 2000–10 the NCS was 41% **higher** than the UKCS
- For the period 2006–10 the NCS was 5.9% **lower** than the UKCS

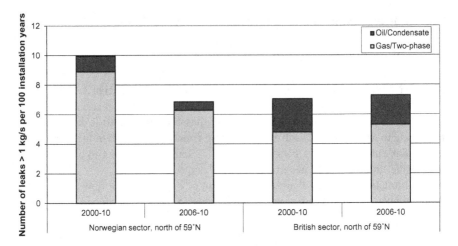

Fig. 6.8 Trend, leaks, normalised against installation year, fixed and floating production installations

6.4.2 Detailed Comparison

A detailed comparison of leak frequencies between the UK and Norwegian sectors was carried out some years ago, using data from the period 2000–04. This was a period when the frequency of leaks was considerably higher in the Norwegian sector, and thus the applicability of the results is somewhat uncertain. The circumstances of leaks have been surprisingly stable through this period, which implies that the results may also be relevant for later periods (see also Fig. 6.16). The details are presented in Ref. [5], upon which the following summary has been formulated.

At the time of the study two Norwegian oil companies together accounted for 51% of the production facilities on the NCS. When it comes to the distribution of leak scenarios, these two companies were quite different from the other oil companies operating in the Norwegian sector. These two companies experienced 82% of leaks in the period 2001–05. Table 6.7 presents the distribution of circumstances when leaks occur split into these two Norwegian companies and all other companies. The same main categories are used as shown in Fig. 6.16.

It should be noted that a detailed analysis of UK data in the same fashion has never been carried out. This is a summary of a coarse analysis based on the available data, but with limited details available. With respect to 'HC release' data from the UK, the period applied was 1.4.2000–31.3.2004. The following limitations have been placed in order to arrive at a consistent and applicable data set:

- Only releases for the northern North Sea (north of 59°N) are used
- Non-process HC release is disregarded (lube/seal oil, mud, etc.)
- Well fluid leaks as well as stabilised oil leaks are disregarded

Table 6.7 Distribution of leak scenarios per 100 installation years for the two groups of companies

Circumstance of leak	Two Norwegian companies	Other companies
Technical faults	14.0	6.9
Latent errors from intervention	27.3	3.4
Immediate errors from intervention	6.7	1.4
Process control	6.7	1.4
Design error	6.7	0.7
Total	61.3	13.8

- The release categories 'major' and 'significant' are used.

The arguments for choosing the northern North Sea are similar to those used in the RNNP for making comparisons in the same area (see Sect. 6.4.1). The main argument is that installations are similar, as the majority are in the Norwegian sector.

The exclusion of non-process leaks is clear, because only process leaks are analysed for Norwegian installations. It is further obvious that the circumstances surrounding HC leaks are quite different from those associated with auxiliary systems, drilling systems and so on.

Broadly speaking, the category 'significant' corresponds to leak rates in the range 0.1–1 kg/s, whereas 'major' leaks correspond to leak rates exceeding 1 kg/s. However, some additional criteria apply in the UK system, relating to the duration of the leak as well as the total HCs released. This has some effect on the number of leaks in each category and thus on the frequencies that can be calculated.

In the RNNP report [6], the differences between these two approaches was illustrated. When identical criteria were used, the number of applicable gas and two-phase releases increased from three to seven for the same area and the same period. In the present case, however, the release categories 'significant' and 'major' have been used for the UK sector, showing some inconsistencies compared with leaks exceeding 0.1 kg/s for the Norwegian sector.

A total of 103 HC releases were recorded for the northern UK sector in the four-year period. These records include codes for design error, technical failure, operational error and procedural error in addition to short verbal descriptions. The codes were in some cases conflicting, and thus verbal descriptions were in general trusted more than codes. The information was in some cases insufficient in order to conclude about the scenario for the leak.

The leak scenario could be determined for 76 releases, which were used in the subsequent analysis. The information was in some cases relatively sparse, to the extent that the main scenario category could be determined, but not the detailed circumstances. For instance, a technical failure was evident in several circumstances, but the exact reason for it was not clear from the limited description. This resulted in many leaks being categorised as 'other technical failures'. However, this

Table 6.8 Distribution of leak scenarios per 100 installation years for all UK and Norwegian companies

Circumstance of leak	UK installations	Norwegian installations
Technical faults	11.8	11.2
Latent errors from intervention	4.1	15.3
Immediate errors from intervention	1.7	4.1
Process control	0.2	4.8
Design error	0.5	2.7
Total	18.2	38.1

does not imply that the distinction between technical, operational or other scenarios is more uncertain.

It is known from previous studies that there might be large differences compared with the distribution found for the Norwegian sector, particularly that the contribution from operational errors could be much lower. As a result, the assessment, although as objective as possible, has erred somewhat in the direction of assuming the existence of operational errors.

The UK results are compared with Norwegian installations in Table 6.8. Whereas Norwegian data show a majority of manual interventions, an even clearer majority applies to technical faults in the UK sector.

Releases associated with human intervention have roughly the same relationship between immediate releases and delayed releases, due to latent errors, with a clear dominance for the second category. However, whereas the Norwegian contribution is 54%, the corresponding value in the UK sector is 28%. It may be argued that the distribution in the UK sector is quite close to the distribution in the Norwegian sector, when the two large Norwegian companies are omitted from the data set.

The message in Table 6.8 is interesting. The frequency per 100 installation years in the UK sector is slightly less than 50% of the corresponding value for the Norwegian sector. It should be noted that this comparison is based on all leaks exceeding 0.1 kg/s in both sectors, implying that identical criteria are used.

6.4.3 Comparison of Ignited Leaks

On the NCS, no occurrences of ignited HC leaks (above or equal to 0.1 kg/s) have been registered since 1992. The number of HC leaks above 0.1 kg/s since 1992 on the NCS is probably in the range of 500. There is evidence that the number of ignited leaks on the NCS is significantly lower than that on UKCS, where approximately 1.5% of gas and two-phase leaks since 1992 have been ignited.

It seems to be inexplicable why over many years there have been fewer HC leaks on the UKCS compared with Norway, whereas the control of ignition sources (or the 'prevent ignition' barrier function) has been so much better in the Norwegian sector. It was demonstrated in Sect. 6.4.2 that there seems to be more leaks due to human intervention in the Norwegian sector on human performance, and it is strange that this is ineffective at the prevention of leaks, but very effective at the prevention of the ignition of the same leaks.

The technical conditions may be better on Norwegian installations when compared with UK installations. This may be an alternative explanation. The ignitions presented in Table 15.5 may suggest that equipment malfunction after all may be one of the critical factors in the case of the UK ignited HC leaks.

6.5 Comparison on a Worldwide Basis

The IRF has 11 members, i.e. national authorities with a responsibility for health and safety in the offshore upstream oil and gas industry. Its objectives are according to its website:

to drive forward improvements in health and safety in the sector through collaboration in joint programmes, and through sharing information.

The number of member regulators has increased steadily, and thus statistics are not available for all countries for the early years. The current member regulators at the end of 2018 are:

- National Offshore Petroleum Safety and Environmental Management Authority, Australia (NOPSEMA)
- PSA, Norway
- US Bureau of Safety and Environmental Enforcement (BSEE)
- Danish Energy Agency (DEA)
- Agencia Nacional de Energía, Seguridad y Ambiente, Mexico (ASEA)
- New Zealand Department of Labour, (DOL)
- Canada-Newfoundland and Labrador Offshore Petroleum Board, (C-NLOPB) and Canada-Nova Scotia Offshore Petroleum Board, (CNSOPB)
- Brazilian National Petroleum Agency, (ANP).
- HSE, UK
- State Supervision of Mines, the Netherlands, (SSM).

When it comes to HC leaks, IRF statistics have a more limited scope than for instance the scope of the PSA's data collection. The scope of IRF HC leak data is limited to the following (www.irfoffshoresafety.com):

- The major and significant categories are not mutually exclusive. Each incident should be assigned to only one category.
- Include only process-related releases of gas that are being recovered from the reservoir.
- Releases resulting in a fatality or injury will be counted here and as a Fatality or Injury.
- Releases associated with a loss of well control related to production operations should be included.

- Releases associated with losses of well control related to well activities (drilling, completion, workover, and abandonment) should not be included. These incidents should be included in losses of well control under "Operational Incidents".
- Any gas release with a rate less than 0.1 kg/s or an amount less than 1 kg will be excluded.

There is no information from the IRF about reporting practices and completeness. This is a crucial issue when comparing statistics from various countries. The comparisons presented below should thus be considered with considerable scepticism, because of the lack of any indications of reporting reliability. It is also noted that for some countries, some years are missing completely.

Figure 6.9 compares the number of leaks per 100 million barrels of oil equivalent (BOE) gas production for those countries that have reported such data: Australia, Brazil, Canada, Denmark, Mexico, Holland, Norway and the UK. See also comment about Brazil in connection with Fig. 6.10.

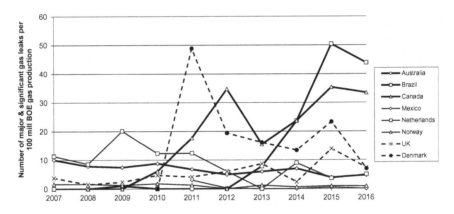

Fig. 6.9 Comparison of the number of gas leaks per 100 mill BOE, 2007–2016

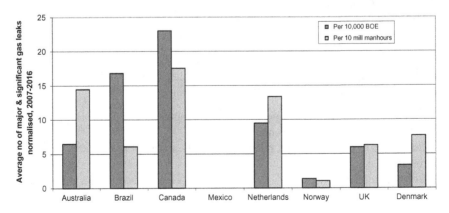

Fig. 6.10 Comparison of the average number of gas leaks per 100 mill BOE in the period 2007–16

Normalisation in relation to production is quite different from how the UK and Norway have been compared in Sect. 6.4. In addition, the number of installations is available, but this is expressed as the total number of installations. This is also different from how the UK and Norway have been compared in Sect. 6.4, where only installations with production and processing equipment are counted.

Figure 6.10 presents the average number of gas leaks in the period 2007–16 for the countries indicated above. The number of leaks is normalised against production as well as manhours. Reporting completeness is not known. Differences in reporting reliability may affect such a comparison significantly, although we know that reporting reliability is high in the UK and Norway. The implication is that countries such as Australia, Brazil, Canada and Holland may have substantial improvement potentials.

It is noteworthy that data are not available for New Zealand and the US. The US has had production operations for a long time, and it is surprising that data on HC leaks are not available. Mexico has reported some data in some few years, but has been omitted in Fig. 6.10 because the reporting is very incomplete.

It has become a recognised fact that data on occupational injuries (including fatalities) does not have a close correlation with major hazard risk precursors (see discussion in Sect. 22.2), such as HC leaks. Nevertheless, FAR values (per 100 million working hours) have been included in the present comparison, in order to provide a more complete picture of the differences between IRF countries. FAR values are available for more countries than those that have HC leak statistics (see Fig. 6.11). The diagram also includes the actual number of fatalities in the period 2007–16.

Data on fatalities and injuries are presented in IRF statistics, but the comparison here is limited to fatalities, as it is considered that the completeness should be the highest in all countries, when it comes to occupational fatalities. The reporting of working hours may, by contrast, be less complete in several countries. Such information is not available.

Multi-fatality accidents will have significant effects on such comparisons. The values in Fig. 6.11 are influenced by the 11 fatalities in the US in 2010, the FPSO Cidade de São Mateus in Brazil in 2015 with nine fatalities, the Abkatun fire in Mexico in 2015 with seven fatalities. The number of fatalities in the US in 2008 is also high, but no information on any multi-fatality accidents has been found. Fatalities in helicopter accidents have been removed from the data, only fatalities on the installations are considered.

The value for Brazil is the highest in Fig. 6.11 and the second highest in Fig. 6.10. Norway has the lowest values in both diagrams.

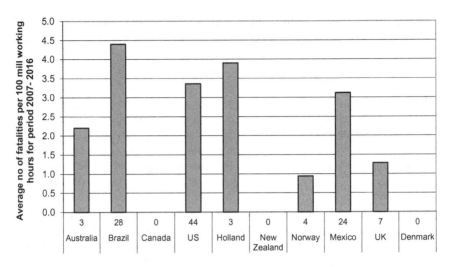

Fig. 6.11 Comparison of the average number of fatalities per 100 million working hours, period 2007–16

6.6 Analysis of the Circumstances and Causes of HC Leaks

6.6.1 MTO Perspective on Leaks

This analysis takes an MTO perspective on HC leaks, with emphasis on the technical, organisational and human (operational) factors that have contributed to the causation of leaks, as emphasised by the categorisation of leaks presented in Sect. 6.6.3.

The analysis takes a pragmatic view that a reduction in the number of leaks on the NCS is primarily a matter of reducing leaks on existing installations. There are few new installations in the Norwegian sector, and current installations are commonly extended with respect to their production periods, through extension with new satellite fields, extension of reserves and so on.

A reduction in the number of HC leaks on existing installations should address the technical, organisational and human factors, as indicated by the categories of leaks presented in Sect. 6.6.3. Technical degradation and design errors are discussed in Sects. 6.6.8 and 6.7.1. With the present perspective it is natural that operational factors will be the main emphasis, as illustrated by the following. If a leak is caused by designing and building process equipment with the wrong type of gasket, this may be seen as a design issue. However, if a wrong gasket is replaced during work on process equipment at an existing offshore installation, the error is in this report considered to be an operational issue.

The replacement of flanges would also have to be considered with respect to risk increase during the replacement period. Large scale cutting and subsequent welding would imply an extensive volume of hot work activity, which would increase risk substantially during the replacement period. It would be doubtful whether the reduced risk due to the redesign of flange connections would compensate for the substantial risk increase due to the extensive hot work during replacement.

6.6.2 Work Process Modelling

Several companies have presented requirements for the planning and execution of manual interventions in the process systems as work process modelling, often using workflow modelling. Figure 6.12 presents a proposed recommended practice for intervention work with a focus on the work steps, based on Norwegian Oil and Gas [7].

A simplified version of the steps in Fig. 6.12 is presented in Fig. 6.13, which illustrates the main steps of the workflow, where emphasis is on verification activities. The following are the main phases as shown in Fig. 6.12:

- Planning
- Put isolation in place
- Carry out work activity
- Reinstatement.

Figure 6.13 shows the main phases. There is a split in three of the phases between the actual work and the verification performed in order to ensure that correct performance has been achieved. The pink boxes therefore focus on:

- The verification of the isolation plan
- The verification of the isolation (the implementation of the isolation plan)
- The verification of the reinstatement (according to the isolation plan).

The following are the main groups of personnel involved in the work process:

- Planning personnel
- Operations responsible
- Executing personnel (mechanics)
- Area technician.

Mechanics are often subcontractor personnel. Previous work and investigation reports may show a trend to focus more attention on the role of the subcontractor personnel (mechanics) than on the production personnel. There may have been too little focus in the past on the planning, isolation and reinstatement phases, compared with the phase where the work activity is carried out.

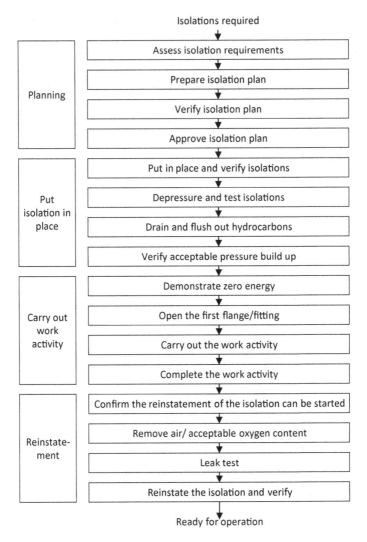

Isolations required

	Assess isolation requirements
	Prepare isolation plan
Planning	Verify isolation plan
	Approve isolation plan

	Put in place and verify isolations
Put isolation in place	Depressure and test isolations
	Drain and flush out hydrocarbons
	Verify acceptable pressure build up

	Demonstrate zero energy
Carry out work activity	Open the first flange/fitting
	Carry out the work activity
	Complete the work activity

	Confirm the reinstatement of the isolation can be started
Reinstate-ment	Remove air/ acceptable oxygen content
	Leak test
	Reinstate the isolation and verify

Ready for operation

Fig. 6.12 Illustration of recommended practice with respect to intervention in process systems

6.6.3 Initiating Events Which May Cause Leaks

The development of the approach to the main circumstances of the scenarios when leaks occur on installations has been documented in Refs. [5, 8], while annual trends are documented by the PSA trough the RNNP project. Vinnem et al. (2007) and Haugen et al. (2011) documented how latent errors have been introduced by the different personnel groups involved in the planning and

Fig. 6.13 Illustration of main
work process steps in manual
intervention in process
systems with emphasis on
verification activities

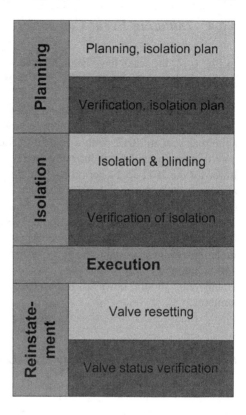

implementation of manual interventions. Latent errors may result from errors made
during planning, if this results in a faulty work instruction, such as opening or
closing the wrong valve. The classification of leaks that has been used in the works
are referred to in the following main categories in Vinnem et al. (2007):

- Technical degradation of system (Category A)
- Human intervention

  - introducing latent error (Category B)
  - causing immediate release (Category C)

- Process disturbance (Category D)
- Inherent design errors (Category E)
- External events (Category F).

The combination of initiating event categories and the work process modelling is
the basis of the analysis of HC leaks from an MTO perspective. The detailed codes
for these six categories are shown in Table 6.9.

6.6.4 Initiating Event Categories

In Fig. 6.14, the leaks have been classified for three periods based on Ref. [9]. In the figure, data in the period 2001–2010 as well as 2008–2010 are compared to the more recent period 2011–2014.

The fraction of leaks due to technical degradation is slightly higher in the period 2011–2014 compared with the two previous periods. This is mainly due to an abnormal year in 2012, with a higher fraction due to technical degradation. The contribution from manual intervention (B and C categories) is correspondingly lower for the 2011–2014 period compared to previous years. In the RNNP project, also the leaks in the period 2015–2017 have been categorized. According to RNNP (2018), the fraction of leaks due to technical degradation was 10–25% in 2015–2017. In the same period, the fraction due to manual intervention (categories B and C) was 50–60%. Based on the above, we notice that the fraction of the leaks in each initiating event category has been surprisingly stable for a long period of time, only with some non-conformities when considering individual years. Results for individual years should, however, in any case be addressed with care due to the low numbers.

6.6.5 Activity Types Involved in Leaks

The previous section addressed the main categories of initiating events. Each main category has several subcategories, according to BORA (Haugen et al. 2007). These subcategories were presented in Table 6.9. The distribution of the main categories is presented in Fig. 6.14. Please consider the subcategories for the categories B and C, i.e. failures during manual intervention, which are presented in Fig. 6.15 including leaks from 2008 to 2011.

B1, B2, B3 and B4 are often grouped together, because they are all associated with the incorrect positioning of valves or connections, which have the same type of barriers. These four subcategories correspond to 22 of the 36 cases of failures due to manual intervention. The maloperation of valves and hoses (B5, B6 and C2) represent seven leaks together, whereas isolation failure or pressurised equipment (C1 and C3) comprise seven leaks.

Of the 11 cases (B2) with the incorrect fitting of flanges or bolts during maintenance, four cases involve errors tightening bolts. These four cases are distributed as follows:

- Bolts not tightened/incorrect tightening: 2 leaks
- Incorrect torque used in tightening: 2 leaks.

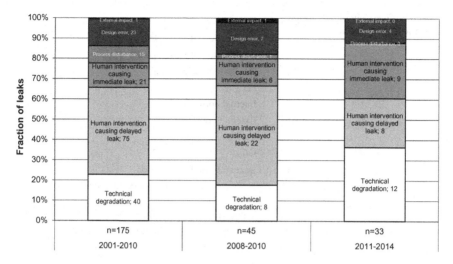

Fig. 6.14 HC leak distribution average for three periods, NCS (Reprinted from Root causes of hydrocarbon leaks on offshore petroleum installations, Journal of Loss Prevention in the Process Industries, vol. 36, 54–62, Copyright 2015, with permission from Elsevier.)

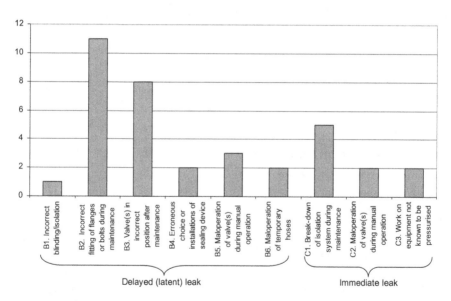

Fig. 6.15 Distribution of subcategories for failures due to human intervention (B&C), leaks NCS, 2008–2011 (n = 36)

Code	Description
Table 6.9 Overview of codes for manual intervention	

Code	Description
A1	Degradation of valve sealing
A2	Degradation of flange gasket
A3	Loss of bolt tensioning
A4	Fatigue
A5	Internal corrosion
A6	External corrosion
A7	Erosion
A8	Other
B1	Incorrect blinding/isolation
B2	Incorrect fitting of flanges or bolts during maintenance
B3	Valve(s) in incorrect position after maintenance
B4	Erroneous choice of installations of sealing device
B5	Maloperation of valve(s) during manual operations
B6	Maloperation of temporary hoses
C1	Break-down of isolation system during maintenance (technical)
C2	Maloperation of valve(s) during manual operation
C3	Work on wrong equipment (not known to be pressurised)
D1	Overpressure
D2	Overflow/over filling
E1	Design related failures
F1	Impact from falling object
F2	Impact from bumping/collision

6.6.6   Time When Leaks Occur

Figure 6.16 shows that the average number of leaks per hour during day shifts (07–18) and early and late night shifts (19–24; 01–06) are 1.0 and 2.2 respectively [10]. The average number of leaks during 19–06 is 2.8 leaks/hour.

For obvious reasons, fewer leaks occur during night shifts. There are few leaks during the period 1900 until 0100, but many leaks during the period 0200 until 0600. This is the period where regulations state that work on process systems only should be carried out if this implies a lower risk, compared with the day shift.

This study shows (see Sect. 6.6.5) that a considerable number of leaks are associated with errors during the implementation of the isolation plan, i.e. when the isolations are put in place. If only C category leaks (immediate release during manual intervention) are considered, then half of the leaks (four out of eight) occurred during night shifts and the other half during day shifts.

Is it likely that leaks could be avoided if the implementation of the isolation plan was carried out during day shifts? This is not necessarily true, if the failure to verify the implementation of the isolation plan is due to a lack of compliance with steering documentation. Such a failure of compliance may occur irrespective of what time of day the work is carried out. On some installations however, it seems that the number of production personnel on the night shift may be low (one person) and therefore there may be a shortage of the right technical staff to perform independent verification at such times. Supervisory personnel could even be called in during night shifts, but reluctance to wake up supervisory personnel in the middle of the night may be an issue on some installations. It may thus be easier to comply with steering documentation verification performance during the day time.

6.6.7   Work Process Phases and Shift Distribution

The leaks in categories B and C are those that occurred in connection with manual interventions, where the work permit system was followed and where the work process phases as presented above are applicable. The structuring of work into phases may be regarded as an essential way to ensure organisationally that the work is carried out in a controlled manner. Work process phases represent a further breakdown of the manual intervention (B and C) categories. Leaks associated with manual intervention were classified into work process phases when errors were made, based on the textual descriptions in the investigation reports. This was possible in 49 of the 96 B and C type leaks. The number of errors shown in Fig. 6.17 is 60, implying that more than one error can be made. The investigation report could be considerably more specific in this regard, which would allow a higher proportion of leaks to be classified into the relevant work process phase.

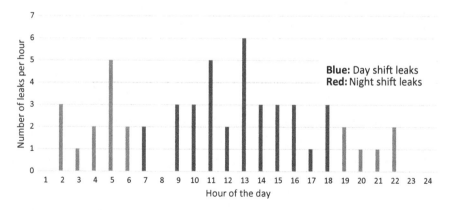

Fig. 6.16 Distribution of times when leaks occur, NCS, all causes, 2008–2011 (n = 53)

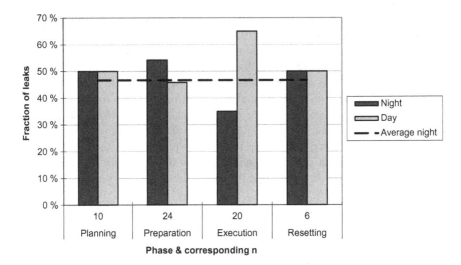

Fig. 6.17 Distribution of errors associated with leaks in work process phases for day and night shift, NCS, 2001–10 (n = 49)

It is further worth analysing whether there are significant differences between work process phases with respect to day and night shifts. This is presented in Fig. 6.17 for the four main work process phases.

For leaks associated with planning, the contributions are equal (50%) from leaks during night and day shifts; the same is the case for the reinstatement work process phase. The two phases that have values that depart most extensively from the average are those that have the highest number of leaks in the preparation and execution phases. These differences may thus be due to random variations.

Figure 6.17 shows that the proportion of leaks during night shifts is highest for the preparation and lowest for the execution phase. This is consistent with the main pattern which we believe is to conduct preparations during night shifts and carry out interventions during day shifts.

6.6.8   Design Weaknesses and Technical Degradation

Figure 6.14 shows that the number of leaks in 2011–14 can be categorised as follows for design and degradation:

- Design error: 4 leaks (12%)
- Technical degradation: 12 leaks (36%).

It should be recognised that the occurrence of leaks can be associated with design issues, such as how a manual valve has to be operated in order to completely close (e.g. the number of turns in one direction followed by one turn back, etc.).

However, many design aspects have to be kept as they are for the rest of the installations' lifetime. To alter the design of process systems is very costly and this would usually involve quite a bit of hot work, implying that the temporary increase in risk during implementation can be substantial.

In the RNNP report, marking process lines and fittings report considered to be a design issue. This is also only partly relevant. When the installation is new, marking is a design aspect, but the need to refresh marking will always occur during operation, and it should therefore also be considered to be an operational issue.

In addition, there are few new installations in the Norwegian sector, whereas the main picture of offshore petroleum production is relatively old installations that often have been subject to, or will experience, lifetime extension. The main efforts relating to the prevention of HC leaks therefore have to focus on leak prevention on existing installations as well as their systems and characteristics. Extensive modifications, such as improved access to valves and instruments are often out of the question, partly due to the increased risk during modification.

The main issue of this report is thus on the prevention of leaks as an operational issue. HC leaks are analysed in a work phase context in Sect. 6.8. Consequently, the current initiative by Norwegian Oil and Gas has so far focused on operational aspects. Improvement in design is also an important topic, which will have to be addressed separately.

6.6.9   Major Hazard Risk Potential

One remarkable feature of the investigations of HC leaks on the NCS is that major hazard potential is in many cases poorly addressed. Most investigations do not recognise the potential for major hazard consequences. In Ref. [11], it has been documented a large difference between what is considered to be the potential consequences in the investigations and the major hazard risk potential reflected in the RNNP classification of the same HC leaks.

It was also documented by Vinnem (2012) that HC leaks have been subject to both investigations by the PSA and company internal investigations. Differences have been identified related to potential consequences. The companies find essentially no major hazard potentials, whereas the PSA has identified significant major hazard potentials for the four leaks investigated.

The actual and potential consequences are taken from the investigation reports, however, in some cases no investigation report is available, and the information is thus taken from the incident reporting system. Only consequences to personnel are recorded.

The weights are taken from the weighting in the RNNP (PSA 2012), which aim at reflecting the potential to cause fatalities, including the potential for leaks to be ignited. These weights essentially express the expected number of fatalities given the occurrence of a leak, which states the different accident chains and outcomes, weighted by the corresponding conditional probability distribution. These weights are based on the analysis of a large number of QRA studies. The weights reflect the special circumstances of the leak scenario, which may increase or decrease the different probabilities involved, for instance in relation to the likelihood of ignition.

The leak rate and ventilation rate together determine the volume of the inflammable gas cloud. Not all HC leaks are gas leaks, but most liquid leaks will flash off substantial amounts of gas, and thus to a large extent behave as a gas leak, at least as far as the possibility for ignition is concerned.

None of the leaks ignited and with exception of one case, there were no injuries. In one case, there was a first aid injury. It can also be seen that the potential consequences in most cases were considered by the investigation committee to be no injury at all. This implies that the investigations did not consider any possibility for the ignition of the leaks.

The weights used in the RNNP based on QRA studies imply the conditional expected number of fatalities, given a HC leak has occurred.

When we look at the correlation in more detail, there is no obvious correlation between the weight of the leaks and what the investigations considered the potential consequences to be (see Fig. 6.18). A linear trend line is also included, demonstrating that the correlation is weak ($R^2 = 0.001$).

The reason why companies may disregard major hazard potential is that the potential consequences are considered in relation to insignificantly changed circumstances. Such circumstances are defined as events that have at least a 50% probability of occurring. Since barrier elements have high reliability, none of the

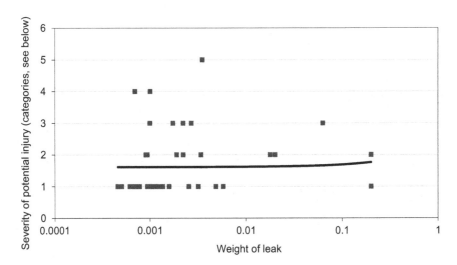

Fig. 6.18 Correlation between the weight of leak and classification of severity potential according to investigations, with a linear trend line (Reprinted from Journal of Risk and Reliability, https:// doi.org/10.1177/1748006x12468670, Vinnem, Use of accident precursor event investigations in the understanding of major hazard risk potential in the Norwegian offshore industry, copyright © 2012 by (Sage Publications) Reprinted by Permission of SAGE.)

barrier failures will have a 50% occurrence probability. This practice should be reconsidered, as discussed in Vinnem (2012).

It should also be noted that the IRIS report 2011/156 ('Learning of incidents in Statoil') (now Equinor) [12], documented that there is inadequate learning from investigations in order to prevent further HC leaks. Further, the arguments are not related to the failure to address major hazard potential, but this relationship should not be ruled out.

6.7 HC Leaks Due to Technical Degradation

6.7.1 Age of Installation with Degradation Failure

Figure 6.19 shows the age distribution of installations at the time of leaks, for leaks due to degradation failure only. It may be argued that maintenance 'friendliness' has improved since the early days, but modern installations are still more compact. We have therefore chosen to assume that age is mainly—if at all—a factor related to technical degradation.

It should be noted that two of the installations in Fig. 6.19. experienced leaks in consecutive years, which influences the trends significantly (because of the limited number of these leaks). This is the case for one installation with leaks at the age of 19 and 20 years, as well as another installation with leaks at the age of 29 and 30 years. The diagram would look quite different, essentially with random failures, without these four leaks.

The overall conclusion is that there is a very weak basis for the hypothesis that degradation failures are closely correlated with age. This has actually been considered in a few other studies, with the same observation; the correlation is impossible to substantiate.

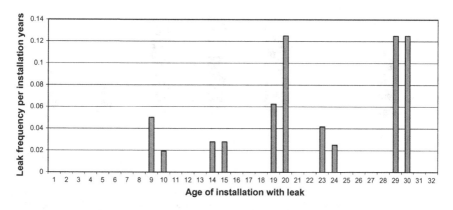

Fig. 6.19 Distribution of installation age when a leak due to degradation failure occurred (n = 13)

This is significantly different from onshore process plants, where corrosion under isolation often is one of the most frequent causes of leaks. Age is in this context one of the obvious factors.

The lack of correlation with age confirms that the preventative maintenance of safety-critical equipment (such as process piping and fittings) should be implemented as intended. If there had been a clear correlation between degradation failures and age, this might have implied that process equipment was not replaced until it was severely degraded.

6.8   HC Leaks Due to Human Intervention

6.8.1   Overview of Work Flow Phases

Figure 6.14 presents an overview of the different steps and phases of a manual intervention in process systems (Norwegian Oil and Gas 2018). The four main phases are planning, isolation, execution of the work task and reinstatement. As emphasized in Fig. 6.13, three of the main phases have a verification activity. The classification of leaks into work flow phases is presented in the following sub-section. Each of the main phases is thereafter discussed separately.

6.8.2   Classification of Leaks During Work Process Phases

Figure 6.20 presents an overview of the errors during various work flow phases, for leaks due to manual interventions (B and C categories; see Sect. 6.6.1). The data set is limited to 32 B and C leaks in the period 2008–11, where information was sufficient to determine the relevant work flow phase.

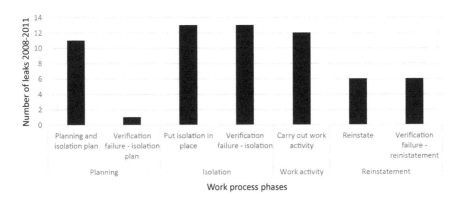

Fig. 6.20 Distribution of errors during leaks due to manual intervention (B and C categories), NCS, 2008–11, for each work process phase separately (n = 32)

It should be noted that several errors were made during planning, but only one of these could have been picked up during the verification of the isolation plan (see further discussion in Sect. 6.8.3).

Further, during the isolation phase, in order for a leak to occur, there has to be an error during isolation implementation together with an error during verification (see further discussion in Sect. 6.8.5). There are thus 13 leaks with failures during both the isolation implementation and the corresponding verification, including cases where no verification was made.

The separate verification of actual intervention work has not been considered. This has been considered during resetting. A failure due to resetting for this phase also requires that failures occur during resetting and they are not discovered during verification, including leak testing. This is discussed further in Sect. 6.8.7.

6.8.3   Personnel Groups Involved in Leaks

Figure 6.21 presents the distribution of personnel categories involved in the errors that caused leaks. This analysis is limited to those 49 leaks with sufficient details available in order to allow the personnel groups involved in the errors to be identified. In some cases, errors were made by more than one personnel category.

Personnel associated with production operations dominate maintenance personnel with respect to those groups involved in causing leaks.

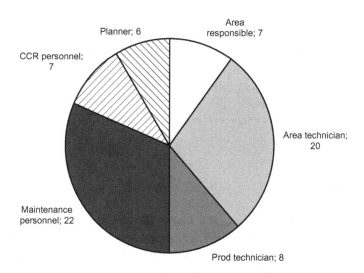

Fig. 6.21 Distribution of leaks by involved personnel groups, NCS, 2001–10 (n = 49) (Reprinted from Journal of Loss Prevention in the Process Industries, Vol 25/4, Vinnem, On the analysis of hydrocarbon leaks in the Norwegian offshore industry, 9 p, 2012, with permission from Elsevier)

6.8.4   Planning

Planning includes the evaluation of the need for isolation, isolation requirements and the implementation and verification of the isolation plan. The majority of planning errors made are not directly associated with the isolation plan. Only one case demonstrated an error associated with the failure to establish an isolation plan (claimed not to be the practice on the installation in question), but no verification was carried out, and the lack of an isolation plan was therefore not identified.

Errors during planning were in all other cases associated with other aspects of planning rather than the actual isolation plan, where the verification of the isolation plan was not relevant in order to identify the error made during planning. The other failures relating to planning were the following:

- Qualification of tools was not carried out and no system was established for monitoring the communication and decision-making relating to risk. Further, no system was established for monitoring the project, asset integrity and operations or for assessing the different levels of management and for decision-making.
- Work permits were insufficient in relation to work process requirements with respect to isolation.
- Description of work was not updated according to latest practices.
- Experience from earlier operations was not included and known errors not rectified.
- There was inadequate risk assessment prior to operation—a HAZOP should have been carried out when an alternative method had to be implemented.
- Insufficient communication prior to planning tasks (e.g. work permits not prepared).
- The process plant was restarted after a prior (six days earlier) gas leak without the proper testing of systems and with incorrect assumptions about the fault.
- Error during planning—bolt torque table was not available and too low torque specified.
- Error during planning—not established practice on the installation in question to prepare an isolation plan.
- The manufacturer's drawings were not in accordance with that installed.

6.8.5   Isolation

Isolation implies putting the isolations into place, including leak testing. This is normally followed by the verification and demonstration of zero energy (used by some companies thus far), as emphasized in Fig. 6.12.

It should be noted that the number of cases with errors during the implementation of the isolations is equal to the number of verification errors in Fig. 6.20. This may be somewhat surprising, but is easily explained in the following manner:

- No leak will be associated with isolation if no error is made during the implementation.
- If the verification is carried out effectively, then there is no leak, even if errors were made during the implementation. This always implies that when a leak is associated with isolation, both implementation and verification failed.

Two categories of verification failure are represented in the data bases: either the verification is not carried out, or it is carried out but failed to reveal the error made. Verification in this context covers the verification of implementation of the isolations as well as the verification of reinstatement after the work has been finished. The following categories of leaks were considered separately:

- B1–B4: Failure in blinding, valves, flange or gasket
- B5 and C2: Maloperation of valves
- C1 and C3: Isolation failure or pressurised equipment.

The overall ratio between verification not carried out and verification failed to reveal the error is:

- 2.1:1 (15 versus seven cases).

'Verification not carried out' is often associated with 'silent deviations'—'it is not our practice to carry out verifications' is an expression often found in investigation reports. This is an important finding. For a more thorough discussion on verification errors and 'silent deviations' see Sect. 6.9.3. The isolation phase is the highest contribution to the number of leaks. There are 13 cases (registered as both setting and verification failure) of failure associated with the this phase.

6.8.6   Execution of Intervention

Execution of intervention is the actual replacement, inspection or modification work carried out. This is the phase in which the second highest number of leaks (12 leaks in the period 2008–10) occurred compared with 13 leaks in the isolation phase. The following is a summary of the errors made:

- Unclear work permit caused error
- Work not performed in accordance with the isolation plan
- Plugs removed without reinstatement: two cases
- Performance failure during flange installation: two cases
- Lack of clarity as to who was responsible for the task/lack of competence: two cases.

6.8.7   Reinstatement

Leaks due to errors during reinstatement of the isolations have the lowest contribution (six leaks) as shown in Fig. 6.20. Five leaks were caused by errors during reinstatement, whereas six were caused by failure during the verification of the reinstatement as the verification that failed was not associated with failure during reinstatement.

6.8.8   Phase When Leaks Occur

Figure 6.23 presents the distribution of when (i.e. in which work flow phase) leaks occurred. This figure can be compared with Fig. 6.20 which shows in which phase the errors were made.

Figure 6.20 shows that some of the errors made in the planning phase caused leaks in subsequent phases. Further, planning errors are often made together with other errors, and the times at which leaks occur are then influenced by these other errors. Type of error will also influence when leaks occur, some errors cause immediate leaks, whereas others lead to delayed leaks, either during reinstatement or after start-up.

6.9   Causal Factors

Immediate causes and potential root causes are documented in the majority of investigations, but the thoroughness of these classifications is variable. Over the past few years, a trend seems to be that some companies' investigations are fewer in number, more thorough and involving personnel with professional investigation competence. The majority of studies are not called 'investigations', but rather 'in depth studies' and are mainly carried out by personnel on the installation. This section documents a number of causal factors, some very important, other less important, but still interesting to document, because there is some attention paid to these aspects.

6.9.1   Risk Influencing Factors (RIFs) from Investigations

Most leaks are recently investigated or studied through 'in depth studies' as noted above. This implies that most analysed root causes and barrier performance. In fact, root causes and barrier performance were available in 47 out of the 56 leaks above 0.1 kg/s. This section is based on those 47 leaks. The analysis focuses on the root causes and barrier failures in the 'M and O' sphere, i.e. failures in organisational systems and human errors.

The distinction between root causes and barrier failures is unclear. Root causes are often defined in investigations as those aspects that lead to one or more of the immediate causes. Barriers are in investigations usually defined rather loosely, as something that could have stopped the chain of events, i.e. prevented an accident or reduced its consequences. Investigations often distinguish between barriers that functioned, barriers that failed and barriers that were not in place or in use. The analysis in this section is limited to barriers that did not function or that were not used.

The more precise terms barrier function, barrier systems and barrier elements are not normally used. This implies that root causes and barrier failures are difficult to distinguish. For example, failure to comply with steering documentation, is in investigations (and similar) classified sometimes as a root cause and sometimes as a barrier failure.

The analysis has thus not distinguished between root causes and barrier failures, and has classified all identified factors as RIFs. Some investigations or 'in depth' studies have identified several root causes and barrier failures, whereas there may be only one or two in other documents. When all available factors are summed for the 47 leaks, the total is 159 RIFs, implying that the average number of RIFs per leak is 3.4. Figure 6.24. presents the distribution of root causes and barrier failures identified in all leaks in 2008–11.

Work practice is the RIF that is cited in most investigations (29 out of 47 leaks). Failure to comply with steering documentation (such as procedures) is the second most frequently used, with 20 out of 47 leaks. If we take failure of work practice **or** failure to comply with procedures, this applies to 35 or out of 47 leaks (74%). By contrast, if we take failure of work practice **and** failure to comply with procedures, this applies to 14 out of 47 leaks (30%). Work practice errors and failure to comply with controlling documentation are therefore the most critical RIFs associated with HC leaks.

'Risk assessment', which should be interpreted as the 'failure to perform relevant risk assessments' as well as the 'lack of apprehension of risk', is almost as frequent as failure to comply with steering documentation (20 out of 47 leaks).

If we focus on those RIFs identified in at least 10 cases, the following are included: the failure of experience transfer, lack of/inadequate procedures and error during design/fabrication. Some RIFs may be less influential anticipated in terms of competence, supervision, time pressure and maintenance.

Figure 6.25 compares the RIFs for all leaks (as in Fig. 6.24.) with only those leaks associated with manual interventions (B and C type leaks, see Sect. 6.6.1). The same RIFs are used in Sect. 6.6.1 and Fig. 6.24. The majority of these are relevant for the leaks associated with manual intervention, i.e. B and C-type leaks.

Errors related to work practice and inadequate compliance with steering documentation only become important if the leaks associated with manual intervention are considered. If we look at failure of work practice or failure to comply with procedures, this applies to 25 out of 30 leaks (83%). However, if we analyse failure of work practice and failure to comply with procedures, this applies to 12 out of 30 leaks (40%). This certainly does not represent a surprise; errors related to work practice and inadequate compliance with steering documentation are essential for those leaks associated with manual intervention.

In addition, failure to perform risk assessment (or failure to apprehend risk) and supervision failure are considered to be more important for the leaks associated with manual intervention. In addition lack of competence and work planning failure are more important for leaks associated with manual intervention, but these represent relatively low contributions. By contrast, the leaks associated with design or fabrication failure have increased somewhat. This was not expected, but it may be due to random variation, as there are relatively few cases (nine) in this category.

Work practice errors dominate during the execution of maintenance work (eight out of nine leaks), but also occur frequently during isolation and reinstatement (18 out of 25 leaks). Failure of communication and failure to perform risk assessment or error in apprehension of risk, are more important during execution than they are during isolation and reinstatement. Supervision is also important during the execution of maintenance work. The only RIF that represents a higher proportion of leaks during isolation and reinstatement is failure to comply with steering documentation. This is typically failure to comply with the isolation plan.

6.9.2 Management and Supervision

The following are some of the weaknesses in management and supervision that were found in investigation reports:

- Weaknesses in management (including management of change) systems

  - Warnings of vibration and noise were not responded to; incident report not taken seriously
  - Qualification of the tool was not implemented; a system for monitoring communication and decisions on risk between the project, construction integrity and management and between different levels of management and decision-making forums as not established
  - No safety system was activated; a high threshold for such actions
  - Documentation was not updated after the modification
  - The system was not given priority with respect to modification; work orders from several years ago were not performed
  - Questions asked of management's handling of start-up after an extensive turnaround
  - Insufficient manning level and inadequate quality assurance
  - Inadequate inspection and maintenance program
  - Risk apprehension, awareness and work performance was not in accordance with accepted standard
  - Previous experience was not taken into account, known errors not corrected

- Steering documentation was not adequate to prevent this type of problem
- There was a weak apprehension of risk in relation to the damage a truck could cause
- Work permit was approved without being aware of a specific exception
- Changes were implemented without a HAZOP or SJA being carried out
- There as a lack of understanding of the risk potential on the involved installations
- Three inadequate plugs were supplied and installed several years ago without discovering the error by the suppliers
- Valve was replaced several years ago with an outdated type of valve stuffing box
- Process plant started after one leak without testing thoroughly, assuming that the error had been found
- Inadequate procedures for verification and control
- Work order issued several years ago, but not given priority by management

- Supervision error and weaknesses

  - No supervisor realised the hazard involved with a large un-insulated volume at low pressure
  - Unclear 'ownership' of temporary equipment
  - Supervisors did not follow up work proactively.

6.9.3 Lack of Compliance with Steering Documentation

First, please note that for the 20 leaks where verification failed, shown in Fig. 6.22, 68% of failures were related to deficient performance in terms of verification, whereas 32% were a failure of the verification itself. It is to be expected, although not identified clearly, that many of the failures to conduct specified verification activities are associated with 'silent deviations' from steering documentation. Unacceptable practices have developed over time on some installations, implying that it is considered to be acceptable not to follow steering documentation, and that instead so-called 'qualified evaluations' conclude that simplifications are just as good. Often such simplifications imply that verification activities either are not carried out at all or are carried out in a simplistic manner.

This is recognised explicitly in some investigation reports, with descriptions such as the following:

- 'Silent' deviations documented
- 'Silent' deviation, made similar error just a few days before
- Lack of compliance with work process requirements when flanges were remounted
- 'Silent' deviation.

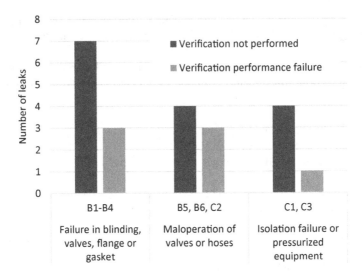

Fig. 6.22 Verification failure, divided into verification not performed and verification performed but failed

Fig. 6.23 Phase in which leaks occur, NCS 2008–11 (n = 30)

It should also be noted that when root causes and barrier failures in investigation reports are recorded (see Sect. 6.9.1), a lack of compliance with steering documentation was reported in 20 out of 47 leaks, i.e. just over 40%. It is therefore reasonable to consider that over one third of all HC leaks in 2008–11 are associated with aspects relating to 'silent deviations'. This implies that eliminating these failures should have high priority in order to reduce the number of HC leaks.

The IRIS report 2011/156 ('Learning from incidents in Statoil') documented that a lack of compliance was essential for the Snorre A subsea gas blowout in 2004 as

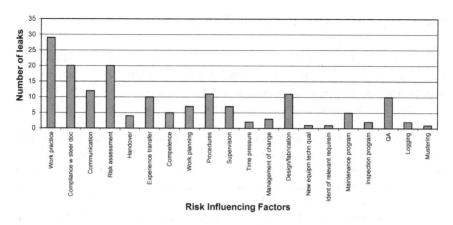

Fig. 6.24 Overview of RIFs from investigations and 'in-depth' studies, NCS, 2008–11 (n = 47)

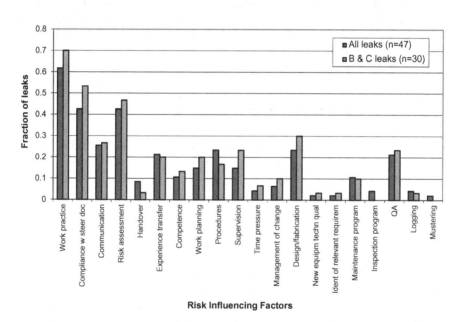

Fig. 6.25 Overview of the RIFs from investigations and 'in-depth' studies, NCS, all leaks and leaks due to manual intervention, 2008–11 (n = 47 and 30)

well as the well incident on Gullfaks C on 19th May 2010. This report also claims that compliance with steering documentation is a general problem within this company, strongly influenced by a complicated and unclear documentation system following the merger of Statoil and Hydro's (now Equinor) system as a consequence of the merger of the two companies in 2007.

6.10   HC Leaks Due to Design Errors

The overview in Fig. 6.14 shows that around 15% of the leaks are associated with design errors. It should in addition be noted that several of the leaks associated with human intervention would also be associated with design weaknesses. The presentations in this chapter should thus not be interpreted to imply that design is not important in the prevention of HC leaks.

However, for existing installations the focus needs to be on the prevention of leaks through operational measures. Fixing design errors or weaknesses may be such an extensive task that it is impossible from an economic point of view. What may be even more important is that design changes usually imply extensive hot work, which implies a significant increase in risk during installation.

The RNNP report in 2011 [4], provided an in-depth analysis of HC leaks and described the actions taken by operators. It was demonstrated that when all leaks were counted where design aspects were involved, this was the highest category. It is therefore important that prevention of leaks also has a strong focus on influencing design standards and practices in order to make new installations less error prone when it comes to manual interventions.

6.11   HC Leaks Due to External Impact

HC leaks due to external impact may be caused by falling or uncontrolled swinging objects, or uncontrolled vehicle movement internally or externally on the installation. Falling objects are frequent on offshore installations, and it could be feared that this represents a significant hazard. However, experience demonstrates that this is not so.

Figure 6.14 shows that only one instance of external impact causing a leak is registered during the period 2001–14 in the Norwegian sector. In fact there is only that one instance even if the period is extended to 1.1.1996–31.12.2017, i.e. the entire data set. This leak occurred on the production deck of a floating production installation (FPSO), when a truck accidentally hit a locking device on a deck valve on one of the tanks. The collision caused the locking device to snap off and the valve to open, resulting in a slow gas release until the valve could be remounted.

6.12   Activities and Systems Involved in the Leaks

The leaks that occurred during the period 2008–2014 have been investigated by Vinnem and Røed [9], and categorized according to which type of activity the leak was associated with, to the extent that this was documented in the available documentation. The following two sections are based upon Vinnem and Røed [9].

6.12.1 The Type of Activities Involved in the Leaks

In some cases, the leak occurred after the activity had been completed, such as when a latent condition causes a delayed leak during or after start-up. Still, the leaks were associated with that activity that caused the latent condition. The distribution is shown in Fig. 6.26.

The figure shows that almost one third of the leaks were associated with preventive maintenance, testing, inspection, and/or cleaning, etc. One special case within the preventive maintenance category was the recertification of Pressure Safety Valves (PSVs). This is quite a substantial contribution within the preventive maintenance category, with seven out of a total of 23 leaks (30%).

It should be noted that preventive maintenance is carried out as a risk-reducing measure; so this is a measure that is intended to reduce risk, whereas on multiple occasions it in fact introduced risk. 'Incident' implies that the leak occurred in response to some unwanted event occurring, such as the tripping of a compressor.

The category 'Modifications' was the second highest contribution, with 11 leaks (neglecting the 'not relevant' category). The third highest contributions were corrective maintenance and 'Start-up', with nine leaks each. The latter category, in a couple of cases, was the starting up after the annual shutdown; the majority, however, were the starting up of wells and other equipment. We also see that a few leaks are related to annual shutdowns: 'Revision stops'. The 'not relevant' cases are mainly in Category A—technical degradation, where the leaks are not associated with any operation. Figure 6.27 shows the distribution of the circumstances for preventive and corrective maintenance, as well as for modifications.

The overall distribution of the circumstances for all of the types of activities shows that immediate and delayed leaks contributed to 55–60% of all leaks. The three activities as shown in Fig. 6.5 had contributions of manual intervention equal

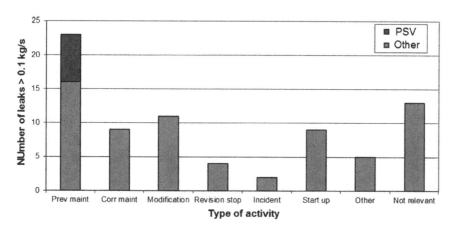

Fig. 6.26 Distribution of the type of activities involved in leaks, 2008–2014, n = 76 (2 leaks with unknown activity)

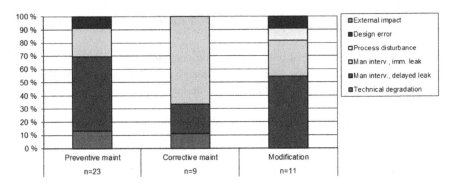

Fig. 6.27 Distribution of the circumstances when leaks occurred in respect of preventive, corrective maintenance, and modifications, 2008–2014, n = 43

to 79%, 89% and 82%, respectively. This implies that the vast majority of leaks associated with maintenance and modification were due to a manual intervention.

6.12.2 The Type of Systems Involved in the Leaks

The systems that were involved in the leaks are presented in Fig. 6.28. The three highest contributions were from gas compression, wells and separation. Metering, oil exports, fuel gas systems, manifolds and gas exports also had significant contributions, whereas gas injection, water treatment, gas lift, and riser/flowlines had quite low contributions.

In this figure, technical failures (Cat. A) and design failures (Cat. E) have been highlighted, since for these fault categories there may be a strong relationship between the fault and the system involved.

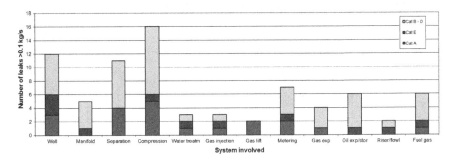

Fig. 6.28 Distribution of the type of systems involved in the leaks, 2008–2014, n = 77 (1 unknown leak)

The number of leaks is highest for the wellhead, manifold, separation and compression areas. This reflects, to some extent, that the numbers of instruments, fittings, bypass piping, etc. are distinctly highest in these areas, and the extent of maintenance operations will also be highest.

When considering technical failures, the most prevalent system is compression, with five leaks. This is not surprising, since this system involves rotating equipment and high pressure. The second and third highest contributions are from separation and well, with four and three leaks, respectively.

6.13 PLOFAM Leak Frequency Model

6.13.1 Introduction

This chapter presents key issues regarding the most updated version of the PLOFAM leak frequency model, issued in December 2018 [13]. The model has been developed to express the expected number of hydrocarbon leaks on offshore installations as input information to QRAs. This expectation value is commonly referred to as a 'leak frequency', and this terminology is also used in the present book, even though the focus is not on historically observed frequencies.

The descriptive text in the following sections is to a large extent directly based on the PLOFAM main report [13]. For further details about the model, reference is made to the PLOFAM report, which is publicly available on the internet.

Before PLOFAM, a frequency model denoted SHLFM was used in the industry in Norway [14]. The last version of SHLFM was solely based on data of leaks occurring at installations on the UKCS in the period 1992–2010.

The PLOFAM model was first issued in 2015 [15] and updated in December 2018 [13]. The main changes from the first revision are the following:

- Data for the period 2015–2017 is included
- The relative leak rate distribution has been reassessed based on leaks on NCS in the period 2001–2017 as opposed to 2001–2015 in the first version. The relative leak rate distributions, including data from the period 1992–2000, have also been assessed.
- The population database used for validation/parametrization has increased from 62 to 109 installations, including all installations that have been in operation on NCS.
- An updated mathematical model for the leak frequency distribution, a reduced rupture fraction for valves and flanges, and updated guidelines for instruments have been included.
- The model for hose leaks is re-assessed giving a reduced leak frequency for leaks giving large released quantities.
- The model is fully aligned with the MISOF ignition probability model [16].

In the PLOFAM report, it is emphasized that the PLOFAM leak frequency model and the MISOF ignition probability model should be used together and not combined with other models. Using the models together ensures consistent interpretation of scenarios having an impact on both models.

6.13.2   Model Characteristics

There are many reasons for leaks occurring at offshore installation process systems, and hence there is a large number of factors that influence the leak frequency. Examples are the components that the process system consists of, the equipment size distribution, the process conditions, the environment around the process system, the maintenance scheme, training of personnel, work culture and time and cost requirements. Many of these factors will be different from installation to installation and some will strongly influence the leak frequency, while others will only to some extent have implications for the leak frequency.

When building a model serving as a tool for prediction of the future leak frequency for topside process leaks in QRAs, it is obvious that all factors influencing the leak frequency cannot be included. Building a model for such a complex phenomenon will be a trade-off between the model complexity, user- friendliness of the model, and the model's ability to predict good overall estimates for single installations. The model should therefore capture the 'most important' contributing factors in topside process leaks in order to reflect the most important differences between the installations.

The reasons for leaks occurring are many, and it is not normally possible to understand all the factors that resulted in an observed leak. However, some failure modes can be understood, and in such cases the aim should be to reflect these known failure modes in the model.

In PLOFAM, the equipment count (for each equipment type) is the only variable included in the model. In addition, known failure modes are reflected in the parametrization of the model. Note that this does not mean that it is concluded that the number of equipment items is the only factor having implications for the leak frequency, but it is judged to be the best single explanatory variable.

6.13.3   Data Basis

The model is based on two sources of data:

- NCS data: 254 incidents recorded at all installations located on the NCS in the period 01.01.2001–31.12.2017
- UKCS data: 4561 incidents at installations on the UKCS recorded in the HCR database in the period Q3 1992–Q1 2015.

6.13.4   Leak Scenarios Covered by the Model

Three main leak scenarios for modelling in QRAs are defined in PLOFAM: process leak, producing well leak and gas lift well leak.

Incidents occurring during well interventions/operations, such as wireline and coiled tubing, are defined as blowouts or well releases, and are covered by Ref. [17], which is based on the SINTEF Offshore Blowout Database. These incidents are not covered by the PLOFAM model.

Note that the leak frequency for process leaks estimated by the model does also account for leaks that occur in the utility system but are being fed from the process system. This is done by including process leaks fed through utility systems, but not equipment counts from utility systems, as a basis for the model validation. This implies that utility equipment should not be counted as the basis for the estimation of process leak frequencies. Furthermore, the model does not give separate leak frequencies for process releases through utility systems and through the process system. This means that a QRA based on PLOFAM will not reflect the potential location of the leak sources in utility systems. Furthermore, the leak frequency contribution from utility systems will scale with the number of equipment counts for the process system. This contribution will in practice vary somewhat with the system at hand, but this cannot be quantified based on PLOFAM.

Leak points in the well system:

- Producing well/injection well: Topside well release where the inventory between the DHSV and PWV is released during normal production.
- Gas lift well: Topside well release where the inventory between the ASV and the barrier towards the process system is released. In cases where no ASV is present, the entire inventory in the gas lift annulus to the ASCV may be released. It is assumed that the check valve ASCV is functioning, otherwise there is no barrier towards the reservoir.
- Release of hydrocarbon fluid from annuli that are not used for gas lift.

Leak points in the process system:

- Leak point in process system between PWV and topside riser ESDV/storage ESDV. The fuel system is regarded as part of the process system.

Leak points in the utility system:

- Leak point in flare system (low-pressure or high-pressure flare system).
- Excessive releases through flare tips and atmospheric vents that exceed the design specification and pose a fire and explosion hazard to equipment, structures or personnel. Such leaks are denoted vent leaks.
- Leak point in utility systems that is fed by hydrocarbons stemming from the process system. Systems covered by the model are:

a. Open drain system
b. Closed drain system
c. Chemical injection systems
d. Produced water.

For all leak scenarios, 0.1 kg/s is recommended as the general leak rate threshold for estimation of the leak duration (in terms of calculation of both fluid dispersion and fire duration) in a QRA, for all leak scenarios in open areas and leaks in enclosures having a net volume more than 1,000 m^3 and with a ventilation rate of 12 air changes per minute (ach) or higher. The lower leak rate threshold is taken as the basis for the lower boundary with regard to the aggregated amount of hydrocarbons released (10 kg). The model distinguishes between leak scenarios (rate >0.1 kg/s) where the total released amount of hydrocarbons is ≤ 10 and >10 kg. These leaks are classified as marginal leaks and significant leaks, respectively.

In a QRA, the risk in terms of fire- and explosion-load exposure to vulnerable equipment and structures such as safety systems, pressurized equipment, load-carrying structures and main safety functions, associated with marginal leaks can normally be neglected. However, the risk to personnel associated with marginal leaks should not be neglected.

6.13.5   Equipment Types Covered by the Model

PLOFAM is a statistical model, based on the assumption that the leak frequency is proportional to the number of each type of equipment. In total, 20 different equipment types are covered by the model, including Gas lift well and Production well, which belong to the well system. The other equipment types included in the model are the most common process equipment types at offshore installations. Equipment types included in the model are as follows:

- Air-cooled heat exchanger
- Atmospheric vessel (vessels with atmospheric pressure)
- Centrifugal compressor
- Centrifugal pump
- Compact flange
- Filter
- Flexible pipe (permanently installed hose)
- Hose (temporary hoses)
- Instrument
- Pig trap (pig launchers and pig receivers)
- Plate heat exchanger
- Process vessel (pressurized process vessels)
- Reciprocating compressor

- Reciprocating pump
- Shell and tube side heat exchanger (includes equipment where the hydrocarbon is on the shell side and/or tube side of the heat exchanger)
- Standard flange (includes all flange types, except compact flanges)
- Steel pipe (process steel pipe)
- Valve (includes all types of valves)
- Gas lift well (wellhead with gas lift)
- Producing well (wellhead with or without gas lift).

6.13.6 Discussion

As mentioned above, the PLOFAM model is based on an assumption that the number of leaks is proportional to the amount of equipment on the installation. This assumption can be questioned. For example, in Sect. 6.8, it is emphasized that a high proportion of historical leaks are related to manual intervention on process equipment and not to technical failures. However, it may be argued that there can be an indirect relationship between the amount of equipment on the installation and the number of work operations.

It must be kept in mind, however, that any model is a simplification of the real world. If, for example, it is suggested to add equipment to reduce the number of operations needed to prepare for work on hydrocarbon inventories, the leak frequency calculated according to the PLOFAM model will increase. However, since many historical leaks are related to work on hydrocarbon inventories, it may be a good idea to include the above equipment. The above emphasizes that no model should be used mechanistically without reflection, since any model is a simplification of the real world.

References

1. Lord Cullen (The Hon) DW (1990) The public inquiry into the piper alpha disaster. HMSO, London
2. HSE (1995) RIDDOR—reporting of injuries, diseases and dangerous occurrences regulations 1995. (www.hse.gov.uk/riddor)
3. PSA (2018) Trends in risk level on the norwegian continental shelf, main report, (in Norwegian only, English summary report) Petroleum Safety Authority, Stavanger, 26 Apr 2018
4. PSA (2011) Trends in risk level on the norwegian continental shelf, main report, (in Norwegian only, English summery report), Petroleum Safety Authority, Stavanger, 27 Apr 2011
5. Vinnem JE, Seljelid J, Haugen S, Husebø T (2007) Analysis of hydrocarbon leaks on offshore installations. In: Presented at ESREL2007, Stavanger, 25–27 June 2007

6. PSA (2012) Trends in risk level on the norwegian continental shelf, main report, (in Norwegian only, English summery report), Petroleum Safety Authority, Stavanger, 25 Apr 2012

7. Norwegian Oil and Gas (2018) Recommended practice for isolation when working on hydrocarbon systems, revised 28 Feb 2018. Available on Norwegian oil and gas webpage

8. Haugen S, Vinnem JE, Seljelid J (2011) Analysis of causes of hydrocarbon leaks from process plants. In: Presented at SPE European health, safety and environmental conference in oil and gas exploration and production held in Vienna, Austria, 22–24 Feb 2011

9. Vinnem JE, Røed W (2015) Root causes of hydrocarbon leaks on offshore petroleum installations. J Loss Prev Process Ind 36, 54–62

10. Røed W, Vinnem JE, Nistov A (2012) Causes and contributing factors to hydrocarbon leaks on norwegian offshore installations, SPE-156846. In: SPE/APPEA international conference on health, safety and environment in oil and gas exploration and production, Perth, Australia, 11–13 Sept 2012

11. Vinnem JE (2012) Use of accident precursor event investigations in the understanding of major hazard risk potential in the Norwegian offshore industry. In: Proceedings of the institution of mechanical engineers, Part O: Journal of risk and reliability, vol 227, pp 66–79

12. Austnes-Underhaug R et al (2011) Learning from incidents in statoil (in Norwegian only), IRIS Report 2011/156, 21 Sept 2011 (available from www.statoil.com)

13. Lloyd's Register Consulting (2018) Process leak for offshore installations frequency assessment model—PLOFAM(2), Report no: 107566/R1, Rev: Final, Date: 06 Dec 2018

14. DNV (2009) Offshore QRA—standardised hydrocarbon leak frequencies, Report no: 2009–1768, Rev: 1, 16 Jan 2009

15. Lloyd's Register Consulting (2016) Process leak for offshore installations frequency assessment model—PLOFAM, Report no: 105586/R1, Rev: Final B, Date: 18 Mar 2016

16. Lloyd's Register Consulting (2018) Modelling of ignition sources on offshore oil and gas facilities—MISOF, November 2018, Report no: 107566/R2, Rev: Final

17. Lloyd's Register Consulting (2017) Blowout and well release frequencies based on SINTEF offshore blowout database 2017, 20 April 2018, Report no: 19101001-8/2018/R3 Rev: Final

Chapter 7
Fire Risk Modelling

7.1 Overview

7.1.1 Cases with Opposite Results

It may be illustrative to consider the vast difference between two incidents in the North Sea that started in almost identical ways, within approximately 24 h, but ended very differently. These two events were presented in Sects. 4.14 and 4.15:

- Explosion and fire on Brent A on 5 July 1988 (Sect. 4.14)
- Explosion and escalating fire on Piper A on 6 July 1988 (Sect. 4.14).

The Piper A accident ended with a lost installation and is very well known due to the catastrophic development of the accident. The Brent A incident on the day before Piper A is virtually not known at all, except those involved, due to the benign consequences.

Both events occurred on mature installations, and started with a gas leak from a flanged connection. The gas leaks were both ignited and a gas explosion resulted with approximately the same peak overpressures. The protective systems onboard both installations were essentially the same. The main characteristics of the two events may be summarised as shown below:

	Piper A, 6 July 1988	Brent A, 5 July 1988
Leak source	Blind flange	Gasket in a flange
Ignition source	Ignition by unknown source	Probably by deluge
Explosion overpressure	Approximately 0.3 bar	Approximately 0.3–0.4 bar
Deluge	Not initiated, fire water supply unavailable	Started (inadvertently) on gas detection
Subsequent fire		Only few minutes fire duration

(continued)

© Springer-Verlag London Ltd., part of Springer Nature 2020
J.-E. Vinnem and W. Røed, *Offshore Risk Assessment Vol. 1*,
Springer Series in Reliability Engineering,
https://doi.org/10.1007/978-1-4471-7444-8_7

(continued)

	Piper A, 6 July 1988	Brent A, 5 July 1988
	Escalating immediately to other areas	
Final consequences	Total loss	Significant damage to gas processing module
Consequences to personnel	167 fatalities	2 injuries

The main difference according to this simplified summary was the activation of the deluge system, but it is slightly more complicated than this. The Brent A gas leak occurred on the highest level far away from the accommodation area, whereas the Piper A gas leak occurred on a lower level below and next to the accommodation block.

7.1.2   Types of Fire Loads

This section gives an overview of accident scenarios that may result in fire loads on primary and secondary structures, (i.e. support structure, main deck structure and module structure). Two tables are presented, Table 7.1 for fixed and Table 7.2 for floating installations.

These tables show that uncontrolled hydrocarbon flow is the main reason for critical fire loads on the structure. It may be noted that dropped objects may also contribute, but only as a result of rupture of hydrocarbon containing equipment. It may be noted that falling objects have not caused hydrocarbon leaks in the Norwegian sector in the period 1996 until 2012, i.e. the period when detailed information about the leaks is available.

Structural failure and collision impact may under certain conditions lead to fires that may affect the structure. It should be noted that the failures that may lead to uncontrolled fire directly, usually dominate over those that depend on escalating sequences. The Mumbai High North fire in 2005 (Sect. 4.18) is an example of external impact from a vessel which caused a riser fire which escalated to an uncontrolled fire.

7.1.3   Structural Fire Impact

Calculating the fire loads on a structure and estimating responses involve these steps:

1. Calculation of release of hydrocarbon

Table 7.1 Overview of types of fire scenarios that may lead to structural fire loading, production installations

Type of accident	Conditions which may lead to structural effects	Structural elements typically affected	Criticality
Blowout	Difficult to control and combat, especially if ignited. Burning blowouts are usually of very long duration, thus very critical	Usually deck structure May also be support structure, if burning on sea level	Usually highest contribution, if wellheads are on platform
Riser failure	Release may have extensive duration, if long pipeline is connected, and no isolation is possible	Deck structure, if riser is inside shaft. Also likely in later phases, when no shaft exists	Usually significant contribution, unless riser is specially protected, or subsea isolation is provided
		Likely if leak is below deck, and no shaft. Also likely if leak is subsea or in splash zone, thus causing burning on sea level	
Pipeline failure	Release may have extensive duration, if long pipeline is connected, and no isolation is possible Release needs to be close to structure and ignited on sea level, in order to be critical	Likely if leak is below deck, and no shaft. Also likely if leak is subsea or in splash zone, causing burning on sea level. Deck structure is likely in later phases	Usually insignificant contribution, because ignition is unlikely
Process equipment failure	Fires may have long duration through escalation to uncontrolled fire	Usually main deck structure	Dependent on likelihood of escalation
Dropped object	Through fire caused by rupture of hydrocarbon equipment	Usually main deck structure	Insignificant contribution

2. Calculation of fire loads
3. Calculation of structural time-temperature distribution
4. Calculation of structural response to temperature distribution.

For each of these steps, simplified methods or more comprehensive simulation tools may be used.

Table 7.2 Overview of types of fire scenarios that may lead to structural fire loading, floating installation

Type of accident	Conditions which may lead to structural effects	Structural elements typically affected	Criticality
Blowout	Dry wellhead: Difficult to control and combat, especially if ignited. Burning blowouts are usually of very long duration, thus very critical	Usually deck structure May also be marine structure, if burning on sea level	Usually highest contribution, if wellheads are on platform
	Subsea wellhead: Difficult to control and combat from platform. Ignited blowout will burn on sea, thus very critical	Usually marine structure, but may also be deck structure	Burning on sea level is less likely, compared with wells on deck. Unit may also reposition
Riser failure	Release may have extensive duration, if long pipeline is connected, and no isolation is possible	Marine structure, as leak is usually subsea	Usually significant contribution, unless subsea isolation is provided
Pipeline failure	Release may have extensive duration, if long pipeline is connected, and no isolation is possible Release needs to be close to structure and ignited on sea level, in order to be critical	Marine structure, as fire is usually on sea level	Usually insignificant contribution, because ignition is unlikely
Process equipment failure	Fires may have long duration through escalation to uncontrolled fire	Usually main deck structure	Significance is dependent on likelihood of escalation
Dropped object	Through fire caused by rupture of hydrocarbon equipment	Usually main deck structure	Insignificant contribution

7.1.4 Fire and Explosion Loads on People

The assessment of the effects of fire explosion on people is parallel with the structural effects analysis, although in some respects considerably simpler.

7.2 Topside Fire Consequence Analysis

The following section provides a brief overview of some of the important parameters that are used in the fire consequence analysis, including an overview of the fire types and their characteristic heat loads. For more detailed introduction (see for instance Ref. [1] or Ref. [2]).

7.2.1 Mechanisms of Fire

Despite the fact that a fire originates from combustion reactions, the process of fire may be dependent on forces or factors that are not directly involved in combustion. The overall process rate may be dependent on, or driven by, a step in the process other than the combustion reactions. It is therefore convenient to separate fires into different types:

- Ventilation controlled fires in enclosures (closed or partly closed)
- Fuel controlled fires in enclosures
- Pool fires in the open
- Jet fires
- Fires in running liquids
- Fire balls (BLEVE)
- Gas fires (premixed, diffuse).

All these scenarios are relevant in the case of offshore installations. Other types of fire that may occur are fires in electrical equipment and fire in equipment in the accommodation. These 'non-hydrocarbon' fires are not discussed. The fire loads are generally lower for these fires, but the smoke production may be a problem.

7.2.1.1 Combustion Reactions

All fires involve combustion in the gas phase except for smouldering combustion where the combustion reactions take place on the surface of the fuel.

Both exothermic and endothermic reactions take place during the combustion process. The overall reaction or process, is considered to be one reaction involving total, or 100%, combustion or oxidation of the fuel.

The combustion reaction rate is dependent of the temperature in the combustion zone, the concentration of the reactants (oxygen and fuel) and the combustion products produced in the zone. If the concentration of the fuel is either too high or too low, or the temperature is too low, the reaction will not take place.

The dependence of the reaction rate on these factors is expressed by the parameters ignition temperature, flammability limits, and oxygen index. Other important properties are the flash point and the self ignition temperature. These properties can be found in a number of text books for different hydrocarbon products.

7.2.1.2   Heat Transfer to Object Within the Flames

The main heat transfer from a flame to an object in the flame occurs by radiation. In some situations, however, the convective heat transfer equals the radiative contribution. The radiative heat transfer can be expressed as:

$$q = \varphi \varepsilon \sigma \left(T_f^4 - T_o^4 \right) \quad \text{kW/m}^2 \qquad (7.1)$$

where

φ configuration factor
ε emissivity/absorption factor for the system
σ Stefan Boltzmann's constant (56.7×10^{-12} kW/m^2 K)
T_f flame temperature, K
T_o temperature of the object, K.

If the object is totally enveloped in the flames we assume that φ is 1.0. If the thickness of the flames around the object is equal to or above 1, we assume that $\varepsilon = 1$. In general it is considered that in a pool fire the diameter of the flame must be larger than 3 m to create flame thickness equal to 1.

The temperature within the flame will vary from 500 °C at the top to 1600 °C locally in the flames. The value of 1100–1200 °C is often used to calculate the radiative heat transfer, but this value has been considered as conservative. Measurements from large scale tests have revealed that even higher temperatures may be generated in certain conditions, during a jet fire as well as a liquid fire. The following approach may be used for practical estimates:

1. The average flame temperature, where the flames totally envelop the objects, may be taken as 800 °C.
2. The product of φ and ε in Eq. 7.2 vary from 0.7–0.9, depending on the flame thickness and tightness around the object.

Using a value of $\varphi \varepsilon = 0.7$ and $T_f = 800$ °C, produces a radiant heat transfer of 52 kW/m^2 which may be regarded as a typical average value.

Consider a region of the flames equal to 2/3 of the total height of the visible flames. The convective heat transfer is high in this region in the beginning of the fire, if the pool diameter is large and flames stagnate on the object. The radiative heat is proportional to the temperature difference to the power four. The convective heat transfer is on the other hand linearly dependent on the temperature difference.

7.2.1.3 Heat Transfer to an Object in a Jet Fire

If an object is in a jet fire, the heat transfer to the object will be far greater than if the object were enveloped by flames controlled by natural convection. This is due to the very high gas velocities which result in 70–90% of heat transfer by convection. In a jet fire the convective heat loads have locally been measured up to 350 kW/m^2 and average values of 300 kW/m^2 are possible.

Heat transfer from a pool fire in the open will normally have a maximum value of 150 kW/m^2. A jet fire will of course be far more local than a pool fire. In the BFETS programme [3], the total incident heat flux onto targets was in general up to 200 kW/m^2, but local maxima could be up to 350–400 kW/m^2 under certain conditions. This occurred for jet fires as well as pool fires in semi-enclosed areas.

The total heat transfer to an object within the flames of a jet fire can be estimated by calculating the radiation heat transfer ($\varepsilon \approx 1$) and assuming this to be 30% of the total.

7.2.2 Fire Balls

Fire balls may be the result of a BLEVE (Boiling Liquid Expanding Vapour Explosion). If a tank ruptures, the gas under pressure mixed with the condensed phase escapes immediately. The flash evaporation of the gas phase entrains liquids in the form of a fog which, when ignited, burns with the look of a ball or 'nuclear bomb mushroom'. The factors of interest are:

1. The hydrocarbon fractions released before combustion
2. The size of the fire ball
3. The duration of the fire ball
4. The radiation intensity from the fire ball.

The fraction participating in the combustion is suggested to be 30%. Hasegawa and Sato [4], have set up the following correlations of size and duration:

$$D_{max} = 5.28\, M_f^{0.277} \qquad (7.2)$$

$$\tau = 1.1\, M_f^{0.0966} \qquad (7.3)$$

where

D_{max} maximum diameter of the ball, m
T duration of the combustion, s
M_f mass of the involved fluid, kg.

7.2.3 Gas Fires

Gas fires and explosions are related phenomena, but it is important to distinguish between them. A gas fire will occur if the following three conditions exist at the same time:

- there is a mixture of gas and air within the flammability limits,
- there is a normal sized source of ignition energy, and
- the enclosure does not contain any flame front accelerating factors.

Flame front accelerating factors can be objects in the enclosure or any narrow passage way to neighbouring enclosures. If the flame front acceleration is sufficiently strong, a deflagration will develop (see Chap. 8).

The gas fire takes place with a flame front that propagates through the mixture at a speed of 0.5–2 m/s and the flow is laminar. The heat and overpressure loads from a gas fire are relatively small.

7.2.4 Air Consumption in Fire

The following figures are given in literature as the air requirements for burning the following fuels:

- Propane: 16 kg air/kg fuel
- Gasoline: 16 kg air/kg fuel
- Methane: 15 kg air/kg fuel.

These are theoretical values for complete combustion in diffusion flames. The air consumption required for jet fires is much higher than for an ordinary diffusion fire. In order for the jet to be maintained, excess air over that needed for the combustion reaction is needed. The jet fire air consumption is 400% the 'normal' consumption, including both the combustion and excess air. For methane jet fires this means 60 kg air/kg fuel.

7.2.5 Choice of Calculation Models

There is a very wide range of available models which may be chosen for calculation of fire dimensions and loads. There are a number of simple, hand calculation models available, mainly based on empirical data. At the other extreme, there are several computational fluid dynamics (CFD) software packages available which enable very sophisticated calculations to be performed.

Using simplified tools is often a good starting point, in order to get some feel for how severe the fire loads are likely to be. If severe loads can occur, then it is wise to use more sophisticated tools (presumably with less uncertainty), to understand the situation better and thereby have a more precise basis for the engineering of protective measures.

An overview of simplified models is found in the Fire Calculation Handbook [2]. Some of the CFD–packages are briefly mentioned in Appendix A.

7.2.6 Analysis of Topside Fire Events

The main fire types are jet fire, diffuse gas fire, pool fire, and fire on sea. Table 7.3 summarises the main characteristics that need to be determined for these fire types.

7.2.7 Fire Simulations

Detailed analysis of fire loads would call for a number of fire simulations with CFD-tools (see Appendix A) for detailed production of fire loads. Table 7.4 presents the cases for CFD fire simulation for a large QRA study of a large production installation, reflecting the actual number of cases considered in the QRA study.

Results from the simulations may be viewed graphically for various parts and angles, with respect to various parameters, most typically the radiation flux at various levels.

The fire simulations for each of these cases are used for the calculation of the following probabilities:

- Probability of immediate fatalities
- Probability of fatalities during escape from area
- Probability of impairment of escape main safety function
- Probability of impairment of shelter area main safety function
- Probability of impairment of evacuation main safety function
- Probability of impairment of critical rooms main safety function
- Probability of impairment of main load-bearing structure main safety function
- Probability of impairment of escalation main safety function.

A typical summary of radiation flux levels from the different simulation cases for three assumed lifeboat stations is shown in Table 7.5 (illustrative values, not from

Table 7.3 Fire load characteristics

Jet fire	Diffuse gas fire	Pool fire	Fire on sea
• Hole size	• Release rate	• Pool size	• Spreading
• Release velocity	• Air supply	• Air supply	• Wind direction
• Direction	• Duration of leak	• Wind direction	• Wind speed
• Duration of leak	• Air supply	• Duration of fire *vs* leak	• Pool breakup

Table 7.4 Fire simulation cases for QRA of a large production installation

Case	Module/area	Type of release	Direction	Mass flow (kg/s)	Wind direction	Wind speed (m/s)
11	M11	Jet fire	South	20	East	4
12		Jet fire	South	20	East	10
13		Jet fire	South	20	South-southeast	12
14	M12	Jet fire	North	10	North	2
15		Jet fire	North	10	East-northeast	2
16		Jet fire	North	10	North-northeast	10
17		Jet fire	North	20	North-northeast	10
18		Jet fire	North	20	North-northwest	10
19	M15 (well area)	Jet fire	East	10	East	2
20		Jet fire	East	10	South	8
21		Jet fire	North	5	East	8
22		Jet fire	North	10	East	8
23		Jet fire	North	10	East-northeast	8
24	Process area	Jet fire	North	10	East	6
25		Jet fire	North	5	North	8
26		Jet fire	North	5	North	8
27		Jet fire	North	10	East	6
28		Jet fire	South	10	East	2
29		Jet fire	South	10	East	10
30	Sea level	Pool fire	–	30	East	2
31		Pool fire	–	30	East	8
32		Pool fire	–	30	North-northwest	6
33		Gas plume	–	30	East	0.1
34		Gas plume	–	30	East	2
35		Gas plume	–	30	North	0.5
36		Gas plume	–	30	North-northwest	0.1
37		Gas plume	–	30	North-northwest	2
38	Drill floor	Jet fire	West	25	East	8
39		Jet fire	West	25	South	12

the study). A typical summary of immediate fatality probabilities in the areas where the accidents originate is shown in Table 7.6 (illustrative values, not from the study). These values reflect the radiation levels in the various modules, and would also reflect some evaluations based on the radiation levels.

Table 7.5 Summary of radiation levels for three assumed lifeboat stations

Case	Lifeboat East			Lifeboat North			Lifeboat West		
	4 kW	15 kW	50 kW	4 kW	15 kW	50 kW	4 kW	15 kW	50 kW
11	x						x	x	
12							x	x	
13				x			x	x	X

Table 7.6 Summary of probability of immediate fatality in area where accident originates

Fire in area	Type of event	Leak rate category		
		Minor (0.1–1 kg/s)	Moderate (1–10 kg/s)	Large (>10 kg/s)
M11	Process fire	0.05	0.30	0.70
	Riser fire	0.20	0.50	0.80
M15	Process fire	0.10	0.40	0.75
	Blowout	0.20	0.65	0.85

7.3 Fire on Sea

A full rupture of a pressure (or storage) vessel containing a large amount of crude oil (e.g. from a separator) will cause the contents to be released more or less instantaneously on the platform. Due to the fact that the deck in process area s on offshore platforms usually comprises gratings, almost all the released oil will flow to the sea forming an oil slick which will rapidly increase in size. There are drain systems installed in order to prevent this scenario, but it is most likely that a significant proportion of the oil will flow to sea, in a major rupture, due to limited capacity of the drain system.

The lighter fractions of the oil will evaporate or flash off as the oil slick spreads on the sea surface. A release from a vessel containing the lighter as well as the heavier fractions will obviously produce a higher amount of gas. Fire on sea level may also occur due to blowout from a well (especially a subsea production well) or a pipeline leak.

It has been considered that very large pool fires (diameter > 40 m) would likely tend to split up into several flame plumes with smaller base diameter and, thus also attain significantly reduced flame heights compared to the height of a non-split-up flame. However, this has never been verified.

Immediate ignition of most crude oil releases on the sea surface will result in an oil slick fire in which the burning intensity will be in the range 0.03–0.08 $kg/s/m^2$, depending on the fraction of light components contained in the oil. It appears that the maximum flame heights will be in the range of 25–50 m. Since the distance from the sea surface to the underside of the cellar deck or the Main Support Frame (MSF) of a typical fixed installation is usually in the range 25–30 m, most oil slick fires arising from a major release of crude oil will impinge on the MSF or underside of the deck. The most severe will engulf the deck and modules on top in flames and smoke. Floating installations have lower air gaps, which implies that heat loads will be higher.

It has traditionally been considered that heat loads from a sea level fire are lower than from a pool fire on deck. The lower burning intensities will result in a flame tip temperature of 500–600 °C, resulting in heat loads of 30–50 kW/m^2. The higher burning intensities will result in the upper part of the flame plume producing heat leads in the range 50–100 kW/m^2 at the MSF and deck, and heat loads of 75–150 kW/m^2 on parts of the platform legs. In large scale tests [3], flame temperatures up to 1250 °C have been measured.

The lack of actual experience data is one of the complicating factors in relation to fire on sea. Very few instances of fire on sea are publicly known. The Ekofisk field experienced one case around 1980, this was a very small amount and short duration, but caused severe panic on the installations. There are also rumours about other cases around the mid-1990s, in the North Sea and in Africa, but these are unconfirmed.

Another aspect of fire on sea, is the possible extensive cost impact if extensive passive fire protection of the structure is to be applied in order to protect against fire on sea level, this is further illustrated in Sect. 7.3.4.

Immediate ignition of an oil spill on the sea level is often caused by running or falling burning liquid droplets from the installation.

7.3.1 Delayed Ignition of an Instantaneous Release

Prior to ignition, the oil will spread on the sea surface due to changing spreading and retarding forces. Figure 7.1 shows the diameter of an unignited oil slick as a function of the elapsed time from the start of the release. This relates to an instant area release of 100 m^3 of stabilised crude oil ignoring any effects of wind waves or currents. No immediate gas flashing or evaporation of light components is taken into account. A significant amount of gas flashing and evaporation would result in a smaller oil slick due to the reduced volume of oil.

From Fig. 7.1 it can be seen that the diameter of the oil slick increases rapidly in the first 30 s after release, after which time the growth rate is slower. For example after 1, 2, 3 and 4 min, the oil slick diameter is approximately 65, 95, 120 and 140 m respectively. After 6 h the oil slick diameter is approximately 370 m. It is evident that if this oil slick is ignited after some minutes, huge flames may theoretically occur.

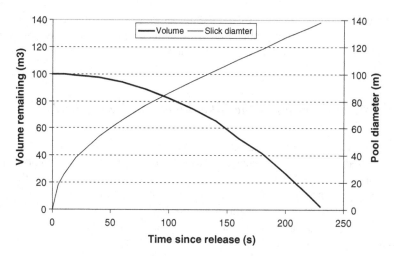

Fig. 7.1 Oil slick diameter as a function of time, provided there is no ignition of the oil (instantaneous release of 100 m³ of stabilised crude oil)

7.3.2   Ignition Probability of an Instantaneous Release

Due to the oil becoming less volatile with time, the ignition probability will correspondingly also decrease. The time to ignite an oil slick when subject to radiant heat will be longer if the oil has been on the water for some time and the thickness has decreased.

With heat radiation above 10 kW/m² the effects of delay and film thickness on time to ignition is hardly discernable. As time proceeds, light oil fractions evaporate, the oil cools down, the slick thickness decreases, and may to a certain extent be mixed with water. In this situation the possibility of igniting the slick decreases rapidly. Crude oil will thus attain ignition properties similar to those of diesel oil which cannot be ignited without the addition of a primer or without the presence of a significant heat source. An instantaneous release of fresh, unstabilised crude oil is easier to ignite the earlier the ignition source is introduced. The ignition probability of the stabilised crude oil in which all major light fractions are evaporated, is rather low. The flash point of such stabilised crude oil is as high as 125 °C.

The temperature of the oil at the point of release may be as high as 90–100 °C, but it will soon cool down by heat transfer to the sea and the ambient air. A major heat source has to be present to ignite a slick of stabilised crude oil some time after release. Such a heat source is unlikely to be present in most offshore situations except when there is already a significant fire on the platform.

On the other hand, if the ignition source is introduced approximately at the time of the release of fresh crude oil (i.e. immediate ignition), the oil is much more readily ignitable. Fresh, unstabilised crude oil has a flash point of 1–10 °C. The fire point however, may be as high as 60–70 °C, which usually is below the temperature at which oil is processed. Taking all these factors into consideration it may be concluded that the probability of immediate ignition of an instantaneous release resulting in a sustained fire is rather low.

7.3.3 What Determines the Likelihood of Fire on Sea?

Fire on sea is not a well documented event. In fact, very few events have explicitly been reported in the public domain. Blowouts are the main type of events where such occurrences have been known to occur. Even for blowouts, there are relatively few cases where there has been fire on the sea. It is therefore essential to establish the circumstances which may result in fire on sea.

SINTEF [5] has carried out a research project for several years examining fire on sea. This has included a series of large scale tests on Spitsbergen in 1994 as well as small scale laboratory tests in 1996–1997. A brief summary of some of the main findings from that work is presented in the following, related to the following main aspects:

- Flow of oil to the sea
- Oil composition
- Environmental conditions.

These aspects will determine whether ignition of an oil slick is likely or not, and whether sustained fire can be expected. The aim of the project was to study the fire behaviour and thermal impact from fires on the sea surface, as input to management of safety in relation to fire on the sea level.

Experiments with fresh crude oil and oil water emulsions were carried out during the test programme. The maximum amount of oil in one experiment was 8 m^3.

7.3.3.1 Flow of Oil to Sea

The full scale tests showed that the manner in which the oil reaches the surface of the sea is important. We may distinguish between oil which reaches the sea in a 'plunge' (i.e. free fall) from a 'run', where the oil runs down elements of the structure. The importance for fire on sea is as follows:

- Oil will 'plunge' if it falls freely from the platform. Even if it is burning when falling from the platform, it may be quenched through plunging into the sea.
- Oil may 'run' down the jacket legs, bracings risers etc., and may still be burning when it reaches the sea surface. Quenching is less likely under these circumstances.
- Oil released subsea can be regarded as 'plunged' oil, but is probably significantly more 'weathered' (emulsified). Ignition would then be dependent on whether a strong external source of ignition is in the area.

7.3.3.2 Oil Composition

Evaporation of the lightest fractions above the liquid surface is the main mechanism that determines the sustained burning ability. The laboratory tests carried out by SINTEF confirmed that a boundary layer still persisted in high wind conditions, if sufficient amounts of the lighter fractions are contained in the crude oil.

The other implication of this is that large differences in behaviour were identified between the different oils. Some oils had considerably more evaporation than others. Obviously condensates ignited most easily. SINTEF suggested that the flammability of the oil is dependent on the extent of flammable components with boiling points below 150 °C.

7.3.3.3 Environmental Conditions

Because of the obvious limitations, the testing of ignitability and sustained burning under different environmental conditions has been one of the most difficult aspects to study. The large scale tests at Spitsbergen were carried out in wind speeds up to 5–10 m/s. Air movement velocities up to 25 m/s were used in the laboratory tests. For the oils with the highest evaporation rates, there was a limited effect from the highest wind speeds. It should be noted that the laboratory air movement velocity of 25 m/s was measured only about 1 m above the surface, whereas official wind speed is measured 10 m above the surface. A wind speed of 28–30 m/s at a point 1 m above the surface would probably correspond to a wind speed of 55–58 knots. This is close to a hurricane which begins at 64 knots.

Wave motions have not been studied in these tests, due to natural causes. The large scale tests were carried out partly in a basin (in the sea ice!) and partly in a sheltered lagoon. The possibility of significant wave actions in both scenarios was non-existent.

The overall effects of wave action on the burning of an oil slick will be most extensive for the thinnest slicks that are close to the critical minimum thickness for ignition. The minimum film thickness for sustained ignition will increase with the time as the oil is weathered. Non-breaking waves will introduce minor fluctuations in oil film thickness and this will affect a proportion of the slick area at any time. If the oil film is thick, this will not have any appreciable effect on the ignition properties other than to slightly enhanced the dilution of the vapour cloud above the slick. Breaking waves will disrupt the thinner areas of an oil slick. Although this will only affect a small proportion of the slick area at any one time, the thickness will not be reinstated. The cumulative effect of breaking waves will be to break up the sheen. This will effectively prevent flame spreading and sustained burning.

There is still considerable uncertainty about the effects of high wave actions, especially breaking waves and extensive sea spray, which would be quite common at high wind speeds. It is claimed that the evaporation rates will not be significantly changed in the high wind state but this remains an unsupported assumption.

However, there is a general agreement that intense burning and sustained fire is virtually impossible in high wind scenarios. This applies to both "weathered" and 'fresh' oil spills (i.e. where little or no weathering has taken place).

An indication of the results under different wind conditions is indicated in Fig. 7.2, which shows that the chance of ignition is virtually independent of the heat transfer at low wind speeds, but much more dependent on the heat transfer at high wind speeds [6].

Table 7.7 [5] presents a summary of the results from the tests carried out by SINTEF. It should be noted that the effects included are limited to wind effects, and are not the combined effects of wind and waves.

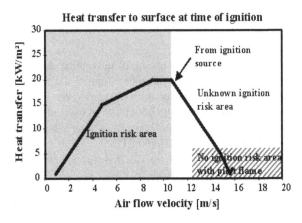

Fig. 7.2 Heat transfer to the surface at the time of ignition in the laboratory tests (courtesy of SINTEF)

Table 7.7 Conditions for ignition and sustained burning for fire on the sea surface influenced by weather

Conditions	Type of oil		
	Condensate	Light crude	Crude oil
Minimum slick thickness (mm)	0.5	1	1–3
Maximum evaporative loss in (%)	>30	<30	5–10
Maximum water-in-oil emulsions (%)	Unknown	Unknown	<25
Maximum wind conditions (m/s)	25	20–25	15

7.3.4 Loads from Sea Level Fire

The most severe fire loads from fire on sea level may have extensive cost implications in relation to the amount of passive fire protection that is needed to protect the structure. It is therefore important to consider the following parameters:

- Flame dimensions
- Fire duration (i.e. burn rates and inventory)
- Fire loads
- Smoke.

Consider, as an illustration, a floating production installation which was installed in the Norwegian sector some years ago. The following structural members were protected against sea level fire with passive fire protection:

- Support structure (columns)
- Deck support structure (beams).

The assessment of dimensioning loads was based on a conservative approach, which resulted in high fire loads, as follows:

- Columns

    - 250 kW/m^2, from sea level up to 15 m
    - 200 kW/m^2, 15 m upwards

- Main structural beams (support for topside modules)

    - 200 kW/m^2

- All members protected for 2 h fire duration.

The total surface with protection was about 13,000 m^2. If we in current prices assume an average price of 600 USD per m^2, then the cost of passive fire protection of the structure is about 8 million USD. But even this may be a low value, if the protection has to be renewed on location. The cost of offshore reapplication would be several times higher.

7.3.4.1 Flame Dimensions

There are three phases of oil slick spreading [7]; gravity driven, viscous force limited and surface force limited spread. The diameter of a burning pool on the sea is dependent on the scenario that has created the burning pool, including a number of factors. All of the following need to be known in order to calculate the likely extent of a burning pool:

- Type of liquid, including its density
- Whether the spill is continuous or instantaneous
- Volume or leakage rate

- Time of ignition in relation to the start of the spill.

The laboratory tests [6] demonstrated that upwind flame spread may occur, but not for the heaviest crude oils. The upwind flame spread may also be limited by strong wind. The crosswind flame spread velocity is also of the same order as the upwind velocity.

7.3.4.2   Fire Duration

The duration is first and foremost dependent on whether it is a continuous release or not. The fire will only last as long as the film thickness exceeds the minimum thickness.

7.3.4.3   Fire Loads

The large scale tests gave surprisingly high heat fluxes. Temperatures of certain regions inside the flames reached 1300 °C or more. The very hot regions are located where the turbulent eddies of the flames entrain sufficient amounts of air in order to allow complete combustion, and where the flames are optically thick.

Soot screening will trap some of the heat inside the fire plume, increasing the temperature and the heat load on an engulfed object. Therefore the fire load on objects inside a fire plume is of the same order as those found by experiments on land. An object enveloped in the flames may experience a fluctuating heat load, with peak values of approximately 400 kW/m^2. Average values may be in the order of 200–300 kW/m^2.

In high winds, heat exposure will mainly be on objects close to the surface, as long as no significant wind blocking from constructions occurs.

7.3.4.4   Smoke

This aspect does not affect structural elements, but rather has an impact on personnel which may be very severe. This may be seen from Fig. 7.3, which shows a sea level fire test using 8 m^3 of stabilised crude oil. The spill was contained in a 10 m diameter ring when it was ignited, after which it was allowed to flow freely, increasing in diameter to almost 50 m, and burning for some 15 min. The height of the flames reached more than 60 m, but was pulsating with a period of some few seconds. The extent of the black smoke may be seen from the picture.

Fig. 7.3 Fire on sea level
tests at Spitsbergen, 1994,
photographed by
J. E. Vinnem

7.4   Analysis of Smoke Effects

7.4.1   Methods for Prediction of Smoke Behaviour

Smoke is one of the major hazards to personnel in fires, especially in oil fires. The reduced visibility due to thick black smoke is the first threat to people who want to escape from, or fight, a fire. Knowledge of smoke production, smoke flow and impact of smoke on people and facilities is available from literature, from laboratory tests, and from experience of real fires such as the fire on Piper Alpha platform [8]. A CFD based computer code is needed for realistic modelling of smoke production and dispersion [9].

The smoke generation in enclosed hydrocarbon fires is mainly governed by the properties of the fuel and the ventilation conditions. The nominal ventilation factor α the air-to-fuel ratio defines certain regimes of a fire from below stoichiometric conditions, characterised by oxygen starvation, through stoichiometric fires, to well ventilated fires.

7.4.1.1   Mass Flow Rate

An important fire scenario for offshore platforms is the enclosed fire. The estimated smoke production in this case is different from an open fire, since the combustion products can be recirculated into the flames. The enclosure itself may then be considered as a combustion chamber, and the mass and heat balance for the enclosure are the basis for estimating temperatures and determining smoke production rates and concentration of soot and gases.

A rough estimate of the smoke production rate in an enclosure may be obtained by multiplying the fuel burning rate by the amount of air needed for complete combustion. For most hydrocarbon fuels this is about 15 times the fuel burning rate. This number can be noted, k_a, the mass air-to-fuel stoichiometric ratio. In fires where the air supply is sufficient for complete combustion, the total mass rate of smoke can be calculated when the fuel burning rate is known. The total smoke production rate is:

$$\dot{m}_{tot} = \dot{m}_f + k_a \cdot m = (1 + k_a)\,\dot{m}_f \approx 16\,\dot{m}_f \tag{7.4}$$

where

$\dot{m}_{tot}$ total mass flow of smoke from the flames
$\dot{m}_f$ fuel burning rate
k_a mass air-to-fuel stoichiometric ratio.

This gives a first step estimate which gives an idea of the amount of smoke that is produced at high temperatures. The smoke produced by the fire is then mixed with entrained air, a process which goes on inside the fire module if excess air is available, or outside the module when the smoke escapes. This mixing process dilutes the smoke, leading to a reduction in temperature, soot concentration, and the concentration of toxic combustion products. The oxygen concentration in the smoke however, increases by mixing with the fresh air. The prediction of mass production rate of smoke is determined by the fuel burning rate, and the ventilation of the fire enclosure.

7.4.1.2   Temperature

The temperature of the smoke leaving a fire enclosure is a result of the burning conditions. If the 'worst case' fire severity is assumed, the maximum smoke temperature will be up to ≈ 1200 °C. A temperature significantly lower than this can be expected in most cases. It is not obvious that a 'worst case' fire development inside the fire enclosure leads to the worst smoke hazard to people on a platform. A very hot smoke plume developing from an opening in a fire compartment will have strong buoyancy and will tend to rise rapidly. In some cases this will be favourable, since smoke can flow above the platform. A cooler smoke plume will be more likely to follow the flow field of air around the platform and may conflict with escape routes and lifeboat stations at lower levels.

7.4.1.3 Soot Production

Soot production rates are given as soot yield, expressed as a ratio of soot production rate against fuel burning rate. The soot yield varies with fuel type, geometrical configuration of the fuel within the fire enclosure, and the air-to-fuel ratio of combustion.

Traditional building materials and furniture have a soot yield typically in the order of 1–2%. Liquid and gaseous hydrocarbon fuels usually have characteristic soot yields significantly above this level.

7.4.1.4 Carbon Monoxide

The concentration of carbon monoxide (CO) in the compartment and at the outlet is typically less than 0.1% for well ventilated fires. This is characterised by an air-to-fuel ratio above 50.

The concentration of CO in fires with limited air supply goes up to about 2% when the air-to-fuel ratio becomes close to stoichiometric (15 kg/kg). CO concentration above 0.5% can be found in fires with air-to-fuel ratio up to three times the stoichiometric ratio. For ventilation controlled fires and fires with oxygen starvation, CO concentrations of up to 35% can theoretically occur. In practice, a CO concentration of 5% at air-to-fuel ratios about half the stoichiometric ratio, has been measured in diffusion flames in poorly ventilated enclosures. This may be considered as a realistic maximum CO concentration in an offshore fire.

7.4.1.5 Carbon Dioxide

Carbon dioxide concentration varies with air-to-fuel ratio and a maximum concentration of 14% has been measured at the outlet opening in experiments by SINTEF [2].

7.4.1.6 Oxygen

Oxygen concentration at the outlet opening varies with air-to-fuel ratio from close to zero at the stoichiometric ratio, up to the concentration of ambient air at large for situations where excess air is available.

7.4.2 Smoke Flow and Dispersion

Smoke from fires in naturally ventilated areas will escape through openings and louvered walls. The hot gases are lighter than air and thus, buoyancy will create a smoke plume. The environment around the platform, characterised by wind direction and velocity, will influence the smoke plume. Both the smoke plume and the wind interact with the platform itself, creating areas with increased flow velocity and recirculating zones. Important questions to answer are:

- Can smoke be transported downwards after leaving an opening from the fire area, and be dragged below the platform?
- Can smoke flow upstream the wind direction?
- Will smoke enter the air intake for the accommodation areas ventilation system?
- Will smoke penetrate into a pressurised living quarter?
- How fast will smoke infiltrate a living quarter if ventilation systems are shut down?
- Which areas will be threatened by smoke in specified fire scenarios?
- What is the major impact of smoke on people?

The hazard of smoke is characterised by three factors:

- Reduced mobility
- Pain and injury to the personnel due to temperature of the smoke
- Incapacitation or death due to toxic or irritating components in the smoke.

The relative importance of these factors can be found by comparing the threshold values with the actual or predicted exposure in a fire scenario.

When soot concentration is determined by modelling of soot production rates and dispersion, this can be converted to a length of vision. Soot particles block the passage of light and visibility is determined by the intensity of the light source, the soot concentration, and the wavelength of the light.

The visibility of an object is determined by the contrast between the object and its background. Light emitting signs, for example, are two to four times easier to see than light reflecting signs.

7.5 Structural Response to Fire

7.5.1 Manual Methods

The simplest methods are directly based on results from fire tests, while the most sophisticated computer models calculate the temperature increase in a structural member based on a given temperature exposure curve and the thermal properties of the materials which are also temperature dependent.

7.5.2   Uninsulated Steel

When unprotected steel is exposed to fire, the temperature will in most cases rise to a critical level within minutes. A critical level is one at which the structural member is unable to fulfil its load bearing function. However, if the steel section is heavy and the temperature level is moderate, the structure may be able to carry adequate loads for a sufficient time without further fire protection. The temperature of an unprotected steel section may be predicted through an iterative process using Eq. 7.5 [2]:

$$\Delta T_s = \frac{h_{cr}}{c_{ps}\rho_s}\frac{A_i}{V_s}\left(T_g - T_s\right)\Delta t \tag{7.5}$$

where

ΔT_s temperature increase in steel section (°C)
T_s temperature in steel section at time t (°C)
h_{cr} $h_c + h_r$, coefficient of heat transfer (includes both convective and radiant heat transfer) (W/m²°C)
c_{ps} specific heat of steel (J/kg°C)
ρ_s density of steel (kg/m³)
A_i/V_s section factor of steel section (m⁻¹), where
A_i area of the inner surface of the insulation material per unit length of the member (m²/m)
V_s volume of the member per unit length (m²)
T_g gas temperature at time t (°C)
Δ_t time interval (s).

This equation is based on the assumption of quasi-steady-state, one-dimensional heat transfer with the steel considered as heat sink. The heat supply to the section is considered to be instantaneously distributed to give a uniform temperature due to the high thermal conductivity of steel.

When the steel section is large, there may be considerable temperature gradients in the cross-sections. In such cases suitable computer codes should be used to predict the temperature response more accurately.

7.5.3   Insulated Steel

There are few methods suitable for the hand calculation of the temperature response of steel with passive fire protection. Those methods that are available are not applicable to all types of fire protection materials. These methods are simple iterative equations where the material properties of both steel and insulation materials are given as constants. However, the thermal properties of both steel and insulation materials may vary considerably with temperature, and the results from such calculations should be carefully checked with test results.

7.5.3.1   Simple Calculation Methods

Simple methods for the hand calculation of temperatures in an exposed insulated steel structure are limited to a one-dimensional approach. The calculation of the steel temperature increase ΔT_s of a member insulated with dry materials during a time interval Δt follows from Eq. 7.6 [2]:

$$\Delta T_s = \frac{\frac{\lambda_{ci}}{\delta_i} A_i}{c_{ps}\rho_s V_s} \left[\frac{1}{1+\frac{2\varepsilon}{3}}\right](T_g - T_s)\Delta t - \left(e^{\frac{\varepsilon}{5}} - 1\right)\Delta T_g \tag{7.6}$$

$$\varepsilon = \frac{c_{pi}\delta_i\rho_i A_i}{2c_{ps}\rho_s V_s} \tag{7.7}$$

where

λ_{ci}	thermal conductivity for insulation (W/m°C)
δ_i	thickness of insulation (m)
c_{pi}	specific heat of insulation (J/kg°C)
ρ_s	density of steel (kg/m^3)
c_{ps}	specific heat of steel (J/kg°C)
ρ_i	density of insulation (kg/m^3)
A_i/V_s	section factor for steel member (m^{-1})
T_g	ambient gas temperature at time t (°C)
T_s	steel temperature at time t (°C)
ΔT_g	increase of the ambient temperature during the time interval Δt (°C)
Δt	time interval (s).

These equations are valid for steel members insulated with dry materials. For wet materials the calculation of the steel temperature increase ΔT_s is based on the same equations with the following modifications:

a. Before reaching a steel temperature of 100 °C, Eq. 7.6 is used.
b. At a steel temperature of 100 °C, a delay in the steel temperature rise, Δt_v is introduced. Methods for predicting the time delay are presented in Ref. [2].

When using hand calculation of temperatures in insulated steel structures exposed to fire, care should be taken when selecting the material data for the insulation material. First, the thermal conductivity varies with temperature. If the thermal conductivity is described with one single value, a relevant average value should be used. Secondly, all materials have an upper validity temperature, and calculations should not be undertaken outside this level. Thirdly, it is a good principle to try to validate calculations with test results.

7.5.3.2 Computer Calculations

There are many non-linear computer codes for temperature analysis of structural elements. Most computer codes can solve one and two-dimensional problems although only a few codes are capable of solving three-dimensional problems. SUPER-TEMPCALC, and TASEF-2, are two codes which may be run on personal computers.

Both codes are based on the finite element method (FEM), and cover one and two-dimensional heat flow. The computer codes can analyse a cross-section of a linear structural element exposed to an arbitrary time/temperature curve. The cross-section can consist of different materials, and the thermal properties (λ, c_p) can be given as non-linear functions of temperature. The heat exposure is modelled either as a time/temperature-curve with a given convection and emissivity-value at the surface, or as a prescribed time–radiation relationship. Hollow sections can also be analysed.

The analysed cross-section is divided into an element mesh, and the governing transient heat conduction equation is solved at the boundaries of each element.

Figure 7.4 shows an element mesh for an insulated I-section steel column. The column is subjected to uniform heat exposure, and due to the two symmetry lines, only a 1/4 of the cross-section has to be analysed.

Results from a calculation should not be accepted unless the validity of the input-parameters is documented. Preferably the validity of the computer code should have been verified through comparison with appropriate test results.

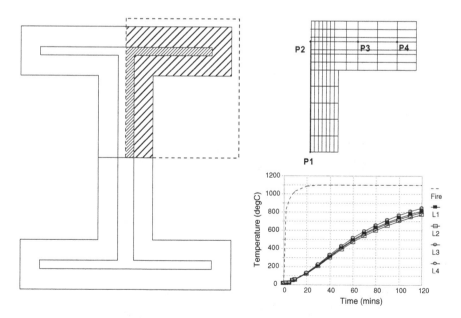

Fig. 7.4 Element mesh for an insulated I section steel column

When using manual methods, care should be taken when choosing the thermal parameters λ and c_p for the materials. For certain materials (gypsum and mineral wool) particular care should be taken into account, when using λ and c_p curves outside certain ranges. When gypsum has reached a temperature of 600–650 °C the chemically bound water is released and evaporated, and a gypsum board will lose its strength and may fall off the protected structure. In calculations it is suitable to simulate this by a rapid increase in the conductivity curve at e.g. 600 °C.

Results from a temperature analysis with FEM computer codes may be presented as time/temperature curves for the individual nodal points, or as isothermal charts at given times.

7.6 Risk Reducing Measures

7.6.1 Overview

Measures to reduce risk in relation to the fire hazard have been the focus of the technical safety discipline for many years. There are several papers and reports published on these topics, and there is no need to go very deeply into these subjects. A full discussion of all relevant aspects is outside the scope of this book, which is focused on risk quantification.

The ISO 13702 standard 'Control and Mitigation of Fires and Explosions on Offshore Production Installations—Requirements and Guidelines' was published in 1999 and updated in 2015 [10]. The following sections of the ISO 13702 standard are relevant to the fire risk reduction:

6. Installation layout
7. Emergency shutdown systems and blowdown
8. Control of ignition
9. Control of spills
10. Emergency power systems
11. Fire and gas detection systems
12. Active fire protection
13. Passive fire protection
15. Responses to fires and explosions
16. Inspection, testing and maintenance.

If the logic of a typical event tree is followed (see, for instance, Fig. 6.10), then the risk reducing measures may be structured as follows:

• Leak prevention	• Welded connections
	• Flange types with reduced leak probability
• Leak detection	• Gas detection
	• Fire detection

(continued)

(continued)

	• Emergency shutdown system • Blowdown system
• Ignition prevention	• Hot work procedures • 'Ex'-protected equipment • Maintenance of electrical equipment
• Escalation prevention	• Installation layout • Segregation of areas • Active fire protection • Passive fire protection

7.6.2 Recent R&D Experience

The most extensive R&D effort has been performed through the so-called 'Blast and Fire Engineering for Topside Systems' programme (BFETS) [3]. This programme has included both fire and explosion research, the explosion related experience is discussed in Sect. 9.1.2. With respect to the fire related research, some of the main issues have been:

- Realistic fire loads have been established for jet fires as well as pool fires. Some of these fire loads are somewhat higher than anticipated.
- Pool fires in ventilated enclosures have been found to give almost as high heat loads as jet fires.
- Use of active fire protection, for instance deluge water systems, has been found to be even more effective than anticipated, especially in ventilated enclosures.

7.7 Dimensioning of Structural Fire Protection

According to the facilities regulations [11], the design loads/actions that are part of the realization of the main safety functions should, as a minimum, be designed in such a way that dimensioning loads with an annual probability of 10^{-4} should not result in the impairment of a main safety function. Each hazard should be considered separately. This means that fire scenarios that may result in the impairment of safety functions should not have a frequency higher than 10^{-4}. Such fires are referred to as dimensioning fires, or more generally, as dimensioning accidental loads, abbreviated as DiAL.

The loads (for example, fires) that an installation is designed to withstand are sometimes abbreviated as DeAL. A company may include additional robustness when design accidental loads (DeAL) are decided upon. However, the DeAL should, as a minimum, include the DiAL.

The NORSOK S–001 standard [12] can be used to demonstrate compliance with the regulations. According to this standard, the design fire for load-bearing structures and fire divisions should as a minimum include:

- the dimensioning fire
- the worst credible process fire, and
- fire class requirements, e.g. H-class.

These aspects are elaborated on in the following using a case illustration.

7.7.1 Case Illustration

The case chosen to illustrate the dimensioning of structural passive fire protection (PFP) concern the upgrading of the deck structure of an existing installation. The deck structure in question is a module support structure consisting of trusses with box section members. There is a limited amount of process equipment installed inside the structure, which is relatively open.

The case study will demonstrate the calculations and risk evaluations necessary to determine whether the most severely affected parts of the structure can be effectively protected by PFP and how this may be efficiently accomplished.

7.7.2 Dimensioning Fire

The previous requirements for design against fire in the Norwegian regulations for Fire and Explosion protection of offshore installations [13] and the regulations on structural design and protection [14] have been replaced by requirements in NORSOK N–001 [15] and N–003 [16]. The requirements are the same, in the sense that adequate fire protection of load-bearing structures is required in relation to the dimensioning fire. Therefore, it is important to establish what a dimensioning fire is and what are the associated fire loads.

It should be noted that whenever PFP is used, it shall also have the capability to withstand the dimensioning explosion load, to protect against a scenario where fire follows an explosion. PFP therefore also needs to have explosion resistance. This is also addressed in a later section. The main characteristics of a dimensioning fire are:

- Heat loads
- Dimensions of fire
- Duration of fire.

The use of constant values for these three parameters will sometimes be too simplistic. The following example may illustrate this. If the contents of a liquid filled vessel (stabilised crude oil) spilled onto the deck is not taken care of by the drain system, a liquid pool on the deck of, say, 20 m diameter may result. The resulting fire will then have the same diameter, and a height of about the same value. More realistically, the height will be limited by the space between decks. The time averaged heat loads will depend on the ventilation rate and air supply to the flames, but may typically be 150 kW/m². If the initial spill is 50 m³, the diameter is 20 m, and the average burning rate is 3 mm/min, then the average thickness of the pool is 0.159 m, and the duration of the fire is slightly above 53 min.

In theory, the diameter and height of flames should be constant during the duration of the fire. In practice, however, experience shows that the size of the pool will shrink, and the flame height reduce (unless it is limited by other constraints, such as distance between decks). In a full scale test with crude oil burning on the sea surface, the size of the unconstrained pool on the sea varied as shown in Fig. 7.5. The pool in that test was physically limited to 10 m, before ignition, after which it was unconstrained. The behaviour in the initial phase must be considered in the light of these constraints.

As a start, it will often be reasonable to assure that the fire dimensions and fire loads will be constant. In that case, the only variable to consider is the duration of the fire.

7.7.3 Fire Duration Distribution

If we make the approximation that all variables except duration are constant, we end up with a 'one-dimensional problem'. Thus the assessment of the integrity of a structure subject to a dimensioning fire is limited to an analysis of fire response under the applicable loading as a function of time. The critical aspect will be to determine whether failure occurs before the fire has subsided. It is recognised that it is impossible to define the fire 'Design Accidental Event', based on the theoretical

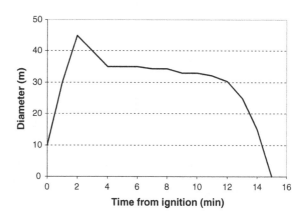

Fig. 7.5 Diameter of burning pool on sea surface as function of time since ignition

definition of DAE. Various practical considerations are therefore used, involving consideration of the following factors:

- Volumes of ESD segments
- Depressurisation capacities and times
- PFP applied on equipment and structures.

Figure 7.6 presents the exceedance frequency function for the cellar deck (inside the support structure) as a function of the duration of the fire, for the case study installation referred to in Sect. 7.7.1.

The frequency of all fires (exceeding 0 min) is 7.3×10^{-3} per year, whereas the annual frequency of fires that exceed 20 min duration is 10^{-3}. The duration that corresponds to 10^{-4} exceedance frequency is 60 min. In the case study, the 10^{-4} exceedance frequency was used as design basis for structural design.

The same distribution is presented in Fig. 7.7, where the conditional exceedance probability is presented for the same fire durations. The conditional probabilities are based on occurrence of fire within cellar deck.

It is shown that there is about 70% probability that fires will last longer than 6 min. The probability that the fire duration exceeds 15 min is 50%, whereas the probability that the fire exceeds 40 min is 20%.

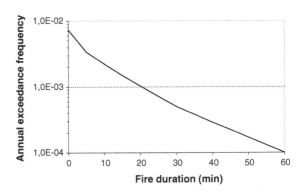

Fig. 7.6 Exceedance frequency distribution for fire duration on cellar deck

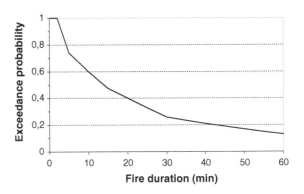

Fig. 7.7 Fire duration distribution on cellar deck

There is no accepted way to interpret these curves in order to define the duration of the dimensioning fire. If the median value in the conditional distribution is chosen, then this would imply a duration of the dimensioning fire of 15 min.

7.7.4 Definition of Dimensioning Fire

Equation 7.8 presents the definition of a dimensioning fire according to the previous NPD regulations for load-bearing structures [14]. A dimensioning fire is defined by the following equation:

$$p(system\,failure) = P(system\,failure\,|\,fire) \cdot P(fire) \leq 10^{-4} \qquad (7.8)$$

This definition is applied to the exceedance probability curve in Fig. 7.8. The fire duration corresponding to a 10^{-4} exceedance probability, is approximately 37 min. The figure gives the frequency of fire as well as the probability of failure for the fire durations shown, as well as the resulting exceedance frequency function.

7.7.5 Worst Credible Process Fire (WCPF)

Worst credible process fire (WCPF) is a term that has been used by the industry in recent years. Some years ago, the petroleum safety authorities realised that there was a need to understand how the operators defined the worst fire in the process

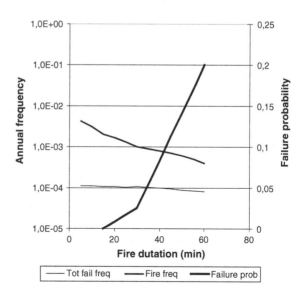

Fig. 7.8 Occurrence frequencies for fire and system failure, and resulting exceedance frequency for system failure

area that the installation was designed to withstand. They initiated a project that resulted in a report (in Norwegian only) with a title that can be translated into 'How is the worst credible process fire defined' [17].

The background for the WCPF term was that it is not possible/realistic/expedient to design an installation to withstand a worst case scenario. And what is a 'worst case' scenario, after all? It is always possible to imagine something even more severe. The WCPF was introduced to define a 'credible worst case scenario': a severe scenario where some safety systems are assumed to work as intended. In 2018, the WCPF was defined in NORSOK S–001 [12]:

> The worst credible process fire (WCPF) in a fire area is derived from an ignited leak in the ESD-segment including possible escalation that will give the worst exposure of main load-bearing structures and fire divisions with regard to duration (not related to the time needed for safe evacuation) and heat load distribution. ESD valves and emergency depressurization valves limiting the supply of fuel can be assumed to function. With respect to liquid spills, the open drain or grating in the area can be taken into account. The load from the WCPF shall be covered by the design accidental load for main bearing structures and fire divisions. The identified WCPF shall not escalate to hydrocarbon pipeline risers or to wells (flow from reservoir) to avoid impairment of the main load- bearing structures.

7.7.6   USFOS® Modelling

The program USFOS® is tailored for progressive collapse analysis of jackets and topside structures under extreme environmental and accidental loads such as ship collision, fire and explosions. The program is based on general continuum mechanics principles and accounting for non-linear geometry and material effects include several special features facilitating efficient non-linear static or dynamic analysis. The basic idea behind the formulation is to use one finite element to model each structural component and still obtain a realistic representation of the non-linear material and geometry effects.

Due to the choice of displacement interpolation functions, USFOS® predicts the exact buckling load of an axially compressed member with arbitrary end conditions and nodal forces. Non-linear material behaviour is accounted for by introducing yield hinges at the beam ends and mid-section. The plasticity model is formulated in the force space based on bounding surface theory. The thermal load effects presently accounted for include thermal expansion and reduction of yield stress and E-modulus at elevated temperatures. The effect of temperature loading on an elastic element is to produce forces related to axial expansion and bowing due to tem-perature gradients over the cross-section. Thermal expansion and reduction of E-modulus yields consistent nodal forces. Within the boundary surface concept both the yield and bounding surface contracts (at different rates) for increasing temperatures reflect the degradation of cross-sectional capacity.

7.7.6.1 Application in Risk Reduction Context

A comprehensive USFOS® analysis has been performed in order to provide further insight into the fire performance of the module support frame and the possible effects of installing PFP on selected members. Three locations are considered, referred to as the following cases:

- South-west corner
- South-east corner
- Frame structure around top of shaft.

The South East corner is particularly critical, because this part of the MSF supports a heavy module. The fire loads on the MSF are as defined below:

• South-west corner	Diffuse gas fire, 18 m diameter sphere, truncated by distance between decks
• South-east corner	Condensate pool fire, 24 m diameter sphere, flame height restricted by distance between decks
• Frame structure around top of shaft	Diffuse gas fire, 18 m diameter sphere, truncated by distance between decks

The PSA regulations require that active fire protection (water spray systems) is not considered when PFP is calculated. On the other hand, in the QRA, the combined effect of both active and passive systems may be considered. In the present cases, the loads as shown in Table 7.8 are utilised as typical loads.

These values are presented in the NORSOK S–001 standard [12]. It should be noted that the USFOS FIRE® package has a module that may be used for calculation of the detailed fire loads as a function of fuel supply ventilation and time.

A detailed and specific analysis will be especially important when the modules are either considerably under- or over-ventilated. The typical values shown in Table 7.8 are otherwise usable (somewhat conservative) when ventilation is around the stoichiometric ratio (15–16 kg air per kg hydrocarbon fuel), typically in the range 10–20 kg air per kg hydrocarbon fuel. It should be noted that the following failure criterion was used in the analysis:

- Failure of support structure: considered to occur when relative movements exceed 0.3 metres in any direction.

Table 7.8 Typical fire loads according to fire type and fire water application

	Jet/liquid spray fire	Pool fire
Local peak heat load	350 kW/m^2 for leak rates m > 2 kg/s. 250 kW/m^2 for leak rates m > 0.1 kg/s	250 kW/m^2 for burning rate m > 2 kg/s. 150 kW/m^2 for burning rate m > 0.1 kg/s
Global average heat load	100 kW/m^2	100 kW/m^2

Please note that such movement may not necessarily imply collapse, but is likely to cause secondary rupture of piping, vessels, etc. Such mechanisms may supply extra fuel to the fire, and fire water systems may fail.

7.7.6.2   USFOS® Results

The results from the USFOS® analysis are presented in Table 7.9 for the three areas considered. The last case also includes the effects of providing PFP on selected trusses in the South-east corner. The details are discussed in a later subsection.

The last calculation, reflects the condition where PFP has been applied to the truss members beneath module support points. Under one module support two members are insulated while under the other supports only one of the members is insulated. This point, although it is the most highly loaded, is not likely to be subjected to high fire loads (only far field radiation), whereas the first support point would be engulfed by flame in most fire scenarios.

7.7.7   QRA Modelling

The initial failure model of the deck used in the QRA is shown in Fig. 7.9. It should be noted that:

- The curve is the overall average failure function for all decks and areas.
- The performance of the module support frame members is likely to be in the lower probability range.

Table 7.9 Main loads from USFOS® analysis

Area	Time to failure (min)	Deluge accounted for	Comments
South-west corner	7	No	Fire size conservative for area
	10	Yes	Conservative fire size Model slightly conservative
	>150	Yes	Realistic fire size Most optimistic location
Top of column	16	No	Standard fire size used
South-east corner	22	Yes	Calculated with realistic fire size and heat load
	45	Yes	PFP on selected members

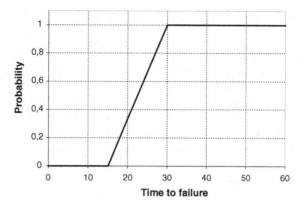

Fig. 7.9 Base case failure probability model used for decks and areas in QRA

This is a simplistic model which does not pretend to be accurate in any sense, but useful for the QRA modelling. The implications of the model are as follows:

- Structural failure (to an extent that it gives escalation of the fire from one deck to that above) is impossible up to 15 min.
- Escalation of the fire through deck failure is certain after 30 min.
- It is assumed (as a simplified model) that the failure probability increases linearly between 15 and 30 min.

The definition of the curve may be expressed mathematically as follows:

$$P_f = \begin{cases} 0, & t < 15 \\ \frac{t}{15} - 1, & 15 \le t \le 30 \\ 1, & t > 30 \end{cases} \tag{7.9}$$

where

t duration of fire (in minutes).

It is important to note that the USFOS® studies seen in the context of the QRA may be characterised as follows:

- They provide 'snapshots' i.e., they provide a single deterministic calculation of one case out of a myriad of possible cases.
- They provide a specific time to failure in defined areas under given fire and structural loading.
- Since the USFOS® studies are normally conducted for the most critical areas in the most severe circumstances, they are inclined to define the lowest starting point of the failure curve.

For the evaluation of USFOS® results with respect to the QRA, the following should be noted:

- For the ALARP analysis, active and passive fire protection can be considered in combination, even though the regulations do not allow active fire fighting to be accounted for, when passive fire protection is dimensioned.

Following the results from the USFOS® studies, the failure model in Fig. 7.9 was revised as shown in Fig. 7.10.

This failure curve is similar to that defined by Eq. 7.9, except that the failure probability starts to increase above zero after 5 min rather than 15 min. The failure probability is less than 1.0 until the fire duration reaches 30 min, at which the probability of failure is 1, as with the previous model.

Lastly, Fig. 7.11 presents the failure function assumed to model the situation when the structure was protected with Passive Fire Protection in the most vulnerable areas.

The implication of Fig. 7.11 is that even in the most critical areas the structure will not fail for 20 min. In this case there would be only limited variations between the different parts of the structure.

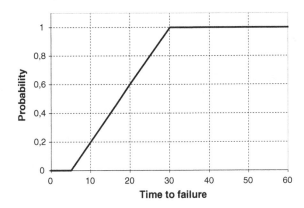

Fig. 7.10 Revised failure probability model to reflect vulnerability of structure when unprotected

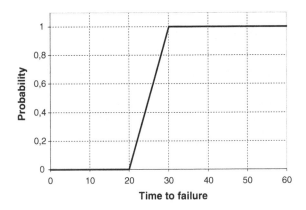

Fig. 7.11 Structural failure model after improvement of structural fire protection

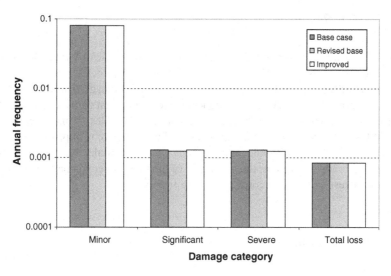

Fig. 7.12 Overall results for material damage risk

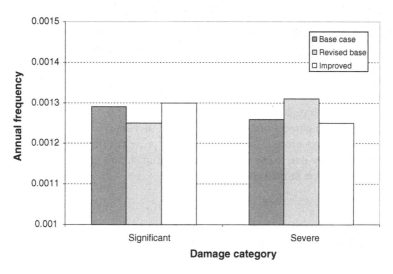

Fig. 7.13 Material damage results only for significant and severe damage

These two modelling cases (represented by the failure models in Figs. 7.10 and 7.11) were selected to represent a conservative approach to the ALARP demonstration, in that a limited protection by PFP in one area could not improve the situation more extensively than is implied by these two diagrams, probably less extensively.

7.7.8 QRA Results

Figures 7.12 and 7.13 present the results for four categories of risk of material damage to the assets.

It can be observed that the influence on the material damage risk is very small, only in the order of 3–4% change of frequencies for significant damage (one module damaged) and severe damage (damage to two or more modules). This is interpreted in the following way: When the fire protection of the structure is improved, some of the scenarios that earlier would escalate to a second module, will be contained within one module. Therefore, the frequency of the most severe damage is reduced, while the lowest category is correspondingly increased.

7.7.9 Observations

The case study discussed in this section has shown several aspects of the use of advanced non-linear structural analysis in combination with QRA. The main observations that may be drawn from this case are the following:

- The non-linear analysis will enable reflection of the differences in structural sensitivity according to which structural members are affected.
- The analysis with the USFOS® (and associated software) allowed significant improvements to be identified from a very limited application of PFP on the structural members. Thus a realistic case for improvement could be defined and implemented. More extensive PFP application would not have been practical, neither from a technical nor economical point of view.
- The limited effect on risk level s from the improvement of structural fire protection is mainly due to the fact that no accommodation facilities are provided on the installation in question.

7.8 Blast and Fire Design Guidance

The Steel Construction Institute, in collaboration with British Gas and Shell, and supported by a number of sponsors, issued in 1992 the document 'Interim Guidance Notes for the Design and Protection of Topside Structures against Explosion and Fire' [18]. The guidance document has been undergoing revision for a number of years and is now referred to as 'Fire and Explosion Guidance', published by Oil and Gas UK [19].

References

1. Lees FP (2004) Lees' loss prevention in the process industries, 3rd edn. Butterworth–Heinemann, Oxford
2. SINTEF (1992) Handbook for fire calculations and fire risk assessment in the process industries. SINTEF/Scandpower, 1992 Trondheim, Norway
3. SCI (1998) Blast and fire engineering for topside systems, Phase 2. Ascot, SCI. Report No. 253
4. Hasegawa K, Sato K (1977) A study of the blast wave from deflagrative explosions. Fire Saf J 5(3–4):265–274
5. Opstad K, Guénette C (1999) Fire on the sea surface, ignitability and sustainability under various environmental conditions. In: 6th international symposium on fire safety science (IAFSS), 5–9 July 1999, Poitiers, France
6. SINTEF (2005) New knowledge about offshore fires (in Norwegian only) SINTEF NBL, 14 February 2005, Trondheim, Norway. Report No. NBL A04148
7. Fay LA (1969) The spread of oil slicks on a calm sea. Department of Mechanical Engineering, Massachusetts Institute of Technology, Boston, May 1969
8. Lord Cullen (The Hon) (1990) The public inquiry into the piper alpha disaster. HMSO, London
9. Holen J, Magnussen BF (1990) KAMELEON FIRE E–3D—A field model for enclosed pool fires. Trondheim, SINTEF. Report No. STF15 F90010
10. ISO (2015) Control and mitigation of fires and explosions on offshore production installations —requirements and guidelines. International Standards Organisation, Geneva, ISO13702, 2015
11. Petroleum Safety Authorities Norway (2017) Regulations relating to design and outfitting of facilities, etc. in the petroleum activities (the facilities regulations). Last amended 18 Dec 2017
12. Standards Norway (2018) Technical safety, S–001, 5th edn. Standards Norway, Oslo, June 2018
13. NPD (1992) Regulations relating to explosion and fire protection of installations in the petroleum activities. Norwegian Petroleum Directorate, Stavanger, Feb 1992
14. NPD (1992) Regulations concerning load bearing structures. Norwegian Petroleum Directorate, Stavanger, Feb 1992
15. Standard Norway (2012) Integrity of offshore structures, N-001, Rev. 8, Standards Norway, Oslo, Sept 2012
16. Standard Norway (2017) Actions and action effects, N–003, 3rd edn. Standards Norway, Oslo, Jan 2017
17. SINTEF (2017) Hvordan defineres "worst credible process fire"? Report issued by SINTEF (in Norwegian only), 08 Feb 2017
18. SCI (1992) Interim guidance notes for the design and protection of topside structures against explosion and fire, Document 53. Steel Construction Institute
19. Oil and Gas UK (2018) Fire and explosion guidance, Oil and Gas UK, Mar 2018

.

Chapter 8
Explosion Risk Modelling

8.1 Overview

8.1.1 Introduction

Explosions on offshore installations have come very much into the focus in the last 20 years because new insight revealed that previous knowledge about blast loads was obsolete. Even worse, the blast loads that result from the latest tests are so high that they cannot be designed against, in many cases.

There is a lot of focus on what are realistic blast loads, how they may be determined, and what is the most appropriate approach to design against blast loads.

Calculation and assessment of blast loads are thus important subjects for this book. The physical laws and detailed calculation methods are outside the scope of this book because it would be a book in itself. A brief introduction to some of the main concepts is provided.

8.1.2 Explosion Loads on Structure

The calculation of blast loads on a structure and its response follows a similar series of steps to that used in fire analysis:

1. Calculation of HC releases
2. Calculation of explosion overpressure loads as a function of time and ignition location
3. Calculation of structural response to the time dependent overpressure loads
4. Evaluation of secondary blast effects, such as missiles, etc.

© Springer-Verlag London Ltd., part of Springer Nature 2020
J.-E. Vinnem and W. Røed, *Offshore Risk Assessment Vol. 1*,
Springer Series in Reliability Engineering,
https://doi.org/10.1007/978-1-4471-7444-8_8

Explosion modelling has been substantially improved as a result of a programme of large-scale tests conducted through the BFETS (Blast and Fire Engineering for Topside Structures) research programme [1].

The knowledge gained through this programme has led to the realisation that loads are likely to be considerably higher than previously thought.

8.1.3 Explosion Loads on People

An assessment of the effects of explosion loads on people is parallel with the structural effects analysis, although in some respects considerably simpler. Some indicative loads were mentioned in Sect. 12.3.3.

8.2 Explosion Frequency

8.2.1 Event Tree Analysis

The frequency of explosion events is usually calculated from an event tree analysis in QRA studies. Consider Fig. 8.1, which is the same as used in Chap. 5 in order to illustrate the simple event tree following a process leak. The conditions in Fig. 8.1 which imply occurrence of explosions, given a medium gas leak, are the following:

- ESD unsuccessful, ignition inside module
- ESD successful, ignition inside module.

The simple event tree assumes that all ignitions of the gas leak lead to explosions. A detailed event tree will differentiate more explicitly between ignition causing an explosion and just causing a fire, depending on whether it is immediate or delayed.

Calculation of event frequencies in the event tree will establish the explosion frequencies for all explosion cases, irrespective of associated blast load.

8.2.2 Historical Frequencies

A study [2] has been conducted of potential explosions in the 25 year period 1973–97, covering the following geographical areas:

- Norwegian sector

  - North Sea
  - Norwegian Sea

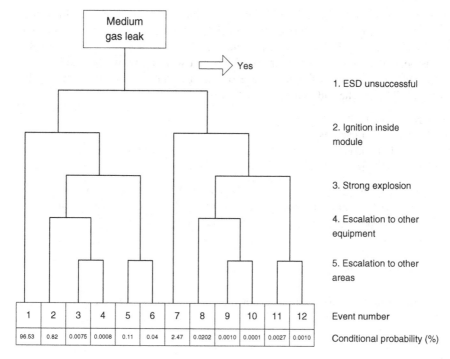

Fig. 8.1 Event tree for medium gas leak

- UK sector
  - North Sea
  - Irish Sea
  - Atlantic Sea (West of Shetland)
- Dutch sector
  - North Sea
- Danish sector
  - North Sea

With respect to explosions, there are registrations in the UK as well as Norwegian sectors. In the Danish sector, one exploration drilling well blowout resulted in an explosion in 1977, but mobile drilling units are excluded from the study referenced. There has also been a significant explosion on a production installation (in 2001) in the period after the study period.

In the Dutch sector, no explosions have been registered in the WOAD® database, neither on fixed nor floating installations. However, there are confirmed reports of one accident in the Dutch sector in August 1995, whereby a small explosion occurred in an open area ignited by the hot oil heating system, with minor damage. Details of the case are not known, and the event is not registered in either WOAD® or Lloyds' List database. The fire is certain to have occurred, but the explosion preceding it is not confirmed, due to the lack of sources. Conservatively, it has been assumed that it was an explosion, but with a low overpressure.

34 relevant explosion incidents have occurred in the relevant areas during the period 1973–97. The following distribution is found:

- UK: 16 incidents
- Norway: 17 incidents
- Holland: 1 incident
- Denmark: 0 incident

It is likely that there is some extent of underreporting from the sectors outside the Norwegian sector, due to the fact that the WOAD® database has closer connections to Norwegian authorities than to British, Dutch and Danish.

8.2.2.1 Blast Load Categorisation

The following categories were used for the classification of overpressure:

- <0.2 bar
- 0.2–1 bar
- 1–2 bar
- >2 bar

The following are additional aspects that may need to be considered in relation to gas explosions and the responses to blast loads:

- Local maxima in relation to average maximum overpressure on a panel
- Duration of pressure peaks.

Both these two aspects are important when responses and effects are being evaluated, but the descriptions were too vague (even after consulting with the companies) in order to allow any such considerations in relation to the 34 incidents.

8.2.2.2 Blast Occurrences

Table 8.1 presents an overview of the 34 blast occurrences and an assessment of the applicability under various assumptions. Clear indications were found that improvements had been made over the years, due to the following:

- All the significant explosions occurred on installations that were installed off-shore before 1980.

Table 8.1 Overview of blast load classification

Sector/Assumptions	Overpressure classification (bar)			
	<0.2	0.2–1	1–2	>2
UK sector				
• All events	7	7	2	0
• All events, but excluding events during shutdown for maintenance	7	6	1	0
• All events, but excluding events during shutdown for maintenance and irrelevant due to non-representative solutions	7	6	1	0
• All events, but excluding events during shutdown for maintenance and events that are ignited outside classified areas	7	5	1	0
Norwegian sector				
• All events	16	1	0	0
• All events, but excluding events during shutdown for maintenance	16	1	0	0
• All events, but excluding events during shutdown for maintenance and irrelevant due to non-representative solutions	15	1	0	0
Dutch sector	1	0	0	0
Danish sector	0	0	0	0
Sum all sectors				
• All events	24	8	2	0
• All events, but excluding events during shutdown for maintenance	24	7	1	0
• All events, but excluding events during shutdown for maintenance and irrelevant due to non-representative solutions	23	7	1	0
• All events, but excluding events during shutdown for maintenance and events that are ignited outside classified areas	23	6	1	0

- Where comparison could be made between 'older' (pre-1980) and 'newer' (post-1980[1]) installations for the insignificant explosions (< 0.2 bar), there was almost a 2:1 ratio between them.
- There is a clear downwards trend in the number of explosions per year for the 'old' installations, as time passes. Such a trend may be observed for significant as well as insignificant explosions (i.e. above and below 0.2 bar).

It is thus concluded that quite considerable conservatism is implied by using the 25-year period with all installations included. An adjustment of the overall frequencies has therefore been made. Unadjusted values are also presented in Ref. [2].

Comparison was also made between Norwegian and UK installations, but the numbers were too low in order to draw firm conclusions.

---

[1]It may seem surprising to call post-1980 installed facilities as 'new', but this reflects the time of the original study, about 20 year ago.

8.2.2.3 Calculated Frequencies

Frequencies have been normalised 'per platform year' as well as a 'per explosion area year'. Figure 8.2 is limited to results for explosion areas and shows adjusted exceedance frequencies. These frequencies have been based on different interpretations of the data, whereby only events relevant to normal operation are used in one case, whereas all events are used in another interpretation. Also an upper 50% prediction limit is presented, based on only the so-called 'relevant events'. The exceedance diagram below uses the lower limits in the overpressure ranges as basis for the plotting.

It was considered that there is some extent of conservatism in the adjusted frequencies. It is therefore recommended in the report [2] that the 50% upper prediction limits are used as the upper limits against which results from QRA may be compared. The calculated frequencies of failure of explosion barriers and escalation are as follows:

- Frequency of explosion barrier failure: 5.7×10^{-5} per explosion area year
- Frequency of escalation: 2.9×10^{-5} per explosion area year

The values are based on the events relevant to normal production, and a conservative interpretation of the data regarding damage.

A final note on the validity of the results relates to explosion accidents after the study was performed. There are no known cases in the UK or Norway which would imply that the results are invalid. However, the blast loads in the Macondo accident (see also Chap. 5) might have influenced the results significantly. The exact loads

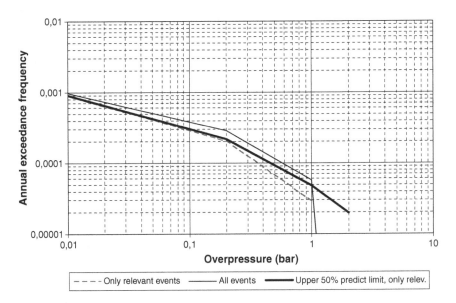

Fig. 8.2 Exceedance frequencies per explosion area

are however, not known to anybody, because the rig sank after about 36 h. The only possibility to establish the most likely range of overpressure would have been a detailed inspection of the damage, including the distinction of explosion versus fire damage. This has for obvious reasons never been made.

8.3   Explosion Consequence Analysis

8.3.1   Types of Explosion Loads

Explosion loads can range from less than 100 mbar overpressure to many bars overpressure. The loads may cause the following range of effects on structures:

- Direct catastrophic failure.
- Considerable damage (to tertiary structures) which may be further extended by the ensuing fire.
- Little or no damage (structurally), but cause critical failure of safety systems thereby preventing control of the ensuing fire.
- Damage to passive fire protection, thereby reducing the survivability of structural members.
- Damage to process equipment thereby causing immediate escalation of the accident.

It is worthwhile noting that the Piper Alpha accident was of the last type. Very few events of the first type have occurred on offshore installations, but there have been several on onshore petrochemical and chemical plants. The best known accident of this type is the explosion at the Flixborough plant in the UK in 1974.

This phenomenon has several names: 'Gas explosion', 'gaseous explosion', 'unconfined vapour cloud explosion', 'vapour cloud explosion' or 'fuel–air explosion'. The term 'gas explosion' is used in the following.

An explosion is defined as an event leading to a rapid increase of pressure. This pressure increase may arise from many different causes; nuclear reactions, loss of containment in high pressure vessels, high explosives, metal water vapour explosions, run-away reactions, combustion of dust, mist or gas (including vapours) in air or in other oxidisers.

The burning of gas, liquids, or solids in which fuel is oxidised involves heat release and often light emission. Combustion of methane (CH_4) in air can be described by the chemical equation:

$$CH_4 + 2(O_2 + 3.76\,N_2) \rightarrow CO_2 + 2H_2O + 2(3.76\,N_2) + \text{Energy} \qquad (8.1)$$

The chemical products from complete combustion of a HC fuel are mainly CO_2 and H_2O vapour. The combustion process will result in increased temperature, due to the transformation of chemically bound energy into heat. It should be emphasised that the equation above constitutes a great simplification of the real combustion process.

The combustion of gaseous fuels in air may develop in two different modes. The most common is fire, where fuel and oxygen are mixed during the combustion process. In the other case fuel and air (or another oxidiser) are premixed and the fuel concentration must be within the flammability limits for ignition to occur. In general the premixed situation allows the fuel to burn faster i.e., more fuel is consumed per unit time. The premixed fuel may also burn as a fire, if ignited prior to building up a cloud of any size.

8.3.2 Gas Explosion

A gas explosion is a process where combustion of a premixed gas cloud (i.e. fuel–air) causes a rapid increase of pressure. Gas explosions can occur inside process equipment or pipes, in buildings or offshore modules, in open process areas, or in unconfined areas.

The consequences of a gas explosion will depend on the environment in which the gas cloud is contained or which the gas cloud engulfs. Therefore it is natural to classify a gas explosion from the environment in which the explosion takes place. There are in general three categories of explosions:

- Confined gas explosions within closed rooms, vessels, pipes, channels or tunnels.
- Partly confined gas explosions in compartments, buildings or offshore modules.
- Unconfined gas explosions in open area in process plants and other unconfined areas.

It should be pointed out that these terms are not precise, and in an accidental event it may be hard to classify the explosion. As an example, an unconfined explosion in a process plant may also involve partly confined explosions in compartments into which the gas cloud has leaked.

Confined gas explosions are explosions within tanks, process equipment, pipes, culverts, sewage systems, closed rooms and underground installations. Confined explosions are also called internal explosions. A typical property of this kind of explosion is that the combustion process does not need to be fast in order to cause serious pressure build-up.

Partly confined explosions occur when fuel is accidentally released inside a building which is partly open. Typical cases are compressor rooms and offshore modules. The building will confine the explosion and the explosion pressure can only be relieved through the explosion vent areas (i.e. open areas in the walls or light relief walls that open quickly at low overpressure), or through failure of the surrounding enclosure.

The term 'unconfined gas explosion' is used to describe explosions in open areas such as process plants. Large scale tests have demonstrated that a truly unconfined, unobstructed gas cloud ignited by a weak ignition source will only produce low overpressures while burning (flash fire). In a process plant there are local areas which are partly confined and obstructed. In the case of a deflagration it is these areas that are causing high explosion pressures. A deflagration has a limited burning velocity, in the range 100–500 m/s.

However, if an unconfined gas cloud detonates, the explosion pressure will be very high, in the order of at least 20 bar, and in principle independent of confinement and obstructions. The detonation front travels as a shock front, followed by a combustion wave. The velocity of the detonation front reaches that of the speed of sound in the hot products, and thus substantially higher than in unburnt mixture. The detonation front velocity may reach 2–3,000 m/s.

Deflagrations are luckily rare events; the author witnessed a deflagration which transitioned to detonation during model tests at a test site in Norway around 1980. The Buncefield explosion and fire occurred outside London on 11th December 2005. Some experts have analysed damages to equipment and structures and concluded that the deflagration in some local spots may have transitioned to deflagration [3].

8.3.3 Blast Wave

A blast wave can be defined as the air wave set in motion by an explosion. The term 'blast wave' includes sonic compression waves, shock waves and rarefaction waves. Figure 8.3 illustrates in principle different types of blast waves. We can have:

1. a shock wave followed by a rarefaction wave,
2. a shock wave followed by a sonic compression wave and then a rarefaction wave,
3. a sonic compression wave and a rarefaction wave.

The type of blast wave depends on how and when the energy is released in the explosion and the distance from the explosion area. For strong explosions Category 1 is typical. Weak explosions initially give Category 3, but the wave can be 'shocking up' and end as Category 1 when it propagates away from the explosion.

Fig. 8.3 Blast waves

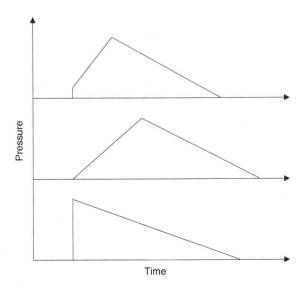

8.3.4 Pressure

Pressure is a type of stress which is exerted uniformly in all directions and is measured as the force exerted per unit area. In fluid dynamics we often use the terms like static pressure, dynamic pressure and stagnation pressure. Static pressure is what we normally call the pressure.

Dynamic pressure is the pressure increase that a moving fluid would have if it was brought to rest by isentropic flow against a pressure gradient. The dynamic pressure can also be expressed by the flow velocity, u and density, ρ.

$$P_{Dyn} = \frac{\rho \cdot u^2}{2} \tag{8.2}$$

Stagnation pressure is the pressure that a moving fluid would have if it was brought to rest by isentropic flow against a pressure gradient. The stagnation pressure is the sum of the static and the dynamic pressures.

$$P_{Stag} = P_{Stat} + P_{Dyn} \tag{8.3}$$

For blast waves and shock waves we use the terms side-on pressure and reflected pressure. The side-on pressure is measured perpendicular to the propagation direction of the wave. Side-on pressure is the static pressure behind the shock wave. The reflected pressure is measured when the wave hits an object like a wall head-on. Since reflection is not an isentropic process there is a difference between stagnation pressure and the reflected pressure. These definitions of side-on and reflected pressures are illustrated in Fig. 8.4.

Fig. 8.4 Side-on pressure
and reflected pressure

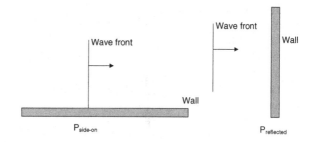

8.3.5 Formation of Explosive Cloud

If the gas cloud formed by a release is outside the flammable concentration range (i.e. the interval between LEL and UEL), or there is no ignition source, combustion will not occur. Subsequently the gas cloud will dilute and disappear. In the case of an immediate ignition a fire will develop. The most dangerous situation, however, will occur if a large flammable premixed fuel–air cloud is formed and ignited. A serious explosion may then result.

8.3.5.1 Jet Release and Evaporating Pool

The released substance can be a gas, an evaporating liquid, or a gas-liquid (two phase) flow. The source will be characterised as a jet release (i.e. gas, two phase) or evaporating liquid, or a diffuse release, (i.e. evaporating pool).

The two sources have quite different characteristics. The jet release will have a high momentum and establish a strong flow field due to additional air entrainment. Recirculation zones may be generated where the gas concentration can reach a combustible cloud. The evaporating pool will act as a diffuse release source, the wind forces and buoyancy will control the dispersion process. The flow velocities will be much lower than for the jet release. If the evaporating liquid forms a dense gas, a layer of combustible gas may be formed at the ground level, or in a lower compartment. Similarly in an open area a dense gas cloud will have the tendency to intrude into confined spaces such as buildings, which may pose serious problems due to high overpressure.

8.3.5.2 Gas Cloud and Ignition

To ignite a gas cloud requires an ignition source with sufficient strength. The minimum ignition energy depends on fuel concentration and the type of fuel (see Fig. 8.5). The minimum energy occurs for a concentration which is close to the stoichiometric mixture (i.e. where the amount of oxygen is the exact value required

Fig. 8.5 Minimum ignition energy

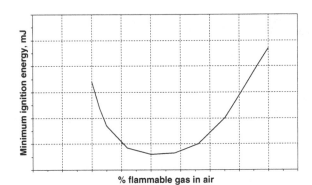

for full oxidation). The curve illustrates the principle, and the scales have therefore been omitted.

As the source of the leak is emptied the release rate will gradually be reduced and the gas concentration in the cloud will decrease. A weak ignition source will 'sit and wait' until the gas cloud has the right composition before it is ignited. In several accidental gas explosions the time from the release until the explosion was 10–20 min. In such cases it is probable (but speculative) that the gas concentration is decreasing at the time of ignition.

When the ignition source is strong the gas cloud will be ignited when the edge of the cloud with flammable concentration reaches the ignition source. If the ignition source is weak, however, the source may fail to ignite the cloud in the early phase of the dispersion process or ignite only a small part of the cloud. Subsequently, a homogeneous large gas cloud may be formed. This cloud reaches a flammable concentration as the pressure reservoir is emptied and a weak ignition source may ignite the cloud. This discussion shows some of the complexities involved in assessing ignition probability and the formation of explosive gas clouds.

8.3.5.3   Ventilation of Compartment

It has been claimed that 'the best building has no walls'. This is particularly true with respect to gas explosion safety. In an open building the natural ventilation will enhance the gas dispersion and if an explosion occurs, the pressure is dissipated through the open areas. If the release rates are small there is no doubt that mechanical ventilation systems can counteract the formation of explosive gas clouds. However, for a massive release, the forced ventilation rate will in general be too low.

A ventilation system may also transport gas from one area to another. This occurred onboard the MODU 'West Vanguard' in 1985 (see Sect. 4.5), when gas from a shallow gas pocket blew out under the platform, was sucked into the ventilation system and 'distributed' around the platform. The subsequent gas explosions also followed the ventilation ducts.

8.3.6   Deflagration

8.3.6.1   Deflagration Waves and Explosion Pressure

A deflagration is a gas explosion where the flame front propagates at subsonic speed (relative to the unburnt gas), immediately ahead of the pressure wave. The propagating velocity can span more than three orders of magnitude in a gas explosion. The mechanism of flame propagation will be quite different at different velocities.

When the cloud is ignited by a weak ignition source (i.e. a spark or a hot surface) the flame starts as a laminar flame. For a laminar flame the basic mechanism of propagation is molecular diffusion of heat and mass. The diffusion of heat and mass into the unburnt gas is relatively slow and the flame will propagate with a velocity of the order of 3–4 m/s. The propagation velocity of the laminar flame depends on the type of fuel and the fuel concentration.

The laminar flame will accelerate and transit into a turbulent deflagration (i.e. turbulent flame) in most accidental explosions, since the flow field ahead of the flame front becomes turbulent. The turbulence is caused by the interaction of the flow field with process equipment, piping, structures etc. The mechanisms generating turbulence ahead of the flame front are discussed below.

One of the mechanisms causing the increased burning rate in turbulent deflagrations is the wrinkling of the flame front by large turbulent eddies. For this combustion regime the increased flame surface area causes the burning rate to increase.

When a flame propagates through a premixed gas cloud there are two mechanisms causing pressure build-up. These are:

1. fast flame propagation
2. burning in a confined volume.

In most accidental explosions a combination of these two effects causes the pressure build-up. The pressure behind the flame (in the burnt gas) will gradually decay away from the flame. This pressure decay will mainly depend on the boundary conditions at the end of the tube (i.e. open or closed tube) and on the flame velocity.

Since the flame front is a subsonic combustion wave, the burning will influence the flow ahead of the flame. The pressure ahead of the flame depends on the flame acceleration and speed. In order to get a shock wave ahead of the flame, a high flame speed is required.

If the explosion happens inside a closed vessel, fast flame propagation is not required to obtain high pressures. A stoichiometric fuel–air cloud in a closed vessel will give up to 8–9 bar when exploding. By opening up part of the vessel wall, relief will be provided and the pressure will be reduced. The reduction will depend mainly on how fast the flame is burning in the vessel and the location and size of the vent area.

8.3.6.2 Flame Acceleration in a Channel Due to Repeated Obstacles

In a partly confined area with obstacles (i.e. process equipment, piping etc.) the flame may accelerate to several hundred metres per second during a gas explosion. The mechanisms causing the increased burning rate in turbulent deflagrations are the wrinkling of the flame front by large eddies and the turbulent transport of heat and mass at the reaction front. This turbulence is mainly caused by the interaction of the flow with destructions such as structures, pipe racks, etc.

Figure 8.6 shows how turbulence is generated in the wake of obstacles in a channel. When the flame consumes the unburnt gas, the products will expand. This expansion can be up to 8–9 times the initial volume. The unburnt gas is thus pushed ahead of the flame and a turbulent flow field may be generated. When the flame propagates into a turbulent flow field, the burning rate will increase dramatically. This increased burning rate will further increase the flow velocity and turbulence ahead of the flame.

The mechanism of flame accelerations due to the pressure of obstacles causing turbulence constitutes a strong positive feedback loop. Flame accelerations may, to some extent, be avoided by venting the hot combustion products. The flow and turbulence in the unburnt mixture ahead of the flame will be reduced. Venting combustion products is a very effective way of minimising the acceleration effect of a member of obstacles.

Venting of unburnt gas ahead of the flame may also contribute to a lower explosion pressure, particularly when the venting directs the flow away from the obstacles. If unburnt gas passes a series of obstacles before it is vented, flame acceleration will most likely occur. This is illustrated later in this chapter (see Fig. 8.9).

This discussion shows that there are two mechanisms governing the pressure build-up in deflagration of partly confined gas clouds, namely:

- Flame acceleration due to enhanced burning arising from turbulence generated by obstacles.
- Pressure relief venting thereby reducing the effect of the feedback mechanism.

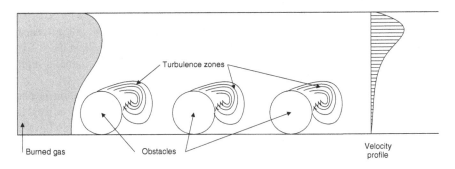

Fig. 8.6 Turbulence generation in a tunnel due to repeated obstacles during a gas explosion

These mechanisms have competing effects. The flame acceleration due to turbulence will increase explosion pressure, while venting will reduce the pressure. It is the balance between these two effects that is governing the pressure build-up. When analysing gas explosions we have to take both of them into account.

8.3.7 Confined/Semi-confined Explosion

Blast waves from explosions in rooms, offshore modules etc., are difficult to calculate and thus several research programmes have been carried out in order to find a realistic calculation model for such explosions.

The blast wave will be affected by equipment etc., in the room although the effect is difficult to quantify. It is, however, possible to place the equipment and other obstacles favourably in order to reduce the maximum overpressure. This must be done in the planning/engineering phase. Some important principles for modules are listed below:

- Venting areas must be placed as near as possible to probable ignition sources.
- In many cases the most probable ignition sources are known. Ventilation should then be placed on as many walls as possible in this area.
- If this is not possible, avoid venting areas only on the smallest wall.
- If this is impossible, then place possible ignition sources near openings.
- Avoid long and narrow rooms with openings only in the ends (cannon). If this geometry is necessary, then place venting areas on at least one sidewall for its entire length.

These principles are also shown in Fig. 8.7. All equipment in the room will produce turbulence which will increase the burning velocity of the gas, thereby increasing the overpressure.

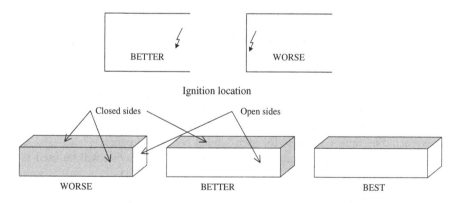

Fig. 8.7 Placing of possible ignition sources and venting areas

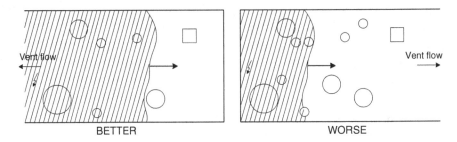

Fig. 8.8 Placing ignition sources

The following principles should be followed in order to reduce the maximum overpressure:

- Try to place the equipment and ventilation areas in a way that the ignition sources will be between the largest equipment and the venting areas.
- The largest equipment should be placed as far as possible from the venting areas.
- The long side of equipment should be parallel to the venting direction.
- Sharp profiles are worse than rounded profiles.
- An increase of the cross-section of equipment in the ventilation direction will give an exponential increase of the overpressure.

These principles are also shown in Figs. 8.8 and 8.9.

Equipment and structural elements will be subjected to drag loading after the overpressure phase, caused by the transient winds behind the blast wave front. The drag force will be highest near the venting areas, and is often important for structural elements; as columns and beams. Further away from the opening the drag force will be considerable less.

Further details about explosion phenomena and calculations may be found in Refs. [4, 5].

8.3.8 Calculation of Explosion Loads

FLACS® is one of the most advanced software packages available for simulation through computational fluid dynamics (CFD) of gas explosion scenarios, through ventilation and dispersion simulations, followed by explosion simulation. Location of ignition sources is input to the software. FLACS® can give pressure-time histories, including drag velocities, for user defined points and panels.

Table 8.2 presents an illustration of results from FLACS® that will be used as input to probabilistic explosion load assessment (see Sect. 8.4).

Fig. 8.9 Layout of
equipment

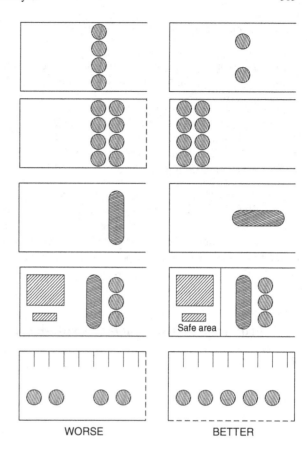

WORSE   BETTER

Table 8.2 Example of FLACS result calculations for cellar deck

Case no	Gas cloud size (%)	Ignition point	Deck El212	Wall T330/ M10	Wall L400/ M19
301	100	0	0.92	2.09	1.83
350	50	0	0.77	1.27	1.09
355	50	5	0.70	1.16	0.99
330	30	0	0.45	0.77	0.68
332	30	2	0.24	0.37	0.35
320	20	0	0.29	0.53	0.46
324	20	4	0.21	0.26	0.30
310	10	0	0.13	0.21	0.19
315	10	5	0.13	0.21	0.20

8.3.9   Explosion Design of Facilities

A Fire and explosion guidance document has been published by Oil and Gas UK, sponsored also by Health and Safety Executive in the UK [6]. The guide has the following parts:

- Part 0 Hazard management (formerly FEHM).
- Part 1 Avoidance and mitigation of explosions.
- Part 2 Avoidance and mitigation of fires.
- Part 3 Detailed design and assessment guidance.

A practical design explosion guide for offshore facilities has been published by Czujko Ref. [7]. This engineering handbook has been developed based on the results of the research up to 2001. The main objective was to develop methods and procedures to improve gas explosion load assessment using computational fluid dynamics (CFD) and structural behaviour using nonlinear finite element methods. A number of practical studies have been documented in the book.

8.4   Probabilistic Approach to Explosion Load Assessment

8.4.1   Basis

For design against blast loads, the dimensioning blast load must be determined. This is usually based on a probability distribution. The assessment of design loads implies that an exceedance function has to be established for each structural element to be designed. This exceedance function may be defined as follows:

The annual frequency of exceeding a specified overpressure load, is a function of the overpressure level.

This exceedance function is established on the basis of uncertainties in the explosion load assessment. Such uncertainties are related to:

- The actual location of the ignition point which may vary considerably and have a strong influence on the resulting explosion overpressure.
- The strength of the ignition source which may vary depending on the type of ignition source.
- The volume of gas cloud.
- The homogeneity of cloud.
- The gas concentration in the cloud relative to a stoichiometric concentration.

8.4.2 Approach to Probabilistic Evaluation

A statistical analysis of the occurrence of all aspects of the event sequence leading up to an explosion is required, in order to establish a probabilistic representation of blast loads. This will include the following aspects:

- location of the leak source
- direction of gas jet
- flow rate of the leak
- wind direction and speed
- performance of barrier elements, in order to limit size and duration of cloud.

Probability distributions need to be defined for all parameters. This is available from environmental data for the wind conditions. Hole size distributions are usually also available from the leak statistics, thus distributions for the flow rate s may be generated. Distributions for the location and the direction of the leak are usually based on geometrical considerations.

These variations will generate input scenarios to dispersion CFD simulations. Most of these parameters may have continuous variations, which in theory could generate an infinite number of dispersion scenarios. Some form of categorisation will have to be performed, in order to limit the number of cases. The basis for the categorisation should as far as possible be scenario dependent, reflecting specific limitations and considerations. If a leak from a flange is considered, there may be variations in the direction corresponding to 360°. A coarse categorisation would split the directions in two; below horizontal and above horizontal. If the flange is close to the ceiling, and there is a vertical truss above and somewhat to the side of the leak source, then such restrictions may limit the free flow of a gas jet and the subsequent dispersion of a gas cloud. The angle, at which separation between the two categories should be made, is where the gas jet will be split to either side of the truss.

The resulting number of dispersion calculations may still be high, even if all parameters are classified into categories. The dispersion simulations will therefore have to be made with a coarse grid, in order to limit the computational time.

When the dispersion simulations have been completed, it will be required to reduce the number of cases for blast load simulations through elimination of cases which are unignitable and categorisation of cloud conditions when they are similar.

One difficult aspect is to identify those dispersion scenarios that will be able to reach ignitable atmospheres, but where the extent of the cloud in space or time is insufficient to give a deflagration with significant blast effects. This has been shown to be important in large scale gas leak dispersion tests carried out at Spade Adam in 1998 Ref. [8].

When all non-relevant dispersion scenarios have been eliminated, explosion simulations should be carried out for the remaining cases, possibly combining some of the cases into broader categories. When blast loads for all of the cases have been simulated, the resulting blast load distributions may be generated from a combi-nation of simulated blast load and scenario probabilities.

The selection of combination of variations and cases may be carried out according to Annex F; Procedure for probabilistic explosion simulation of NORSOK Z-013; Risk and emergency preparedness assessment [9]. This procedure is used extensively in QRA studies for Norwegian installations.

8.4.3 Probabilistic Evaluation

The simulation of uncertainty is dependent on the ability to express the uncertainty as a function of a set of parameters and knowledge about the relationship between the overpressure and the parameter in question. A brief qualitative discussion of some of these aspects is presented below.

This evaluation is based on use of a simplified model which does not take into account all the experience from the BFETS test programme. It is nevertheless useful in order to illustrate the main parameters and their effects.

It should be emphasised that the science of establishing a probability distribution for blast loads is far from well established. What is presented below should be considered an overview of some of the important aspects, rather than a specific recommendation for what should be done.

8.4.3.1 Distribution

The change in overpressure as a function of the basic parameters is not well known and in some cases rather simplified so coarse models and functions have to be used. Once each individual distribution is known (or modelled), the overall distribution may be generated, by either:

- Statistical simulation, or
- Numerical solution.

The consultancy Safetec has an in-house software tool for the generation of such distributions, called SERA (Safetec Explosion Risk Assessor). Other consultants have similar tools, but the following illustration of methodology is based on information from Safetec. Figure 8.10 presents a flowchart for the SERA approach.

It should be noted that SERA Module 1 includes a gas dispersion module and ignition probability model, whereas SERA Module 2 is the module that combines the frequencies of ignition of gas cloud sizes with the resulting explosion over-pressure results. The distribution is then actually created in the second module.

There are various consequence studies that create the input data to SERA. The following are the main input data:

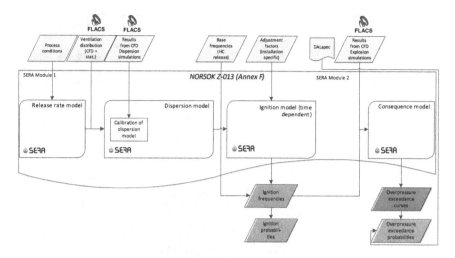

Fig. 8.10 Flowchart for probabilistic explosion analysis using SERA

- Dispersion simulation for various combinations of leak source location and direction, as well as wind direction and speed.
- Ventilation simulations.
- Gas leak frequencies for leak sources and sizes.
- HC release calculations in order to establish leak rates as functions of time.
- Input to ignition probability modelling, using the 'JIP' model outlined in Sect. 15.8.5.
- Reliability/availability data for barrier systems/elements.

SERA Module 1 has a simplified dispersion model which calculates the fraction of ignitable gas and the equivalent stoichiometric fraction as a function of time. The dispersion model is used in a calculation loop in order to calculate results for all leak profiles and ventilation conditions. Figure 8.11 shows an illustration of the output from dispersion calculations, for a relatively high gas leak rate in low wind speed conditions.

SERA Module 2 combines the equivalent gas cloud distributions from Module 1 with explosion overpressures, using linear interpolation for calculation of explosion overpressures that are lacking. Figure 8.12 presents an example of output from SERA, illustrating overpressure distribution for a certain structural element.

8.4.3.2 Gas Leak Sources

The details of the possible gas leak sources are important for the dispersion calculations. These details therefore need to be considered. The following aspects need to be addressed:

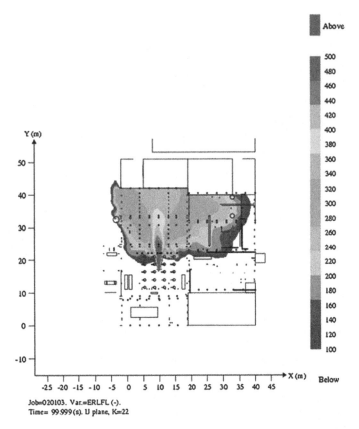

Job=020103. Var.=ERLFL (-).
Time= 99.999(s). U plane, K=22

Fig. 8.11 Illustration of output from dispersion calculation, 5 kg/s gas leak, low wind speed. (Maximum cloud size in diagram corresponds to extension of LEL concentration, central jet from wellheads up to opposite bulkhead is all above UEL concentration.)

- Location of the leak source, in three-dimensional space.
- Gas composition and characteristics, i.e. temperature and specific weight.
- Leak rate.
- Direction of flow from the leak source.
- Unrestricted gas jet or diffuse gas leak.

If all these parameters are allowed to vary, there will be a very large number of combinations, even if the variation in each parameter is restricted to a handful of categories.

8.4.3.3   Ventilation and Dispersion

The ventilation conditions also have considerable influence on the dispersion of a gas leak, and the resulting gas cloud. Most installations have natural ventilation,

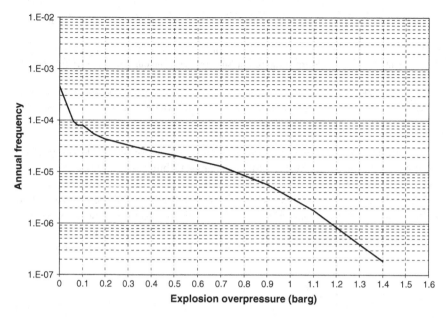

Fig. 8.12 Overpressure frequency curve as output from SERA

implying that the dispersion of a gas leak will be strongly dependent on the wind strength and direction.

The variations in wind conditions are additional to those variations listed above for the characteristics of the gas leak, implying that the number of parameters is even higher.

It is impossible to carry out gas dispersion simulations (using a CFD code) for all possible combinations, even if the variations are limited to large categories. It is therefore required that a set of representative cases is chosen. Experience is probably the only way to determine how such representative cases should be selected. It has been indicated that the number of representative cases should be limited to around 10. This appears very low in relation to the large number of variations that are possible. It would not be unrealistic to expect in the order of 30–50 cases in a detailed study, but only experience will determine how many will be required in order to establish a representative distribution.

8.4.3.4   Ignition Source

The actual location of the ignition point may vary considerably depending on the type of ignition source. This will have a strong influence on the resulting explosion overpressure. There are several types of ignition sources:

• Rotating equipment	These will be major equipment units, with a discrete distribution, related to location of each unit.
• Electrical equipment	There will usually be a high number of possible sources from electrical equipment, such that a continuous function often may be the appropriate description.
• Hot work (such as welding)	Hot work activities are usually possible in most locations, such that a continuous distribution over the area (or volume) would be most representative.

The influence on the overpressure is mainly a function of the location of the ignition in relation to the obstacles that generate turbulence, and thus increase the flame front velocity and the resulting overpressure.

When the possible leak sources are considered, these need to be correlated with the leak and ventilation characteristics. If the leak for instance is close to the floor level with horizontal movement, and the gas is heavier than air, then only ignition sources close to the floor level may be potential initiators of an explosion.

8.4.3.5 Ignition Strength

Theignition strength of the source will depend to a large extent on the type. The strength of the ignition will also influence the maximum overpressure that may be generated and thus the type of source also influences the overpressure generated.

8.4.3.6 Gas Cloud Characteristics

The overpressure is dependent on the gas cloud in several ways:

• Volume of gas cloud	The size of the cloud is dependent on the leak rate, the ventilation, and the ignition time and location. The larger the cloud is, the higher the overpressure will be.
• Homogeneity of cloud	Parts of the cloud may be within explosive limits and other parts may be outside this range. In theory only the part of the cloud inside the explosive limits should participate in an explosion, but it has been shown that the deflagration itself may also cause larger parts of the cloud to participate in the burning.
• Gas concentration	Theoretically, the highest overpressure should result from stoichiometric concentration in the gas cloud, but it has been shown that, actually, the highest pressure results from a concentration somewhat higher than the theoretical value. Most calculations assume stoichiometric concentrations. FLACS simulations are increasingly being performed on the basis of a simulated gas dispersion which more accurately reflects actual conditions.

Virtually all real gas cloud s will be extremely far from homogenous, whereas most of the experimental data are from homogenous, stoichiometric clouds. This is probably one of the most uncertain aspects of transfer of the experimental data to modelling.

The dispersion experiments at Spade Adams [8] have shown that real clouds may be even further away from homogenous clouds than previously assumed. The current practice in the case of inhomogeneous clouds is to determine an equivalent size of a stoichiometric, homogenous cloud as input to the overpressure simulation (according to [9]). This may be a factor which leads to significant conservatism.

8.4.4   Example

Results for an offshore platform based on using SERA are shown in Fig. 8.13. The results from SERA may be used as follows:

- Maximum overpressure is determined by using FLACS simulations. The maximum overpressure is input value to SERA.
- Subjectively, the limit of secondary ruptures of process equipment on the deck was assessed to 0.5 bar (0.05 MPa). This simplification was made to avoid extensive structural response calculations.
- The diagram was used to give a coarse estimate of the conditional probability of escalation of accidental effects in the case of explosion.

8.4.5   Use of Load Function

Section 2.3.6 introduced the need for establishment of a probabilistic description of the blast loads. An introduction to how this may be done using the latest knowledge, is given in the preceding Sects. 8.4.2–8.4.4. This section will discuss the use

Fig. 8.13 Conditional exceedance probability distributions, based on results from SERA

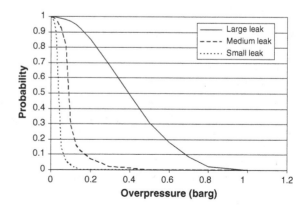

of the probabilistic approach. An exceedance diagram may be used in one of the following ways:

• Dimensioning load assessment under ALS criteria	Compute overall annual exceedance frequency for relevant explosion scenarios, then determine explosion load for ALS check (see Chap. 19).
• QRA (probability of escalation)	If the equipment's resistance to explosion loads is known, then probability of secondary ruptures (escalation) may be determined.

The following would be the procedure (in accordance with the Norwegian regulations for load bearing structures) for determining the design accidental loads for explosion on deck structure:

1. Split the deck structure into sections, with the main emphasis on sections that will be exposed to different explosion scenarios.
2. The system failure probability, 10^{-4} per year, is split into contributions from each section. This distribution is 'ad lib', and an optimisation of the structure may involve redistribution of these contributions several times.
3. For each section of the deck structure, establish the exceedance diagrams for overpressure.
4. For each section, identify the design explosion load that corresponds to the allowed contribution to system failure probability.
5. Check whether an alternative distribution of system failure probability on sections of the deck structure is easier to implement and thus less expensive to provide.

8.4.6 Structural Response Calculations

Because the current knowledge implies that blast loads are higher than previously known, more emphasis has also to be placed on response calculations. The following choices are available:

- Simplified triangular pressure pulse, whereby max load and duration are the required characteristics. Often the maximum overpressure will be limited to pulses that have at least a millisecond duration.
- Detailed pressure–time history.

A relatively advanced structural analysis is required, capable of performing more than just a static analysis. The software packages to be used for such analysis should be able to consider the following:

- Dynamic responses to pressure-time histories (detailed or simplified, triangular)
- Non-linear aspects of the structural response.

The drag forces on structures and equipment also need to be calculated, also this load response calculation should consider dynamic aspects. It should be noted that not very many software packages are able to carry out such analysis. Simplified calculation of drag loads may be done with the formula given in Ref. [10].

8.4.7 Is a Probabilistic Approach the Best Way Forward?

The preceding sections have briefly indicated how a probabilistic approach to assessment of the design explosion loads could be developed. In principle, such an approach is fully feasible, although quite laborious. This is also what may cause some doubt about the realism. Another aspect which is also difficult to implement, is the fact that none of the explosion events that have occurred on production installations (see Sect. 8.2.2) have resulted in particularly high blast loads. It may appear unrealistic to devote extensive resources to simulation of a wide spectrum of results, when experience has not been able to demonstrate any such extensive variations.

The main challenge is the high number of free variables, as already indicated. This implies that the total number of variations will be very high, even if each parameter is restricted to categories. The number of simulations of gas dispersion to be carried out will therefore be quite considerable. The manner in which this subsequently can be reduced to a manageable data set, is far from obvious.

It may therefore be relevant over time, to develop deterministic rules for which cases to use as the design basis. It is likely that considerable experience will be needed as the basis, in order to formulate such rules, and some time will probably be needed before so extensive experience is available.

8.5 Explosion Risk Reduction

8.5.1 Establishing Basis for Design

The design basis for explosion hazards changed markedly in the 2–3 years from 1995/6. Prior to the mid-1980s attention on explosion mitigation was relatively low, even though the Flixborough accident in 1974 in the UK did focus attention on the need to prevent serious explosion accidents. In the mid-1980s an approach to explosion design evolved which may be summarised as follows:

- The worst case conditions were defined based upon insight into the hazard circumstances.
- Loads were simulated for the worst case using what was considered to be appropriate tools.

- Design solutions could be made cost effectively even for the worst case conditions.

This situation has changed considerably in the 1990s, culminating in the so-called 'Spade Adam' test series [1]. It is now realised that worst case conditions will be so severe, that it is impossible to find cost-effective solutions which will protect against the worst case conditions. An alternative approach is thus needed based upon probabilistic modelling.

The new knowledge is such that improvements to both existing and new installations will be needed. It is worthwhile to consider the challenges to engineering work for new installations which include:

- Dimensioning of structure and equipment against blast loads including:

  - Primary structure or hull in the case of an FPSO
  - Deck support structure, tank top for FPSO
  - Module structure
  - Supports for vessels and piping
  - Blast and fire walls.

- The definition of loads including:

  - Peak overpressure for all x, y, z coordinates
  - Panel pressures
  - Impulse or time pressure distributions
  - Drag loads ('Explosion wind') for all/representative x, y, z coordinates.

- Other critical aspects including:

  - Fragments
  - Displacement of structures.

8.5.2 BFETS R&D Experience

The most extensive recent R&D work is the 'Blast and Fire Engineering for Topside Structures' (BFETS) programme [1]. This programme included an extensive series of tests with large scale models having realistic offshore module geometry. The tests were conducted on models having volumes in the range 1600 to 2700 m^3 which were designed to resist explosion overpressures up to 4 barg (i.e. bar overpressure). The parameters that have been studied include the following:

- Confinement
- Congestion
- Ignition location

- Effect of deluge
- Deluge droplet size
- Only homogenous, stoichiometric gas/air mixtures.

The tests were conducted as a joint effort between European oil companies and the UK Health and Safety Executive. The interim results were considered so important by the HSE, that they were released to the industry in order that necessary action could be taken as early as possible. The main results may be summarised as follows:

- The overpressures measured were higher than expected based on previous medium scale tests (e.g. CMR tests).
- Damage to the module occurred in several tests due to exceedance of the design limits.
- The congestion inside the module was shown to be a very important parameter in determining the overpressure.
- Typical for global load characteristics:

  - Determined by the balance between the production of combustion products and their loss due to venting.
  - Important for the design of the main structure.
  - Closer to what was predicted than in the local effects prediction.

- Typical for local load characteristics:

  - Determined by local physical mechanisms such as flow, turbulence, reflections, mixing.
  - Important for design of bulkheads, local structures, and equipment.
  - Sometimes locally quite high.

From these results it may be concluded as follows for local effects and global trends respectively:

- Local effects:

  - Dominated by pressure–time profiles
  - Difficult to predict
  - Short durations, pulse loads may need to be considered
  - Structural response/damage may be less from a pulse loading, if the duration is short.

- Global trends

  - Usually easier to predict
  - Quite good correlation with predictions by the best software tools.

318   8   Explosion Risk Modelling

8.5.3   Main Experience, Mitigation

The local effects of explosion relief were not always as expected. If the extent of explosion relief increases, then it would be expected that there would be a decrease in overpressures. This was not always the case for local conditions although for global effects the trend was as expected.

Smaller sized objects have the larger effect on module congestion, if there are a sufficient number of these objects. The dramatic effect of small sized objects may be illustrated from the test results. When the module congestion was increased from 7.5% blockage (so-called 'low' congestion) to 9.5% blockage (so-called 'high' congestion), the peak overpressure increased by a factor of four. It has been shown in the medium scale tests that cable trays and pipe racks are the most critical small sized equipment for increasing the overpressure. The density of equipment was shown to be most critical for the longest flame paths.

One of the most important potential risk reducing measures is the use of deluge for blast load reduction. For this to occur, deluge needs to be initiated prior to ignition (for instance on detection of a gas leak). The tests have shown that use of deluge is particularly effective in preventing so-called runaway flame accelerations. The tests have, however, also shown that the active use of deluge has given reduction of the peak overpressure in all the tests. The most extensive reductions have occurred for the long flame paths. When the conditions are ideal, quite extensive reductions may occur.

Ideal conditions require droplets from the deluge system to be larger than normal droplets, thus requiring special nozzles. Deluge from standard nozzles however, produces lower overpressure although the extent of the reduction is quite scenario dependent.

The most critical aspect in relation to the use of deluge for overpressure reduction is the need to activate the system prior to ignition. Modelling of ignition has shown that the most likely interval between release and ignition is two to three minutes. Thus to be effective, deluge activation has to be within the first half minute.

8.5.4   Risk Reduction Possibilities

8.5.4.1   Priorities

The general approach to risk reduction is to give priority to the reduction of accident probability over reduction of accident consequences. Probability reduction may often however, be rather difficult to document, due to the following factors:

• Probability reduction may be dependent on operational measures, which may have a limited reliability.

- The effect of the actions on the probability may be qualitatively certain, but the extent of the reduction may be quantitatively unknown.

Consequence reduction is often easier to document and often more reliable if passive measures are adopted. The probability and consequence reduction options in relation to gas explosion are discussed in the following sections.

8.5.4.2   Probability Reduction—Prevent Gas Leaks Through Design

The most obvious action to prevent gas leaks is to reduce the number of potential leak sources, most typically the number of flanges. This is probably easiest to accomplish for a new installation. In the case of an existing installation, it is still technically feasible, but may in itself lead to increased risk, because open flame cutting and welding will most probably be needed in the modification work.

The choice of connection approach therefore needs to be a trade-off between the desire to prevent leaks through the use of all welded connections, and the need to minimise hot work during disconnection (opening) of welded connections.

Other alternatives for reducing the number of gas leaks are: improvement of the quality of the maintenance work in the process area s; selection of higher quality materials for gaskets; the follow-up of minor leaks in order to identify trends and unwanted tendencies at the earliest possible time. Many of these aspects were briefly discussed in Sects. 6.7 and 6.8.

8.5.4.3   Probability Reduction—Prevent Gas Leaks from Operations

It has been demonstrated over more than 15 years that the majority of the HC leaks, at least on the Norwegian Continental shelf, are due to loss of containment during operational/maintenance/inspection activities (see Sect. 6.6.1).

One of the most effective ways to prevent gas leaks will be to improve operational barriers. One approach that may be used in this connection is the BORA approach (see Sect. 15.3.1).

8.5.4.4   Prevent Ignitable Concentration

The next possibility to halt the accident sequence if a gas leak has occurred is to prevent the formation of an ignitable atmosphere. Extensive natural ventilation is one of the obvious actions in order to achieve this. In the design phase good natural ventilation is frequently provided, but sometimes this is reduced during operations by temporary equipment being installed or left temporarily in openings. In other cases, ventilation is purposely reduced, because a need is perceived to improve the working environment (reduce chilling draft).

Increasing the natural ventilation often requires a difficult trade-off between reduced ignition probability and aggravated working environment conditions. Increased ventilation usually implies colder working conditions and possible freezing of equipment. The author still remembers well over 20 years ago, after having initiated opening up of weather cladding on process modules on an old North Sea installation, arriving offshore on the coldest weekend in February. The threat of being thrown over board from persons who had spent many hours out in the cold having to use miscellaneous equipment, including hair dryers, to prevent process instruments from freezing, was probably not real, but it was perceived as quite real.

8.5.4.5 Prevent Ignition

The next option is to prevent an explosive atmosphere from being ignited. Several actions are possible in this regard:

- Reduce the extent of hot work activities. This has been applied successfully on many installations where it has been proven that a wide variety of tasks may be done in a 'cold' fashion i.e., without the use of hot work.
- Improved maintenance of 'Ex-proof' equipment. On many installations there is probably some explosion proof equipment which has improper maintenance routines, most typically this applies to light-fittings.
- Attention should also be given to so-called 'continuous sources' i.e., potential ignition sources that are constantly active, such as a lighted flare.

Prevention of ignition is the last of the probability reducing measures and thus consequence reduction measures are briefly outlined below.

8.5.4.6 Prevent High Turbulence

There are some basic design rules which may help to prevent high turbulence. These rules may be summarised as:

- Optimise the arrangement of equipment
- Avoid extensive multiple pieces of equipment
- Optimise the location of pipe racks relative to likely ignition sources.

The rules were shown in Figs. 8.8 and 8.9 for layout aspects.

8.5.4.7 Prevent High Blockage

The same actions that may contribute to improved ventilation, may also prevent high blockage in the modules, and thus help to prevent increased overpressure. Risk reducing actions may include (see also Figs. 8.8 and 8.9):

- Remove temporary installations which may have been installed during operation and maintenance, containers, new equipment, and weather cladding.
- Arrange vessels in a way which minimises blockage of the most likely path of the flame front.

8.5.4.8 Install Fire and Blast Barriers

Escalation due to explosion may be limited by the provision of fire and blast barriers between modules and areas. There are, however, several problems that may be introduced by such actions:

- More barriers (walls, decks) may cause problems for keeping well ventilated areas.
- Barriers may also restrict explosion relief and introduce more blockage.
- The retrofitting of such barriers may give rise to extensive hot work although this may be avoided by good planning and preparations.

8.5.4.9 Activate Deluge on Gas Leaks

Use of deluge to reduce blast loads has already been mentioned, including the issue of which nozzles to use. This is, however, to some extent a controversial issue. Some operators have claimed that activation of deluge prior to ignition, has in the past been the apparent cause of the ignition itself. It cannot be ruled out that this may be the case in special circumstances although there are indications that the problem is not very large.

The most obvious positive demonstration of the potential advantage of this approach is the fact that many operating companies in the North Sea have over some years an extensive experience with numerous gas leaks which have been 'deluged', without any problem. Even so, there appears to be considerable reluctance to the release of deluge on confined gas detection. The possible effect of deluge may be seen from Table 8.3, with results from BFETS.

The table compares two otherwise identical tests from the BFETS programme [1], especially with respect to the effect of deluge activation prior to ignition. Standard offshore nozzles were applied. It may be seen that the effects on the maximum overpressures is higher than the effect on the average overpressures. This is claimed to be general for the effect of deluge, that the peaks are affected more than the average loads.

Table 8.3 Comparison of BFETS results with and without deluge activation

Parameter	Blast loads without deluge (bar)	Blast loads with deluge (bar)
Maximum recorded	3.73 (+155%)	1.46
Minimum recorded	1.44 (+95%)	0.74
Average recorded	2.38 (+118%)	1.09
Maximum recorded with duration > 1 ms	2.29 (+83%)	1.25
Minimum recorded with duration > 1 ms	1.05 (+46%)	0.72
Average recorded with duration > 1 ms	1.76 (+69%)	1.04

The idealised triangular pressure pulses generated from the maximum and minimum values with at least 1 ms duration are shown in Fig. 8.14.

The triangular pressure pulses may be integrated, in order to show the differences with respect to impulse. The results are shown in Table 8.4.

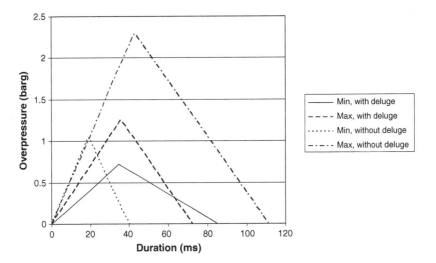

Fig. 8.14 Idealised triangular pressure pulse

Table 8.4 Change in impulse values, according to the use of deluge

Parameter	Min, with deluge	Max, with deluge	Min, without deluge	Max, without deluge
Overpressure (bar)	0.717	1.251	1.051	2.294
Impulse (Ns/m^2)	3047	4504	2102	12732
Change in impulse (%)			−31%	183%

It may be noted that the impulse reduction for the points with the maximum overpressure is 183%, whereas the impulse for the point with the lowest over-pressure actually increases when deluge is applied. This may also be seen from Fig. 8.14, which shows the increased duration for this case.

It may be noted that the minimum overpressure is recorded at the same point physically, whereas the point where the maximum is recorded has moved quite considerably.

Finally it should be noted that other tests have shown that more extensive improvement may be achieved through the use of nozzles that produce larger droplets, implying that more energy than is consumed in the break-up of the larger droplets into many smaller ones.

8.5.4.10   Improve Resistance of Equipment and Structures

The last possibility for consequence reduction is to improve the resistance of equipment and structures to blast effects. The large scale explosion test programme [1] has shown that the resistance of equipment to explosion loads (up to and in some cases above 4 bar overpressure) was better than had been anticipated. To design additional resistance to explosion overpressures is, however, likely to be quite expensive for existing installations.

8.6   Example, Dimensioning Against Blast Load

8.6.1   Introduction

Dimensioning against blast load presents a suitable illustration of the freedom offered by the functional regulations, with respect to accidental loads. Blast loading is suitable to use in this illustration because it may be described by a one-dimensional function. Fire loads are far more complex because there are several degrees of freedom including duration, intensity, radiative, convective and smoke generated loads. The comment should be added that explosion loads are simpler because we make simplifications to make them simpler. Explosion loads would be similarly complex if we took the full pressure–time function into account.

This illustration involves dimensioning the deck structure of a simple unmanned wellhead platform using the regulations for load-bearing structures where applicable.

8.6.1.1 Platform Design

The platform is a simple installation with four wells. The wellheads are located in a wellhead area and first stage separation in the process area. Figure 8.15 presents a schematic overview of the main areas which are placed inside the platform deck structure. The two areas are separated by a blast wall.

The blast wall is considered to separate the two areas, such that explosion in one area should not affect the other area. The blast wall will be designed to resist the most severe loading from both areas.

8.6.1.2 Sources of Blast Loads

For the design against blast loads, the sources of gas explosion need to be identified. For the simple platform in question, the scenarios that may lead to gas explosion are gas leaks from the following sources:

- Leaks from X-mas tree and wellhead or well blowout
- Gas leak from separation
- Subsea gas leak.

For each of these areas, the occurrence of gas explosion is calculatedleak accordingignition the followingleak formulae:

$$f_{gas\ xpl} = f_{gas\ leak} \cdot P(ignition)\ P(xpl|ignition) \tag{8.4}$$

Fig. 8.15 Platform sketch

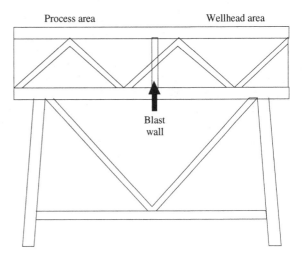

Process area   Wellhead area

Blast
wall

8.6.2 Basis for Dimensioning

According to the applicable Norwegian facilities regulations (Ref. [11] the impairment check shallleak be carriedignition out for blast loads on the deck structure, with the following limitation:

$$\sum P(Deck\ failure\ under\ blast\ loading) \leq 10^{-4} \qquad (8.5)$$

For the leak platformignition in question, with the blast loads as stated above, this equation may be written as:

$$P(Deck\ fails_{WH}) + P(Deck\ fails_{PROC}) + P(Deck\ fails_{SS}) \leq 10^{-4} \qquad (8.6)$$

The optimisation that may be done in the present case is to distribute the failure probabilities among these sources, such that they sum up to the allowable limit as stated above.

8.6.3 Design Capability

An initial estimate of the blast resistance of the deck may be implied by considering other load cases. In the present case, the initial design capability (static loads) is:

0.2 barg (overpressure)

8.6.4 Load Distributions

8.6.4.1 Wellhead Area

Figure 8.16 presents the conditional exceedance probability distribution for the wellhead area. The maximum overpressure has been determined to be 0.5 barg. There is a 90% probability that the overpressure will exceed 0.2 barg, and 50% probability that the overpressure will exceed 0.4 barg.

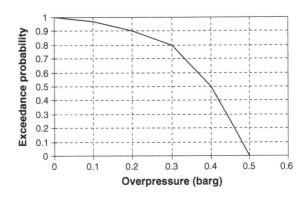

Fig. 8.16 Conditional load distribution, wellhead area

8.6.4.2 Process Area

Figure 8.17 presents the conditional exceedance probability distribution for the process area.

In the process area there are usually many obstructions that produce turbulence and high flame speeds, thus resulting in higher overpressures. The maximum overpressure has been determined to be 1.2 barg. There is an 80% probability that the overpressure will exceed 0.5 barg, and 15% probability that the overpressure will exceed 1.0 barg.

8.6.4.3 Subsea Gas Leaks

Figure 8.18 presents the conditional exceedance probability distribution for subsea leaks.

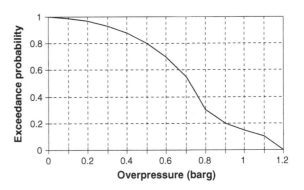

Fig. 8.17 Conditional load distribution, process area

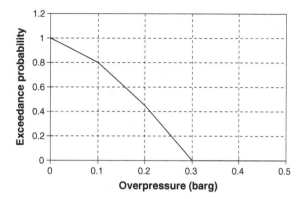

Fig. 8.18 Conditional load distribution, subsea leaks

The gas cloud from subsea sources will be partly outside the platform structure, but also partly inside the platform. Some parts will inevitably have to be inside the platform, in order to find an ignition source. The overpressures are usually low, due to the fact that most of the cloud is outside the structures. The maximum overpressure has been determined to be 0.3 barg. There is a 80% probability that the overpressure will exceed 0.1 barg, and a 45% probability that the overpressure will exceed 0.2 barg.

8.6.5   Gas Explosion Frequency

The frequency of gas explosions are given in Table 8.5, based on event tree analysis (not shown). Just above half the total frequency is caused by explosions in the process area, with almost 40% of the total value due to gas explosions in the wellhead area.

It has been shown already that the explosions in the process area are the most serious in that they are likely to cause the highest overpressures.

By combining the frequencies given in Table 8.5 with the conditional load distributions shown in Figs. 8.16, 8.17 and 8.18, exceedance functions may be generated as shown in Fig. 8.19. Assuming that the wall has the same

Table 8.5 Gas explosion frequencies, wellhead platform

Contribution	Annual frequency
Wellhead area	7.5×10^{-5}
Process area	1.10×10^{-4}
Subsea	2.8×10^{-5}
Total, all leaks	2.13×10^{-4}

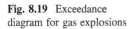

Fig. 8.19 Exceedance
diagram for gas explosions

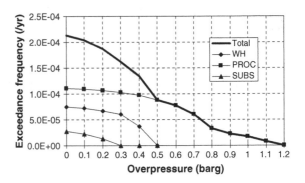

resistance to loads from either side, the pressure with a frequency of exceedance of 10^{-4} per year is 0.45 barg.

This would suggest that the structure would need reinforcement to resist
this higher load, compared to the original strength given as 0.2 barg. With the
overpressures as stated, the following failure frequencies are found:

- Wellhead area: 2.0×10^{-5} per year
- Process area: 8.0×10^{-5} per year
- Subsea leaks: 0

8.6.6   Reinforcement Costs

The regulations give freedom to distribute failure frequency among the areas
in whatever way we want. This may be used for optimisation. In the present
case study, this has been exemplified by assuming that there are differences in
cost according to which side of the wall the protection is applied on.

Figure 8.20 presents the additional costs associated with reinforcement of
the deck structure in the wellhead area for it to be able to resist higher blast
loads. The costs are incremental costs over those associated with achieving
the baseline resistance, 0.2 barg.

Figure 8.20 also presents the additional costs associated with reinforcement of the structure in the process area, to be able to resist higher blast loads.
The costs are incremental costs over those associated with achieving the
baseline resistance of 0.2 barg. The costs are higher for this area, compared to
the costs in the wellhead area.

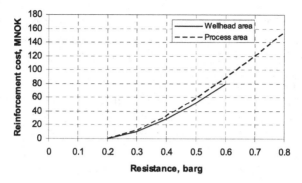

Fig. 8.20 Cost of reinforcement for structure in wellhead and process areas

The overpressure protection needed is such that there is a zero probability that a gas cloud from a subsea leak will give blast loads that need to be considered.

8.6.7 Optimisation

Based upon Fig. 8.20 the total cost of reinforcement if all the deck is reinforced to resist the same overpressure (0.45 barg) is 83 MNOK. The cost for the wellhead area is 38.6 MNOK and the cost for the process area is 44.4 MNOK. An optimisation of cost may be achieved as follows (see also Table 8.6):

- With an assumed level of reinforcement in the wellhead area determine:

  - the failure frequency of structure in this area
  - the cost associated with reinforcement of the structure

- The allowed failure frequency in the process area is thus determined from Eq. 8.6. From this failure frequency, the following can be found for the process area:

  - the required structural reinforcement
  - the cost associated with the reinforcement

Table 8.6 Optimum structural reinforcement of deck in wellhead and process areas

Resistance WH (barg)	Cost WH area (MNOK)	Failure freq WH area	Failure freq proc area	Resistance proc area	Cost proc area (MNOK)	Total cost (MNOK)
0.3	10.00	6.00×10^{-5}	4.00×10^{-5}	0.77	145.56	155.56
0.4	28.28	3.75×10^{-5}	6.25×10^{-5}	0.69	115.78	144.07
0.45	38.62	2.02×10^{-5}	7.98×10^{-5}	0.45	44.43	83.06
0.5	51.96	0.00	1.00×10^{-4}	0.34	20.53	72.49

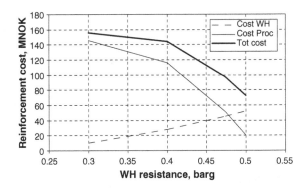

Fig. 8.21 Cost elements of structural reinforcement of deck, wellhead and process areas

- The steps above are repeated for new values of reinforcement in the wellhead area in order to determine the variability of the cost functions.

The values show that the cost increase for the wellhead area is less extensive than the cost increase for the process area. Thus the cost is minimised if the total system failure frequency is taken in the process area.

The differences in cost for reinforcing the different areas are clearly shown in Fig. 8.21. The total cost is thus seen to fall continuously when the structural reinforcement is increased in the wellhead area.

8.7   Case Study; Reduction of Blast Load

This case study is focused on reduction of blast loads within semi-confined deck spaces on a process platform. The criticality of explosion loads was documented through a QRA, which emphasised the need to study possible explosion loads in more detail. In the QRA, explosion loads of different strength were shown to have the following possible effects:

- Rupture of process equipment, thereby causing extended fuel sources which may escalate the accident
- Impairment of safety systems to the extent that prevention of escalation is impossible (Piper Alpha)
- Damage to structure causing direct escalation of the accident.

It was actually shown in the QRA that high overpressures were likely, and thus explosions were the dominating cause of escalation from one deck to another.

The damage mechanisms involved in escalation of this type are very complex, and relatively limited modelling work has been carried out. The physical modelling of damage to equipment caused by blast loads is discussed by Eknes and Moan (Ref. [12]). In the QRA study referred to here, the following coarse assumptions were made:

- Escalation by rupture of other equipment will be caused by overpressure exceeding 0.3 barg.
- Escalation by structural damage to decks and walls is caused by overpressure exceeding 0.5 barg.

These values are quite frequently used. It should be noted, however, that recent R&D has indicated that there may be some conservatism in these estimates.

Because the likelihood of escalation was shown to be quite high, investigation of remedial measures was undertaken, including consideration of the following:

- Removal of equipment not in use
- Partial removal of cladding.

The potential for reduction of overpressure by removal of equipment was found to be rather low. This is therefore not further discussed. The effect of partial removal of the cladding (corrugated light steel plate walls, with air gap close to floor and roof), is discussed in the following text. It should also be noted that removal of cladding usually implies improved natural ventilation and this is also briefly touched upon below.

8.7.1   Layout and Geometry

The platform considered in its original configuration has three decks. The cellar deck and main deck are partially enclosed with cladded walls while the upper deck is open. The deck dimensions are approximately 70 by 50 m. There is normal weather cladding on three sides of the deck, but the north side of the deck is closed off by a series of equipment rooms.

The walls around the process area (except the fire wall) are mainly composed of light corrugated steel plate. There are narrow horizontal (0.3–0.5 m) openings at deck and roof levels. These gaps are intended to provide the natural draft needed for ventilation purposes. It was however, noted that ventilation was not very extensive. A natural solution was therefore to open the cladding to provide increased ventilation and thereby also reduce the explosion overpressure.

8.7.2 Cases and Configurations Analysed

Two types of models were used to consider the effects of removing cladding. The models were based upon the following scenarios:

- Stoichiometric mixture in defined parts of the area
- Realistic gas cloud configuration based upon modelling of the gas dispersion given a defined leak source and environmental conditions.

Stoichiometric clouds are often used in gas explosion studies due to the straightforward definition of cases, and limited calculation resources. It is recognised that such an approach is conservative. The degree of conservatism may be studied in a probabilistic manner.

The case study referred to also includes consideration of the effects of installing a fire wall to subdivide one large fire area into two smaller fire areas. This was in addition to considering the effect of partial removal of wall cladding. The explosion studies that were undertaken for the cellar and main decks were as follows:

- Cellar deck, comparison of cases before and after partial removal of cladding
- Main deck, comparison of overpressure for the following cases:

  - Entire deck prior to segregation by fire wall and removal of cladding
  - Area 1 after segregation by fire wall and partial removal of cladding
  - Area 2 after segregation by fire wall and partial removal of cladding.

8.7.3 Ventilation Results

The effect of partial removal of cladding is usually to improve the ventilation rates which has the following effects (in addition to the reduced overpressure loads):

- Reduced likelihood of ignition, due to more extensive dilution of the gas cloud.
- Reduced effectiveness of automatic gas detection for small gas leaks.
- Increased exposure of personnel to harsh environment.

When considering possible partial removal of cladding, one of the main considerations is to balance the improved ventilation and reduced overpressure against

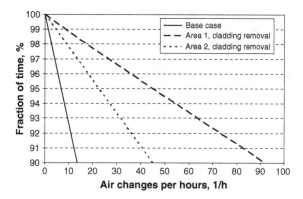

Fig. 8.22 Ventilation improvements on main deck resulting from the removal of cladding

deterioration of working environment conditions. Such deterioration can have an effect on the likelihood of accidents. Improvements in ventilation are nevertheless usually considerably more important.

Figure 8.22 shows the improvement in ventilation rates resulting from the cladding modifications i.e., a higher number of module air changes per hour for the same time fraction.

The diagram was obtained by running CFD calculations of the platform configuration to determine the distribution of ventilation rates for different environmental conditions. It may be added that results for the cellar deck are similar to these for Area 2 on the main deck.

There is no universally used standard defining adequate ventilation. Some operators have stipulated a minimum requirement of 12 air changes per hour, for at least 95% of the time. Increasing the opening of walls ensures that such a standard may be more easily reached.

8.7.4 Explosion Studies

There were three purposes of the explosion studies; determine the need for explosion risk reduction; how such reduction could be achieved; and improve the QRA modelling. The basic approach adopted consisted of the following:

- Explosion (FLACS) studies with stoichiometric methane-air atmosphere

  - Complete filling of areas and modules
  - Partial (50 and 25%) filling of areas and modules.

- FLACS studies with gas clouds resulting from dispersion studies following realistic gas leak scenarios.

The studies were completed with the FLACS version at the time, 'FLACS 1994' (see Appendix A). The experience from the BFETS test programme [1] showed clearly that this version under-predicted the likely blast loads. Selected cases from the initial study were therefore repeated some years later, using the 1998 version of the software. The 1998 version has incorporated the findings from the large scale tests mentioned in Sect. 8.5.2. When comparisons between 1994 and 1998 versions are made, increases are rather scenario dependent, but are typically in the range +50 to 150%.

8.7.5   FLACS Results

The results outlined in the following are based on the 1994 version of the FLACS software. The following observations may be made from the discussion of FLACS results:

- Local overpressure may be up to twice the average pressures on walls, decks, etc.
- The difference between overpressure for the stoichiometric scenario versus a scenario with gas dispersion varies quite considerably. The average pressures to walls, decks and roof are found to be 50% higher in the stoichiometric model whereas the average pressure on small rooms within the area shows an increase of nearly 290%.
- The dispersion based calculation is much to be preferred in terms of realism, but the computing power and time needed for such calculations are very extensive (up to several hours of computing time), thus making it necessary to limit the number of calculations done in this way.
- Calculations based on stoichiometric gas cloud may be adequate in order to study the relative effects of alternatives. The more realistic dispersion based calculations need to be done in order to define dimensioning loads.
- The variations between stoichiometric and gas dispersion models are strongly dependent on local effects.
- The effect of partial cladding removal is most extensive on the main deck and more limited on the cellar deck. One possible explanation is that the size of the area is greater on the cellar deck than on the main deck. It is expected that the effect of explosion relief walls around an area has the highest effect on smaller areas. For larger areas, the turbulence and acceleration increases pressures to a high value, before the pressure wave reaches the relief walls.

8.7.6   Demonstration of Parameter Sensitivities

This section presents another sensitivity study that demonstrates the effect of small changes in parameters. The object of this section is to show how one process module may be separated from a larger process area by the installation of a fire wall. It should be observed that on two sides there is to be a 1 m gap between the module and the rest of the platform. The two other sides of the module are used for ventilation purposes, as they are facing out from the platform.

The results from the following FLACS 94 analyses are presented in Fig. 8.23 and described below:

These different design cases are shown diagrammatically in Fig. 8.24. The process module being considered is shown in the southeast corner of the drawing. The area to the west of this module is an open process area without internal walls.

A. Base case, no fire wall installed, relatively full cladding.
B. Fire wall installed on the module, no change in cladding.
C. Fire wall installed on the neighbouring module, on the other side of the air gap, no change in cladding.
D. As case C, with increased explosion relief through reduced cladding.
E. As case D, with additional wall opened for relief.

8.7.7   Implications for QRA Modelling

Escalation by rupture of nearby equipment is taken to occur at 0.30 barg and escalation through destruction of decks, at 0.50 barg. Reduced maximum over-pressure results in reduced conditional probability of escalation. The conditional probabilities of overpressure exceedance are determined by the SERA software (or similar) and used as an input to the QRA.

Fig. 8.23 Comparison of overpressure results

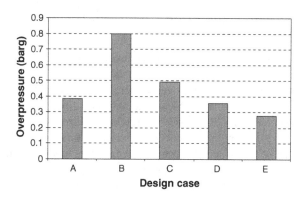

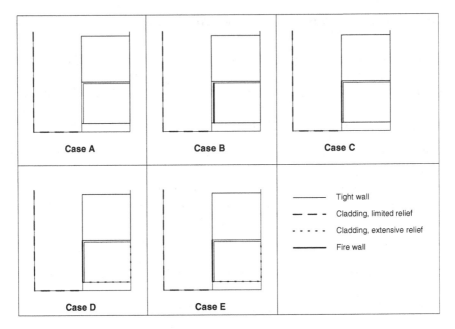

Fig. 8.24 Overview of wall locations

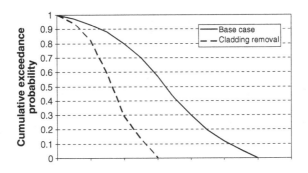

Fig. 8.25 Conditional probability distribution for explosion overpressure

Typical exceedance curves are shown in Fig. 8.25, for the base case before cladding modification, and after modification.

With the reduced overpressure, the probability of escalation to equipment is reduced from 90 to 57%, and the probability of escalation through the deck is reduced from 70 to 10%. It should be noted that the shape of such curves is very dependent on local aspects. These reductions are therefore not universal values, but are very case specific. It is thus seen that reduction of the maximum overpressure by 50% implies the following reductions:

- P(Equipment escalation): 37%
- P(Deck escalation): 86%

These reductions are very large, and suggest that the risk to personnel caused by explosions is drastically reduced by modification to the cladding.

8.7.8 QRA Sensitivity Results

Based on the overpressure results, the following average values were used as input to calculations, for the following decks and areas:

- Cellar deck: 20% reduction in maximum overpressure.
- Main area 1: 50% reduction in maximum overpressure.
- Main area 2: 50% reduction in maximum overpressure.

Table 8.7 presents the changes in risk levels resulting from the reduction in explosion overpressure arising from the cladding modifications. There is a considerable reduction in the potential loss of life (PLL) associated with process system accidents. Such accidents however, make only a limited contribution to the total risk, so the overall reduction in PLL is not particularly large.

The reduction in frequency of accidents causing damage to several modules, or total loss of the platform, is considerable. This is due to the reduced probability of escalation to other areas arising from the reduced explosion overpressures.

8.7.9 Discussion and Evaluation

The risk reduction measure discussed in this case presentation had a total cost of about 4 MNOK. The total risk reduction over the field lifetime is some 0.05 fatalities, implying a cost of 80 MNOK per statistical life saved. This is certainly not a low value, but is in the range of cost per life saved that will be committed to by operating companies.

Table 8.7 Effect on risk parameters from cladding modifications

Risk aspect	Change in risk value (%)
PLL value due to process accidents	40
Total PLL value	5
Total frequency of damage up to one module	2
Total frequency of damage to several modules or total loss	40

A simplified assessment of the risk to assets could focus on the largest consequences, that is, damage to several modules or total loss. If we assume an average accident cost of 1 billion NOK per accident, then the expected value of reduced costs over the field lifetime is about 10 MNOK, taking an NPV approach to cost accumulation. This underlines the fact that the cladding modification is an effective risk reducing measure. In fact, reduction of explosion overpressure is one of the few modifications which has a significant effect on risk to personnel and may be carried out without excessive cost.

It is worthwhile noting that the considerable reduction of the risk level from cladding modifications is strongly influenced by the shape of the curves in Fig. 8.25. These curves have a 'lazy s'-configuration. If these curves had a steeper s-shape i.e., were falling much more rapidly, the effect on the risk level would be less, although most probably, still significant.

References

1. SCI (1998) Blast and fire engineering for topside systems, phase 2. Ascot; SCI, Report no, p 253
2. Vinnem JE (1998) Blast load frequency distribution, assessment of historical frequencies in the North Sea. Preventor, Bryne, Norway; 1998 Nov. Report no.: 19816–04
3. Venart JES (2010) Buncefield: cause and consequences. In: Ale, Papazoglou, Zio (eds) Reliability, risk and safety, Taylor & Francis Group, London, ISBN 978-0-415-60427-7
4. Eckhoff RK (2005) Explosion hazards in the process industries. 1st edn. Gulf Publishing. Company, Houston, 440 pp
5. Bjerketvedt D, Bakke JR, van Wingerden K (1997) Gas explosion handbook. J Hazard Mater 52(1997):1–150
6. Oil and Gas UK (2003) Fire and explosion guidance. Oil and Gas UK, Oct 2003. www.oilandgasuk.co.uk
7. Czujko J (2001) Design of offshore facilities to resist gas explosion hazard. Engineering Handbook, Nowatec.no
8. ERA (1998) Fire and explosion engineering, offshore installations—conference proceedings, 1 Dec 1998, ERA Report 98-0958. Leatherhead, UK
9. Norway Standard (2010) NORSOK Z-013. Risk and emergency preparedness assessment, Standard Norway
10. IOGP (2010) Vulnerability of plant/structure, IOGP Risk Assessment Data Directory, Report No. 434–15, OGP, March 2010
11. PSA (2011) The facilities regulations; Petroleum Safety Authority, Norwegian Pollution Control Authority and the Norwegian Social and Health Directorate, Stavanger
12. Eknes ML and Moan, T. (1996) Modelling of escalation due to explosion. In: Proceedings of the 15th international conference on offshore mechanics and arctic engineering, 16–20 June 1996, Florence, Italy

Chapter 9
Collision Risk Modelling

9.1 Historical Collision Risk

9.1.1 Significant Collisions

According to the WOAD® database, [1] there have been six cases of total loss of a platform since 1980 due to collision or 'contact' (impact by vessel in close attention):

- Two jacket structure s in US Gulf of Mexico have been lost due to collision as the initiating event.
- One jacket structure in Middle East waters has been lost due to collision as the initiating event.
- One jack-up was lost in US Gulf of Mexico during movement, due to listing, structural damage, contact with platform, and finally loss of buoyancy.
- One jack-up structure was lost in the North Sea, due to collision with a pier. This is a non-representative case, involving a small jack-up, which was lost due to severe weather. The jack-up was engaged in tunnel drilling and was standing only a few metres from the waterfront. The jack-up was small and not representative of offshore jack-ups. The accident is disregarded from further discussions.
- One jack-up structure was lost in South American waters (Atlantic coast) due to contact with attending vessel.

It is worthwhile noting that none of these occurrences have taken place in the North Sea although there have been a number of near collapses. There is a story that a North Sea jack-up was impacted by a supply vessel at full speed (making a last minute attempted evasive manoeuvre, resulting in a so-called 'glancing blow'). One of the legs had substantial damage to two of the three vertical members, and between five and ten of the diagonal members. Luckily it was a day with calm

© Springer-Verlag London Ltd., part of Springer Nature 2020
J.-E. Vinnem and W. Røed, *Offshore Risk Assessment Vol. 1*,
Springer Series in Reliability Engineering,
https://doi.org/10.1007/978-1-4471-7444-8_9

weather, and the platform could be removed in a controlled manner with no personnel injuries.

If we turn to fatal accidents due to collision or contact, it becomes clear that the crew of the boats are the ones at risk. Five accidents after 1980 [1] have resulted in two fatalities on a jacket structure (South America, Pacific coast), while the four remaining cases have caused 23 fatalities on the vessels. Two of these accidents occurred in US Gulf of Mexico, one in South America (Pacific), and one in the Middle East.

A structure will have to fulfil a number of other requirements in addition to design requirements against progressive collapse due to an impact load. These additional requirements will often govern the dimensions and the design of the platform. This means that the platform may have quite extensive over-capacity with respect to collision impact, compared to the minimum design limits. This over-capacity is concept dependent and may vary significantly.

9.1.2 Norwegian Platform Collisions

No very serious collisions have occurred in the Norwegian North Sea, but there have been some of medium severity, and some very close calls. The section provides an overview of the significant collisions. All the light contacts between vessels and installations during loading operations and other cases with close proximity, low velocity manoeuvring are excluded, except for shuttle tankers. All other vessels except shuttle tanker will have low impact energy when the speed is low.

It may be noted that whereas most other critical incidents have shown a downward trend during the past 15–20 years, six of the eight significant collisions in the list below have occurred after 1999:

- 1988: Oseberg B jacket, submarine collision
- 1995: Small coastal freighter head–on collision into Norpipe H-7 platform
- 2000: Shuttle tanker impact into FPSO (see Sect. 4.30)
- 2004: Large supply vessel colliding with mobile installation at high speed
- 2005: Large supply vessel colliding with bridge between two Ekofisk installations
- 2006: Shuttle tanker impact into Njord B storage tanker
- 2007: Large supply vessel colliding with Grane production platform at low speed
- 2009: Large well simulation vessel colliding with unmanned Ekofisk installation at high speed.

The second event did not actually occur on the Norwegian Shelf, but in the German sector. The platform, however, was operated from Norway, as a compressor station on the Norpipe line from Ekofisk to Emden.

There have also been various light contacts with supply vessel s and collisions of shuttle tankers into loading buoys while loading offshore. There have also been several cases of drifting vessels and vessels on a collision course, but none of them have ended up impacting on an installation. Finally, there have been several near-misses, which had the potential to result in serious head-on collisions. Several of these have occurred at Haltenbanken in the Norwegian Sea. Collisions involving FPSOs and shuttle tanker s are discussed in Sect. 10.4.

9.1.2.1 Oseberg B Collision

The submarine collision at Oseberg occurred on 6th March 1988, when the West German submarine U27 collided with the Oseberg B steel jacket platform. Personnel on board were evacuated to the flotel 'Polyconfidence', which was linked to the platform with a gangway. A later survey found that a 1.2 m diameter bracing had been dented to a depth of about 20 cm. The repairs which could be safely postponed until the following summer season, had a final cost of some 80 MNOK (1988 value). The submarine was navigating approximately 20 m below the surface. The platform was marked on the map, but no signals from the sonar were received. The submarine sustained damage to the bow, bridge, and navigation equipment, but there were no injuries.

9.1.2.2 Norpipe H-7 Collision

The head–on collision of a merchant vessel with the Norpipe H7–platform, occurred on 30th September 1995 when a small (273 GRT) German cargo ship 'Reint' was en route for Aalborg in Denmark. The fact that the vessel was heading straight for the platform was only discovered at a late stage by personnel on the platform. The platform's standby vessel tried to contact the ship, but with no success. The standby vessel also tried to intercept the vessel, but the ship went on. Then at 10:42 h platform shutdown was initiated before the ship hit in a 'glancing blow' manner on one of the platform's legs. There was only minimal damage to the platform, and neither any injuries nor spills were caused. The ship's mast was smashed. The platform had two large risers one on each side and thus there was the potential for a very serious accident. As it happened however, the ship was not even close to hitting the risers. It was later revealed that the ship was on autopilot when the accident occurred.

9.1.2.3 West Venture Collision

On 7th March 2004, a supply vessel was just a few months into its first contract, impacted directly into a corner column of the mobile installation West Venture in the middle of the night. The vessel had a displacement of about 5,000 tons, and a speed of 7.3 knots, making this a high energy collision.

Two ballast tanks in the column were punctured, but no water ingress occurred, as the holes were above the sea level. The installation experienced a sideways movement of about seven metres, while the BOP was suspended in 300 m riser, but was not in close proximity to other equipment.

The vessel had its bow indented about 1.5 m, and damage to the bulbous bow. 117 persons were on the installation and 14 crew on the vessel, none of whom were injured during the accident. Both the installation and vessel could travel to shore under their own propulsion for repairs.

The direct cause of the accident was that the standby vessel Far Symphony was heading for the installation without deactivation of the autopilot. Due to the autopilot being active, the officer on the bridge was prevented from manoeuvring the vessel in a normal manner. The autopilot did not have the installation as the waypoint, but had not corrected its course for deviations due to wind and waves. The officer in charge had not tested the manoeuvring system prior to entering into the safety zone. The crew was not familiar with the emergency manoeuvring equipment.

9.1.2.4   Ekofisk Centre Collision

This collision occurred on 2nd June 2005, in dense fog, with 100–150 m visibility. The standby vessel steamed at full speed into the bridge between two installations on the Ekofisk field. The displacement of the vessel was 5,600 tons, and the speed was about 6 m/s, implying a very high collision energy level. The hit was on the bridge between two installations, which was hit by the vessel's superstructure. The bridge was left unusable, but the damage was small, comparatively speaking, due to the lucky circumstances of the impact. Had the hit been on the small steel jacket platform at the end of the bridge, this would probably have experienced a total loss. Operations on this installation had been abandoned in the late 1990s, and thus hydrocarbons had been removed and it was not a 'live' installation.

The direct cause of the collision was communication failure during crew change on the bridge immediately prior to arrival at the Ekofisk field.

9.1.2.5   Njord B Shuttle Tanker Collision

The shuttle tanker Navion Hispania collided with Njord B storage tanker on 13th November 2006 during preparations to start off-loading. The shuttle tanker lost power to all but one of the propellers, due to contaminated fuel. The Dynamic Positioning (DP) system should have ensured that at least 50% of the propulsion should be intact, but had an unrevealed fault. There was thus insufficient power available to prevent collision, which occurred with 1.2 m/s shuttle tanker speed. The collision energy was 55 MJ, making this one of the highest energy collisions offshore ever. No injuries or spills occurred, but the potential for damage was present due to the high collision energy.

9.1.2.6 Supply Vessel Collision at Grane

The supply vessel Bourbon Surf (3,120 dwt) was heading for the installation on 18th July 2007. Both the captain and the first mate left the bridge to attend to other tasks, after having passed the safety zone limit around the installation. When they returned, it was too late to avoid the collision, but the speed was reduced to 1 m/s before the impact.

9.1.2.7 Big Orange Ekofisk Collision

The well stimulation vessel was approaching one of the Ekofisk complex installations to start on a job on 6th June 2009. The captain had activated the auto-pilot, and had forgotten to deactivate the auto-pilot when approaching the installation. When attempting to manoeuvre manually, his actions were overridden by the auto-pilot repeatedly. He managed to avoid two installations, but collided with an unmanned water injection platform with high speed, 9.5 knots. Serious damage to the jacket structure and the bridge to the platform, as well as the stimulation vessel, but no injuries occurred.

9.1.3 Attendant Vessel Collisions

The most comprehensive study of attendant vessel collisions in the past was considered to be J. P. Kenny`s study on the protection of offshore installations against impacts [2]. This study was prepared for the UK Department of Energy. Collision incidents recorded by the Department of Energy between 1975 and 1986 were analysed.

Since then, the HSE has issued analysis of all collision incidents recorded in their database, as well as other databases, the most recent report cover the period 1975 to 31 October 2001 [3]. Out of the total 557 collision incidents included in the report, the following distribution on vessel types may be summarised:

- Supply vessels: 353 incidents, 63.4%
- Standby vessels: 87 incidents, 15.6%
- Other attending: 74 incidents, 13.3%
- Passing vessels: 8 incidents, 1.4%
- Unspecified vessels: 35 incidents, 6.3%.

With respect to attendant vessels, the following trend may be reported, when we consider 10 year periods:

- 1975–1984: 218 incidents
- 1985–1994: 211 incidents
- 1995–2001: 85 incidents.

Attendant vessels account for 514 incidents, which is over 96% of the total, if unspecified vessels are disregarded. The falling trend since 1995 is clearly visible, even if the last period is shorter, due to absence of data for the period 2002–2005. The report also presents volume of installation years, which allow calculation of the following frequencies per installation year:

- Supply vessels: 0.010 per installation year
- Standby vessels: 0.0030 per installation year
- Other attending vessels: 0.0038 per installation year.

These frequencies are down almost an order of magnitude from what was reported previously by JP Kenny. If we consider the differences between fixed and mobile installations, we would expect the mobile installations to have higher frequencies, due to two floating structures. The following frequencies for the period 1995–2001 confirm this expectation:

- Fixed installations: 0.026 per installation year
- Mobile installations: 0.102 per installation year.

With respect to causes of collision incidents, detailed and overall distributions are available, the following is the overall distribution on main categories:

- External factors: 82 incidents, 14.7%
- Mechanical control failure: 126 incidents, 22.6%
- Human control failure: 152 incidents, 27.3%
- Watch-keeping failure: 15 incidents, 2.7%
- Unspecified: 182 incidents, 32.7%.

There are also many light collision recorded in the RNNP project for the Norwegian sector, a small number of these collisions have been high speed collisions, as described in Sect. 9.1.2.

The number of high energy collisions with attendant vessels has increased as shown above. Three significant collisions with attendant vessels occurred during the period 2000–2010. There are about 100 installations on NCS, including production and mobile installations. If we conservatively assume that there are two supply vessel visits in average per installation per week, there are about 10,000 supply vessel visits to Norwegian offshore installations each year. Three incidents during 10 years imply a frequency of 3×10^{-5} per year. This is below the 10^{-4} limit according the Norwegian regulations.

9.1.4 What Should Be Design Basis?

There has been some uncertainty in recent years about what should be design basis for structures under Norwegian regulations. The requirements according to the Facilities regulations [4] have not changed, but the signals from PSA have changed somewhat, in the sense that PSA apparently states that passing vessel collision has been given too high weight, and that attendant vessel collision needs to be emphasised more.

What could be the rationale for such recommendations? For Norwegian installations, there has been only one passing vessel collision, the H7 collision in 1995 (Norwegian installation in German sector, see Sect. 9.1.2.2), but there are several anecdotal cases known where passing vessels have missed production installations by just a few meters, apparently unaware. The number of RNNP reported passing vessels on collision course in the Norwegian sector has dropped significantly over the last 15 years, apparently due to Equinor's information campaign towards the North Sea fishing fleet. There are several passing vessel collision cases in the UK sector, most of them before 2000. The requirements for Automatic Identification System (see Sect. 9.3.8) have been applicable since 2004, implying that international vessel navigation is significantly more controlled now compared to 20 years ago.

All of this points to a reduced risk of passing vessel collision in recent years, but can this be interpreted as implying that such collisions should not be designed against?

PSA has in the recent years warned against basing risk assessments only on calculation of frequencies (see also Sect. 2.1). Based on the number of observed incidents and accidents, they argue that also uncertainty and unknown mechanisms should be considered. It is therefore surprising that the lack of passing vessel collisions in the North Sea in the last 15 years should be the basis for omitting passing vessel collision as a design basis.

In the early hours of Thursday 8th November 2018, there was a collision between the Norwegian naval frigate 'KNM Helge Ingstad' and the Maltese registered tanker 'Sola TS' in the Hjeltefjord, outside the Sture crude oil terminal in Øygarden Municipality in Hordaland County, Norway [5]. Both vessels are modern with all available navigational systems, but apparently the Automatic Identification System was turned off on the naval vessel. As a result, the frigate navigational personnel misinterpreted the radar picture completely, and in spite of being warned about the collision hazard, did not take evasive actions in time. The frigate subsequently sank, possibly due to the failure of watertight compartments, luckily without casualties. This accident, although remotely (and unimportantly) connected to oil and gas, is important because it shows that navigational errors may occur under special circumstances, in spite of all the modern navigational systems.

The overall conclusion from this discussion is that passing vessel collision should not be eliminated from the design basis of offshore installations. The risk is probably lower compared to 20 years ago, but it is not negligible and can not be disregarded. PSA's signals over the last few years are to some extent misleading.

9.2 Modelling Overview

9.2.1 Introduction

The initial step in a review of collision risk models will be to specify the different vessel categories which have to be considered. The different types of vessels that may pose a collision hazard for an offshore platform (fixed or floating) are shown in Tables 9.1 and 9.2.

Table 9.1 Categories of external vessels considered in relation to risk modelling

Traffic category	Vessel category	Remarks
Merchant	Merchant ships	Commercial traffic passing the area
Naval traffic	Surface vessels	Both warships and submarines
	Submerged vessels	Submerged submarines
Fishing vessels	Fishing vessels	Divided into vessels in transit and vessels operating in the area
Offshore traffic	Standby boats	Vessels going to and from other fields
	Supply vessels	Vessels going to and from other fields
	Offshore tankers	Vessels going to and from other fields
	Tugs	Towing of drilling rigs, flotels, etc.

Table 9.2 Categories of colliding field related traffic considered in risk modelling

Traffic category	Vessel category	Remarks
Offshore traffic	Standby boats	Dedicated standby boats
	Supply vessels	Visiting supply vessels
	Working vessels	Special service/support as diving vessels, anchor handling, well servicing, etc.
	Offshore tankers	Shuttle tankers visiting the field
Floating units	Storage vessels	Dedicated floating units at the field
	Flotels	
	Drilling units	
	Crane barges	

Each of the vessel categories is presented in the following sections with an evaluation of relevant traffic patterns and vessel behaviour as a basis for the discussion of relevant collision probability models. A broader presentation of models may be found in Haugen [6]. The models presented in the following are:

- Passing vessel collision model
- Collision model for drifting floating units.

9.2.2 Merchant Vessels

Merchant vessels are frequently found to represent the greatest hazard, for several reasons:

- They may be large and thus possess considerable impact energy.
- They are usually travelling at high speed and therefore almost any vessel size will represent a considerable impact energy.
- The traffic may be very dense in some areas.
- These vessels are among the 'traditional' users of the ocean, as opposed to the vessels and installations associated with the oil and gas exploration and exploitation. The tradition of the maritime world is that the oceans are free with a minimum of restrictions, and merchant vessel masters are generally not very willing to accept limitations on their operation.

The last aspect may have changed for the better during the last 20 years or so. Traffic restrictions are more common, and monitoring systems imply that vessel operations become more transparent. This may have contributed to a change whereby the merchant vessel fleet will accept restrictions more easily.

There are few occurrences of collision by merchant vessels. This is obviously a good thing, but implies at the same time that there is limited experience data for risk analysis purposes. Consequently, there is considerable uncertainty about the risk estimates.

Navigation failure of passing vessels is often called 'powered' collisions i.e., vessel steaming towards the installation on a collision course. These are often further split into:

• Blind vessels	Ships with ineffective watch-keeping, because the radar is inoperable or ships being incorrectly operated, in conditions of poor visibility
• Errant vessels	Ships where no effective watch is being kept due to the watch-keeper's absence, distraction, incompetence or any other form of incapacitation. The watch must be ineffective for a significant period (at least twenty minutes) for a vessel to be classified as errant

It may be assumed that blind or errant vessels should be possible to avoid with current navigational aids and monitoring systems. The list in Sect. 9.1.2 suggests that the number of events has actually increased rather than decreased.

9.2.3 Naval Traffic

Traditionally naval vessels were never considered in risk analyses prior to the collision in 1988 because it was assumed that they would always have excellent control over their navigation. This assumption was to some extent shown to be false by the Oseberg incident. Naval traffic is, however, always a problem because no navy is willing to give out information on movements of either their own or foreign vessels. Projections of the traffic are therefore difficult to obtain and very often have to be based on more or less subjective evaluations. Naval traffic may be divided into two main categories, surface traffic and submerged submarines.

In spite of the collision at Oseberg, naval traffic is not assumed to represent a significant risk contribution for the following reasons:

- The number of vessels is generally relatively low.
- The vessels are considered to be technically very reliable, thus giving a low probability of them drifting out of control.
- The manning level and the standards are generally high, implying that it is less likely that the watch-keeper(s) will not be aware of any platform on his course.

Naval vessels are therefore often disregarded in collisions risk studies due to the perceived low risk.

9.2.4 Fishing Vessels

Fishing vessels vary in size from large factory/freezer ships to smaller vessels operating near the coast. Fishing vessels are divided into two groups, depending on their operational pattern:

- Vessels in transit between the coast and the fishing areas.
- Vessels fishing in an area. The vessels' operation and behaviour during fishing will be complex and varied, but usually at low speed with no preferred heading.

The vessels are usually so small that they represent no serious hazard to the structural integrity of a platform. Typically, a large fishing vessel will have a displacement of around 1000 tons. This implies that the collision energy will be less than 20 MJ in most cases and neither drifting vessels nor vessels under power will normally be able to threaten the integrity of any installation. Risers and umbilicals, if located in an exposed position, may however, be subject to damage from fishing vessels.

9.2.5 Offshore Traffic

9.2.5.1 External Offshore Traffic

Passing offshore vessels, including tankers, supply, standby and work vessels, are in many respects similar to passing merchant vessels. Their level of knowledge about installations should be considerably higher because they operate in the area all the time. The risk assessment approach for passing vessels is therefore applicable also to this vessel category, but the parameter values will have to reflect the relevant facts about this traffic.

9.2.5.2 Field Related Attendant Vessel Traffic

Vessels in this category are travelling to and from the field which is being considered. In addition, they may stay at the field for some time. This means that they have two different operational modes that should be considered.

The majority of attendant vessel collisions occur at low velocities while the vessels are manoeuvring in the vicinity of the platform [2]. These low energy impacts will normally not threaten the integrity of normal manned offshore platforms. An exception may occur for some of the new, very lightweight structures for not normally manned installations, which may not have the same load resistance as a manned structure. It was considered in the 1980s that general design rules usually provided sufficient resistance for low energy impacts. The DNV GL rules [7] imply a collision resistance of 14 MJ, with respect to local damage.

PSA has argued for several years that the average size of supply vessels has increased by a factor of two to three during this 30-year period, but the 14 MJ collision requirements has not been changed. It is questionable if this resistance is adequate for the new, large attendant vessels.

Accident statistics for attendant vessels are often used as a base for a collision frequency assessment. These data will, however, have to be modified for the specific vessel types considered, the installation design, and other specific field data.

Vessels may also collide with the platform while they are approaching it. This can occur if, for example, a supply vessel is heading for the platform, but fails to alter course or slow down due to some equipment failure or human error, as has occurred on NCS during the past 10 years (see Sect. 9.1.2). The point is sometimes made that the use of satellite based navigational systems (GPS, etc.) means that a vessel which fixes its route on the exact position of an installation, will inevitably ram into it, if there is a failure to control course and speed during the final stages. Thus, many operators require supply vessels to enter the route terminal point a minimum of 500 m off the installation (outside the safety zone), in order to eliminate the potential for ramming into it.

Historical data show a significant risk contribution from attendant vessel s approaching the field. These collisions will therefore have to be included in a complete collision risk study.

9.2.5.3 Field Related Shuttle Tanker Traffic

A shuttle tanker will be totally dependent on its computer controlled propulsion (dynamic positioning system) while loading oil. The failure frequency of the tanker will be based on the possibility of breakdown of main engines, control systems or reference systems.

The contribution from shuttle tanker s waiting at the field for loading is usually quite low, since these vessels will be located on the leeward side of the platform to prevent drifting into the installation.

9.2.6 Floating Units

The final category is collision between a floating unit and an installation. The term floating unit includes flotels, drilling rigs, crane vessels, barges, storage tankers, diving vessels as well as shuttle tanker s located at the field while loading. The floating unit may be positioned by mooring lines or a DP-system, or a combination of both. If the positioning system fails, the unit may start drifting. The models for drifting of floating units are mainly based on equipment failure frequencies and geometrical considerations. This is discussed in Safetec [8], and is not covered in the following.

9.3 Passing Traffic

Collisions between vessels and offshore installations are distinctly different from vessel–vessel collisions. When two vessels collide, this is often due to failure in the coordination of passage at a safe distance, due to misunderstanding, communication failure, etc. When a vessel collides with an offshore installation, only the vessel is moving and the installation is stationary. Floating installations can move marginally, but this is nothing like the movement of a vessel manoeuvring under power and seamanship. FPSOs, may be considered something of an exception as they in theory may be able to rotate.

Because of this significant difference, the collision models that apply to vessel–vessel collisions cannot be used for vessel–platform collisions and vice versa. The models presented in the following for powered and drifting collisions are to a large extent based on the 'COLLIDE®' model; see [6]. The original model was further developed by M. Hassel in his PhD thesis, 2014–2017 [9]. The basic approach is presented based on Haugen's work, and the updated model is presented based on Hassel's work. Hassel's work has included revised modelling based on modern navigational equipment and practices as well as modern theories for accident causation.

9.3.1 Introduction

Collisions can occur because the vessel is drifting, (i.e. it is out of control), or it is steaming without its crew being aware that it is heading towards the platform. This can be expressed as:

$$P_{CP} = P_{CPD} + P_{CPP} \tag{9.1}$$

where

P_{CP} probability of passing vessel collisions
P_{CPD} probability of collision due to a passing drifting vessel
P_{CPP} probability of powered collisions.

The probabilities of collision are usually quite low values, such that there is in practice no real numerical validation possible of the probability of collision. In the following the probability notation above is used in the sense of frequencies. The model for drifting vessels is general for all traffic categories and is not discussed here. Powered collisions are considered in the following.

9.3.2 Powered Passing Vessel Collisions–Model Overview

9.3.2.1 Basic Approach

The basic equation for calculating the passing vessel collision frequency for a platform in a specific location may be expressed as follows:

$$P_{CPP} = \sum_{i=1}^{m} \sum_{j=1}^{6} \sum_{k=1}^{n} N_{ijk} \sum_{l=1}^{4} P_{CC,jkl} \, P_{FSIRjkl} \, P_{FPIR,jkl} \tag{9.2}$$

where

P_{CPP} = annual frequency of powered passing vessel collisions.
N_{ijk} = annual number of vessels in vessel category j in size category k travelling in lane i. The risk contribution from each relevant 'lane' is calculated and added together to get the total risk to the platform.

$P_{CC,ijkl}$ = probability that a vessel in vessels category j in size category k in traffic group l travelling in lane i is on a collision course at the point when the vessel can observe the platform, visually or on radar. There are six vessel categories:

- Merchant vessels
- Fishing vessels
- Standby boats
- Supply vessels
- Shuttle tankers
- Naval vessels (including submarines).

$P_{FSIR,jkl}$ = probability that the vessel itself does not initiate some action to avoid a collision with the platform (Failure of Ship Initiated Recovery).

$P_{FPIR,jkl}$ = probability that the platform or the standby vessel does not succeed in initiating avoiding action on the vessel, given that the vessel has not initiated such action itself (Failure of Platform Initiated Recovery).

Equation 9.2 may appear to be based on independence of the individual factors, but this is not the case. These probabilities should rather be considered as conditional probabilities, similar to those used in an event tree. The two failure probabilities are specific to vessel, size, and traffic category, but not dependent on the lane.

The traffic volume, N_{ijk}, is the most straightforward parameter. No modelling is necessary, the only problem with this parameter is that quite a lot of data are needed. The COAST® database [10] provides an extensive, up-to-data shipping route database covering entire UKCS and NCS, in addition to other areas around the world (see also Appendix A). The COLLIDE® software uses data from COAST® as input for calculating the collision frequency for a location.

The probability of collision course, P_{CC}, is the 'geometrical factor', which includes all factors related to the composition and position of the traffic flow.

The two causation factors, P_{FSIR} and P_{FPIR}, deal with the underlying mechanisms when the vessel fails to take actions to avoid a collision. These two factors were quantified in the previous COLLIDE® version, they are not quantified explicitly in the new model, but the functions are still implicitly present in the influence diagram and the BBN model.

9.3.2.2   Model Evaluation

As far as possible, the COLLIDE® approach attempts to describe the situation for the responsible navigator on the vessel. The previous model was based on the use of the radar as the main tool for passing vessels in order to avoid impact with the installations. This leads to a dependency on the sequence of what is observed on the radar, and the navigators' responses.

The obvious requirement is that the vessel is on a collision course with an installation. The second main element is the installation's ability to initiate contact with the errant ship, to take corrective action to avoid an collision. This is mainly achieved by detection and communication, which may lead to interception or emergency move-off. The third main element is the ship itself taking corrective actions, which is heavily influenced by the navigator's actions. The main elements influencing the navigator's responses are training, experience and alertness, along with distractions, the safety culture and the quality of the technical navigation aids available.

In the new COLLIDE® model all relevant factors are reflected in an influence diagram, with the following top factor:

- Successful recovery, with inputs:

  - Ship initiated actions, with inputs:

    Navigator action
    Corrective action made difficult.

  - Emergency move-off performed (mobile), with inputs:

    Successful detection of collision course
    Procedures
    Training and competence
    Installation type.

  - Platform initiated action

    Vessel on collision course
    Successful intervention by helicopter/standby vessel
    Communication with ship
    ARPA radar on installation.

Figure 9.1 presents an overview of the top levels of the influence diagram developed for collision risk. The detailed influence diagram is presented in Ref. [9].

The most comprehensive input in the influence diagram is for the factor 'Navigator performance', with output to 'Navigator action' and the following inputs:

- Navigational method
- Navigator skills
- Task difficulty
- Navigator readiness
- Manoeuvrability
- Bridge ergonomics
- Morale
- Incapacitation
- Cumulated tasks

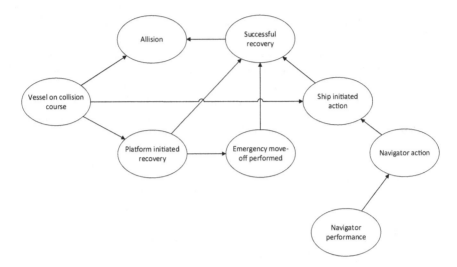

Fig. 9.1 Influence diagram of collision risk from passing vessels towards offshore installation (top levels, based on [9])

- Simultaneous operations
- Situational awareness.

One aspect of the previous as well as the current approach, which may be worrying, is that the approach assumes that navigators operate in a responsible manner, which most navigators do. Yet, for each 999 responsible navigators there may be one who behaves irresponsibly. This person probably accounts for 90% of the collision risk alone. Is the modelling of the approach taken by the 999 persons (which may only be accountable for 10% of the risk) representative of the behaviour of the one irresponsible person? This is mainly a philosophical consideration, as attempting to model the irresponsible person's behaviour would be impossible. As a further illustration of this point consider what was pointed out with respect to the captain of the vessel 'Reint' which collided with the H–7 compression platform in 1995. It was reported that the same person during a period of less than 10 years had been involved in three accidents or incidents involving violation of navigation rules (and his licence was revoked).

The influence diagram is described by Hassel [9] to be most suitable for hazard identification, which should use the full version of the influence diagram (not shown here due to permission issues). The influence diagram is not suitable for quantification.

9.3.2.3 Quantification of Collision Risk

A simplified version of the influence diagram has been developed as a BBN for quantification purposes [9], which shows a quantified risk model with input values quantified by an expert panel. The BBN is outlined in Fig. 9.2, but the values are not shown due to permission issues.

Table 9.3 presents a comparison of values from the previous COLLIDE® version with those calculated by Dr. Hassel [9] with the new model and expert input, for a number of routes with different Closest Point of Approach (CPA), standard deviation and number of vessels per year in route.

There are two cases where the frequency increases somewhat, by a factor of 1.23 and 2.25, for routes M and B respectively. Otherwise the frequencies are reduced, from 2% (Route N) up to a reduction factor of 54 (Route D). More detailed discussion of implications is found in Ref. [9].

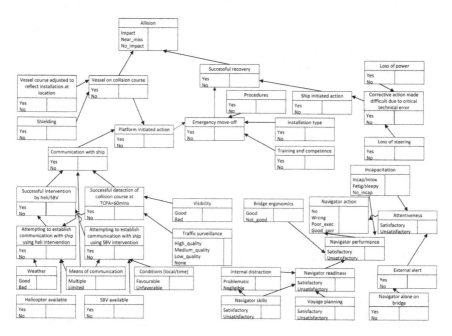

Fig. 9.2 Bayesian belief network for collision risk from passing vessels towards offshore installation simplified from influence diagram (based on [9])

Table 9.3 Comparison of previous and new COLLIDE® versions, probability of impact per installation year (values from Ref. [9])

Route	CPA (nm)	Std Dev	No of vessels in route (/year)	COLLIDE® previous	New BBN model
A	0.5	0.7	118	1.85×10^{-4}	4.89×10^{-5}
B	1.2	0.4	98	4.79×10^{-6}	1.08×10^{-5}
C	4.7	1.5	2434	3.12×10^{-5}	9.40×10^{-7}
D	9.2	1.8	4674	4.52×10^{-9}	8.31×10^{-11}
E	0.9	1.1	2034	1.77×10^{-3}	3.29×10^{-4}
F	1.7	1.2	49	1.12×10^{-4}	5.03×10^{-6}
G	1.7	1.2	67	6.41×10^{-5}	7.25×10^{-6}
H	1.8	1.3	43	6.10×10^{-5}	1.79×10^{-6}
I	1.6	0.8	27	1.83×10^{-5}	2.88×10^{-6}
J	2.2	1.2	39	1.50×10^{-5}	2.44×10^{-6}
K	0.1	1.1	94	1.34×10^{-4}	1.31×10^{-5}
L	3.4	1.6	70	1.55×10^{-5}	3.69×10^{-6}
M	2.3	0.9	155	5.37×10^{-6}	6.61×10^{-6}
N	4.9	0.8	94	2.30×10^{-9}	2.26×10^{-9}
O	5.7	1.0	65	1.88×10^{-7}	1.79×10^{-8}

9.3.3 Traffic Pattern and Volume

The traffic volume is probably the parameter which most directly can be based on observations and which can be treated statistically without applying analytical considerations or engineering judgement. This is therefore the parameter which requires the least effort in terms of modelling.

The database with route data for collision calculations (COAST®) has already been mentioned. A more fundamental source is Lloyd's Maritime Intelligence Unit's port log data, which on a daily basis reports worldwide ship movements. COAST® uses this data as one of its main data sources. Port statistics may also be used in addition to actual observations on the location for which the collision risk is assessed. In all cases it should be noted that traffic patterns and volumes change fairly often.

The lateral sailing patterns in a certain lane are often described using a normal distribution. The width of the lane is often taken as four standard deviations (e.g. COLLIDE®), as shown in Fig. 9.3. There may also be other traffic distributions (see [6]).

One of the basic premises of the model is that merchant vessels travel in relatively well-defined routes or lanes. This has been confirmed in numerous traffic surveys, which have shown that this assumption is reasonable.

The AIS (Automatic Identification System, see also Sect. 9.3.8) gives the best possibilities for mapping of movement data. All vessels that represent a threat to offshore installations are required to have AIS. The waypoints of the routes as well

Fig. 9.3 Route based traffic

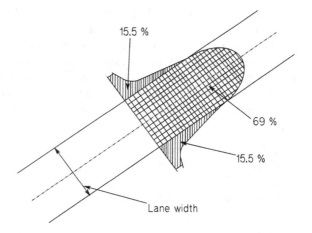

as the lateral distribution of ships in the routes may be documented by the use of AIS data.

Ideally, routes should be defined for each possible, or likely, combination of ports that vessels will travel between. This division would give a traffic picture which would be simple to handle from the point of view of risk calculations because all routes would be clearly defined. It would also improve traceability of the data and improve the possibility of updating the traffic data. Aspects such as 'shuttling' (the fact that some vessels perform repeated journeys over a short time interval) on a route and also ship size distribution would be easy to control with such a route definition, provided sufficient data collection had taken place.

However, this would be inconvenient, as it would lead to a very large number of routes, of which the majority would have very limited traffic. In addition, it is not always necessary to divide the traffic to such a degree of detail because even if two vessels start from different ports they may nevertheless have to pass a common way point which would imply that they will travel along the same route for at least part of the way. These way points are used in the definition of routes.

9.3.4 Probability of Collision Course

The basic condition for a collision is that a vessel is on a collision course. It is assumed that the collision course is reached before the platform can be observed. It is very unlikely that a vessel not already on a collision course will change course in such a way that they are heading towards the platform if they can observe it. Consequently it is the process before the point where the platform can be observed which determines the value of the parameter P_{CC}.

A vessel on a course straight for a platform is a normal occurrence in the North Sea. The statistics from the RNNP have confirmed this beyond doubt. This in itself can therefore not be regarded as a critical situation, if the distance from the platform is sufficiently large. The criticality arises only when the distance between the vessel and the platform is decreasing and the vessel still maintains its course.

A route which passes through open waters, with no particular obstructions, is initially considered. It is then reasonable to assume that the traffic is distributed across the route width in accordance with a normal distribution. Deviations from the straight (or ideal) route between two points may be due to a number of different causes, e.g.:

- different choice of route
- different navigational practice
- inaccuracies in navigation (human related)
- inaccuracies in equipment, charts, etc.
- deviation due to wind, waves, and current.

All these factors are to some extent random, and it can be assumed that they together will produce deviations from the straight line which may be modelled with a normal distribution. This however, only relates to navigation in open sea.

A special group of vessels are those which use the platform as a navigational mark. Some ships divert from the shortest course line to locate an installation for position-fixing purposes. Fishing vessels have been known for using this practice. Statoil (now Equinor) performed 10–15 years ago a massive effort in the North Sea, in order to inform fishing vessel crews about the effect on an installation when the fishing vessels headed directly for an installation for a prolonged period of time, which ultimately could cause a complete precautionary evacuation of all personnel, if the fishing vessel did not reply on radio. The number of fishing vessels with such practice has been significantly reduced.

Dr. Hassel has in his thesis [9] provided a comprehensive study of the effects on navigation when a new offshore installation is installed (see Fig. 9.4). AIS data was used before and after commissioning for seven installations in the Norwegian

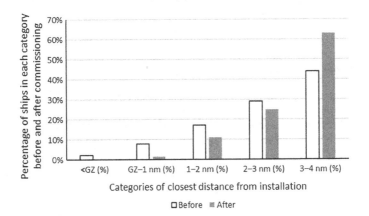

Fig. 9.4 Normalized results for all installations in the survey, before and after commissioning

sector: Goliat, Skarv, Knarr, Gjøa, Valemon, Edvard Grieg and Gudrun, the first four being floating production installations, the remaining steel jacket production installations. All installations are so recent that AIS data were available before and after installation, and all installations are isolated in the sense that they do not belong to a field with several installations located together.

9.3.4.1 Example

An illustration is presented later in the section, for an assumed field installation, referred to as 'XXXX'. For illustrative purposes, one route, and one vessel category are picked out in order to present some of the input data and specific results. The data for the actual route is presented in Table 9.4, Figs. 10.5 and 10.10. It is shown that the majority of the traffic is small vessels i.e., below 1500 dwt. Since the quantitative BBN diagram in the new COLLIDE® model is not available, the example is completed with the previous version of the COLLIDE® model.

For the illustration, the second size category is picked out, vessels 1,500–4,999 dwt, which again is broken down into three subcategories. The distributions for speed categories, flag of registration and shuttling frequency are also shown.

For illustration of other input data, the following may be mentioned (please note that the data is only valid for the stated vessel size category):

- For vessels which pass up to once per year:
  - 80% will know about the platform
  - 10% update their charts regularly
  - 40% will exercise avoidance planning, with a safe distance of 2 nm
  - 15% will use the installation as a fixed navigation point.

Table 9.4 Input data for route Stadt–Tyne/Tees

Merchant	<1500	1500–4999	5000–14999	15000–39999	≥40000
Vessels/year	113	25	0	1	1
Lgt	0.333333	0.333333	0.333333	0.333333	0.333333
Med	0.333333	0.333333	0.333333	0.333333	0.333333
Hvy	0.333333	0.333333	0.333333	0.333333	0.333333
Slow: 5.8 kn	0.100000	0.100000	0.100000	0.100000	0.100000
Ordn.: 9.7 kn	0.550000	0.550000	0.550000	0.550000	0.550000
Fast:13.6 kn	0.350000	0.350000	0.350000	0.350000	0.350000
Flag cat. A	0.650000	0.650000	0.650000	0.650000	0.650000
Flag cat. B	0.100000	0.100000	0.100000	0.100000	0.100000
Flag cat. C	0.250000	0.250000	0.250000	0.250000	0.250000
Pass/year < 1	0.650000	0.650000	0.650000	0.650000	0.650000
1 < P./y. < 4	0.120000	0.120000	0.120000	0.120000	0.120000
4 < Pass/year	0.230000	0.230000	0.230000	0.230000	0.230000

- 140 vessels per year, traffic in both directions, all year seasons
- Gaussian lateral distribution, with 6 nm standard deviation
- 1.35 nm distance to installation XXXX.

The input data have conditional probabilities for several aspects relating to vessels in the different size groups:

- Subdivision of sizes (light (Lgt), medium (Med), heavy (Hvy))
- Speed categories (slow, ordinary (Ordn), fast)
- Flags of registration (Categories A, B, C, see Table 10.5)
- Shuttling categories (< 1 pass/year, 1–4 passes/year, >4 passes/year)

The values in Table 9.4 are conditional probabilities which in each size category and for each aspect should sum to 1.0. The values are usually different for the different size categories, which is not the case for the values in Table 9.4.

The width of the platform is 18 m, almost perpendicularly on the lane heading (see Fig. 9.14). In a normalised Gaussian distribution, the following data is used for further calculation:

- 1.35 nm ~ 0.225 standard deviations
- 1.35 nm + 18 m ~ 0.22662 standard deviations.

For a Gaussian distribution, the following value may be found:

$$F(0.22662) - F(0.225) = 0.00063$$

This implies that the conditional probability of a vessel hitting the platform is 0.063%, when on a course line as stated, and with the standard deviation as given.

9.3.5 Probability of Failure of Ship Initiated Recovery

So far the situation before the vessel can actually observe the platform, either visually or on radar, has been considered. Up to this point, the vessel has therefore been totally reliant on charts and navigational equipment to determine its position relative to the platform.

A new situation arises once the platform is within the radar coverage zone of the vessel or the platform can be observed visually. At this stage, the situation changes from being based on the crew's previous knowledge of the area to a situation where any person, provided he or she can use radar, can detect the platform.

In this zone the previous division of the vessels into those who know the platform and those who do not, and those who plan their voyage in detail to avoid it and those who do not, starts to lose significance because all vessels now have a possibility of obtaining the same information and executing the same actions. This does not imply that the different vessel categories will necessarily act in the same way even after they have entered this zone, but all will have the same opportunities.

The outer limit of this zone will depend on the radar range of the vessels and possibly the visibility.

This range may vary considerably with the vessel size and the installed equipment. For small and medium sized vessels, it may, however, not be feasible to operate with a radar range greater than 10–12 nm because the radar antenna cannot be positioned sufficiently high to give reliable radar coverage outside this area. On larger vessels, which presumably will have a higher superstructure and the antenna higher above sea level, greater radar range may be used, although it is assumed that 12 nm is sufficient radar range also for these vessels. This assumption may be conservative for large vessels.

Captains were asked in a questionnaire survey [6] when they would change their course if they discovered that they were heading for a platform. The answers ranged from 'immediately' to 3–5 nm away from the platform. According to Haugen [6] this suggested that the 'normal' period for recovery to take place would be from the moment the platform was discovered until perhaps 3–5 nm away.

The ship initiated recovery is therefore divided into two 'phases', of which the initial normal recovery has been called 'early recovery'. Recovery during this period is in accordance with normal practice and does not represent an extraordinary situation. However, if recovery does not take place during this period, the implication is that the watch-keeping on the vessel is ineffective or non-existent. There are however, still 20 min (assuming a distance of 4 nm and an average speed of 12 knots) to recover from the situation, and the possibility that the situation is corrected in this period must also be considered, called 'late recovery'. Three main modes of failure of ship initiated recovery may be noted together with some of the likely causes:

- No reaction by the watch-keeper on the bridge (watch-keeping failure)

  - Absent from bridge
  - Present but absorbed in other matters
  - Present but incapacitated
  - Present but asleep
  - Present but incapacitated from alcohol
  - Ineffective radar use (bad visibility only).

- Erroneous action by the watch-keeper on the bridge
- Equipment failure.

The second item is probably less important, because the required action is obvious. Equipment failures leading to collision are also a low probability scenario and can probably be disregarded, with the exception of radar failure.

The six modes of watch-keeping failure listed above are mainly human failures. It is also possible that the navigator continues on the collision course on purpose. Excluding terrorists and suicidal persons, this may be due to other obstructions, platforms or vessels, which cause the navigator to choose to continue closer towards the platform before changing its course. Rather general failure models are often used for the probability of the different failure modes. More detailed and explicit models are therefore needed in this area.

Accident statistics clearly show that the probability of collisions between two ships and grounding or stranding increase when the visibility is reduced. The same trend can also be expected for collisions with platforms. The visibility conditions therefore constitute the most obvious factor which needs to be taken into account.

The visibility conditions may be accounted for by developing separate event trees for good and bad visibility conditions. The equipment on the vessel and how it is used will be of importance. If the vessel has only one, perhaps erratic, radar, the probability of collision in bad visibility will obviously increase unless the vessel takes other precautionary measures e.g., reducing the speed and thereby increasing the time available for discovering the platform and performing avoidance actions. On the other hand, there is no point in modern, sophisticated equipment if it is not used properly.

The efficiency of the radar is also affected by installation of RACON on the platform. A platform with RACON will return a radar signal which is easier to detect and identify than the signal from a platform without RACON.

The manning level on the ship is also important. There will be a considerable difference between a ship with only one person on the bridge compared to a vessel with two. This has an effect upon a number of aspects, but mainly it affects the probability of falling asleep. The manning level in general is largely related to ship size and type, and this may therefore be used as a factor in the evaluation of the risk level. But the number of personnel on the bridge may not be so dependent on size and type. Manning levels are also correlated with the so-called 'flag effect', which is considered as shown in Table 9.5.

Another factor which also may be considered is the general activity which takes place in an area. If a platform is located in an area where there are many other platforms and where there are many other limitations on navigation it is likely that the watch-keeper has a higher awareness level and is generally more alert compared to a situation where the platform is in an isolated location and where there are no other restrictions on vessel travel. Typical values for ship initiated recovery are shown in Table 9.6.

The values for platform initiated recovery and ship initiated recovery were originally derived in a coarse manner in the late 1980-ties, without a study that involved modern MTO or HRA analysis techniques. The values for are thus in need of reassessment and revision.

Table 9.5 Flag effect categories

Category	Countries considered to belong to category
A	Denmark, France, Germany, Holland, Italy, Japan, Norway, Spain, Sweden, UK, US
B	Cyprus, Greece, Liberia, Panama
C	Other flags

Table 9.6 Typical probabilities for recovery failure

Recovery failure mode	Failure probabilities for recovery failure	
	By ship	From platform
Alcohol	1.1×10^{-4}	0.72
Asleep	2.3×10^{-3}	0.2
Accident	1.2×10^{-5}	0.75
Absent	9.5×10^{-4}	0.012
Distracted	9.5×10^{-4}	0.01
Radar	9.8×10^{-3}	0.18

9.3.6 Probability of Failure of Platform Initiated Recovery

The final stage of collision avoidance is called platform initiated recovery. The probability of failure of this is easier to handle than ship-initiated recovery. The incoming vessel is not regarded as a threat to the platform until it reaches the 20 min limit. We can in this case also assume that there will be limited differences between the planning, non-planning and unknown vessels. The ability to perform Platform Initiated Recovery is mainly based on whether the actions including the following are performed in time:

- Identification of the vessel as a possible threat
- Attempt to call the vessel on VHF
- Position the standby vessel alongside the vessel
- Undertaken correct avoidance action by the standby vessel.

As noted earlier, the platform initiated recovery and ship-initiated recovery are strongly correlated. The probability of platform initiated action will therefore be highly dependent upon the reason for the failure of the ship initiated recovery [6].

Typical values for ship initiated and platform initiated recovery are presented in Table 9.6. It should be noted that the failure probabilities of platform initiated recovery are conditional probabilities i.e., given that failure of ship initiated action has already failed.

In the Norwegian sector, a majority of the installations are being monitored with respect to vessels on potential collision course, from a ship traffic surveillance centre. There are two such centres, one offshore for the southern Norwegian North Sea, and one onshore, at Statoil's (now Equinor) Bergen operations office, for the Northern Norwegian North Sea and Norwegian Sea. The latter has the highest number of installations being surveyed, and probably also the highest reporting reliability. The radar signals from the installations are transmitted on fibre cable to the centre, where they are combined into a common presentation of large areas. This means that the radar operators in the traffic centre may follow the vessels for long distances, especially in the North Sea. With AIS, the identification system, they will also know the names and other details of the vessels.

9.3.7 Example Results

The results for Installation XXXX and Route 6 are shown in Table 9.7, for different subgroups of size category 1,500–4,999 dwt, and different speed categories.

It should be noted that the previous version of COLLIDE® has been known to produce too high results, the new version gives results [9] with significantly lower collision frequency.

9.3.8 COAST®

The COAST® database [10] is a route database that was started for the North Sea, and was first released March 1996. It has since undergone continuous updating to ensure that any changes in the traffic pattern and density are detected. An increased

Table 9.7 COLLIDE® results for Route 6

Coll.typ Traffic cat	Vessel	Weight (DWT)	Speed (knots)	Coll. Energy (MJ)	Coll. Freq. (n/year)	Cum. Freq. (n/year)
Head on Route # 1	Merchant	3800–4999	13.61	128.01	6.5E–006	6.5E–006
Head on Route # 1	Merchant	2600–3799	13.61	94.32	6.2E–006	1.3E–005
Head on Route # 1	Merchant	3800–4999	9.72	65.31	1.0E–005	2.3E–005
Head on Route # 1	Merchant	1500–2599	3.61	57.94	6.0E–006	2.9E–005
Head on Route # 1	Merchant	2600–3799	9.72	48.13	9.7E–006	3.9E–005
Head on Route # 1	Merchant	1500–2599	9.72	29.56	9.4E–006	4.8E–005
Head on Route # 1	Merchant	3800–4999	5.83	23.51	1.9E–006	5.0E–005
Head on Route # 1	Merchant	2600–3799	5.83	17.33	1.8E–006	5.2E–005
Head on Route # 1	Merchant	1500–2599	5.83	10.64	1.7E–006	5.3E–005

Energy categories: (MJ)

	Categ.1	Categ.2	Categ.3	Categ.4	Categ.5	
Traffic category	0–15	15–50	50–100	100–200	>= 200	Total
Route # 1	1.7E–006	2.3E–005	2.2E–005	6.5E–006	0.0E+000	5.3E–005

number of offshore installations and changes in port activity might have large effects on the traffic pattern in an area. The COAST® database utilises MapInfo® as the GIS (Geographical Information System) platform and digitalised nautical ARCS® charts for presentation of route and vessel information. The database is based on data from several different sources, including:

- Lloyd's Port Log
- Port statistics
- Statistics in ferry traffic
- Data from coastal radars
- Information from operators of ships and offshore installations
- Information from pilots
- Data from offshore radars
- Data from land based and offshore AIS-systems
- Offshore traffic surveys.

The initial release of the COAST® database focused on the UK part of the North Sea. In order to extend the database to Norwegian waters the Norwegian Oil and Gas Association and Safetec started a project in 2001 to map the regular shipping traffic within Norwegian waters (COAST® Norway Database v2002). This project included collection of radar data from fixed offshore installations, data on offshore traffic from operators, collection of voyage plans for larger tankers and extended port statistics. The database currently holds information about regular shipping traffic, including:

- Volume of traffic on each route
- Route standard deviations (defines the width of each route)
- Vessel type distribution on each route (general cargo, bulk, tanker, container, RoRo, supply, standby and ferry)
- Size, flag and age distribution of vessels on each route.

As COAST® holds information about regular shipping traffic the database does not cover non-regular traffic such as fishing vessels, naval vessels, pleasure crafts and traffic related to mobile offshore installations.

Safetec has been granted access to historical AIS data for the North Sea and the Norwegian coastline in order to increase the accuracy of the COAST® database. The AIS system has a significantly larger coverage area compared to a standard radar system, as AIS utilises VHF radio signals instead of radar. The AIS data also holds information that enables tracking of a vessel through a larger area, in addition to providing more detailed information about the ships in each route. Collectively these give wide opportunities related to increasing the accuracy of the database, and thereby establish a more correct picture of the traffic pattern in the North Sea and Norwegian Waters, in addition to easing traffic pattern updates. COAST® is at the time of writing being updated. A plan for regular collection of AIS data has been establish for future updates of the COAST® database.

The COAST® database is unique in its worldwide coverage, at present the following marine areas:

- UK waters
- Irish sea
- Norwegian waters
- The Netherlands
- The Faeroe Islands
- Gulf of Mexico
- Mediterranean
- Straits of Hormuz
- Singapore Strait
- Gulf of Paria (Venezuela).

9.3.9 Traffic Monitoring in the Norwegian Sector

There are two traffic surveillance centres for ship traffic monitoring based on electronic transfer of radar signals, these cover well over 90% of the installations on NCS. One centre is located in the Ekofisk field, covering all installations in the southern Norwegian North Sea. Statoil (now Equinor) has a centre in Bergen, which cover all Statoil installations on NCS, production as well as mobile installations. They also perform traffic surveillance for some other operators.

It has been observed that the surveillance performed by these centres is reliable, efficient and imply a long warning time with ample opportunity to contact the ships in order to make them change course before the installation has to start preparations for shutdown and evacuation [11].

The experience is clearly that where the installation itself or its standby vessel has the responsibility for detection of vessels on collision course, vessels are rarely detected on collision course. As there is no reason to assume that there are fewer vessels on collision course in these areas, the implication is that such surveillance is less reliable. It would be highly preferable if all installations had traffic centres to take care of the ship traffic monitoring.

Virtually no ships at all were detected on collision course before these centres were started up. There were 40 vessels registered annually on collision course on NCS in 2003, whereas the annual value the last three years has been in the range 15–20 per year [12]. These are incidents that satisfy the following criteria:

- Vessels that have not replied to radio message when time to closest point of approach, $t_{cpa} = 25$ min, alternatively if standby vessel or SAR helicopter has been mobilized
- Fishing vessel fishing and pleasure boats are disregarded.

The installations have implemented a practice implying that all personnel are mustered in their lifeboats, if the vessels are contacted and have not replied at t_{cpa} = 50 min, which is required in order for the lifeboats to be ready to launch at t_{cpa} = 25 min. So far, lifeboats have not been launched in any incident.

The clear reduction in annual reports over a period of just over five years is probably due to an effort by Statoil (now Equinor) traffic centre personnel to inform especially fishing vessel crews about the consequence of using offshore installations as way-points in their navigation. The consequence may easily be that all personnel are mustered in their lifeboats until the contact with the fishing vessel is established and a course change is implemented.

9.3.10 Model Validation

The proceeding discussion is to a large extent based on the COLLIDE® approach, which also appears to be most widely accepted in the industry. The main weaknesses of the COLLIDE® and COAST® models are as follows:

- The model for failure of ship initiated recovery is quite detailed and describes explicitly many different failure mechanisms. However, the quantification is based on a very limited data set with generic data and many assumptions. Considerable uncertainty is therefore inherent in the model.
- The quantification of the new COLLIDE® model is based extensively on expert input, which still leaves significant uncertainty.

Earlier, the position of the routes was considered to be a significant weakness of the model. The definition of route terminals (especially the so-called macro positions) is very critical when a route is passing near to an installation. The location of these points is uncertain. Improvements have been made by limiting the use of macro positions significantly, in order to reduce the importance of this aspect. Further, it is common practice to supplement the information from COAST® with radar survey data (if available) and/or AIS-data. This is used to verify the position of the routes, and also to adjust them as necessary. This is therefore not considered to be a key weakness of the model as such, but if the COAST® database is used uncritically, this may introduce considerable uncertainty.

Model validation is discussed in some detail by Dr. Hassel in his thesis [9].

9.4 Attendant Vessel Collision

A Joint Industry Project (JIP) was conducted in order to develop methodology for collision by attendant vessels [13], financed by Equinor, ConocoPhillips, Vår Energi (formerly ENI) and Floatel International, who have provided access to the

report from the work. It should be noted that there are some reservations related to
the proposed approach. The main objectives of the work have been to:

- Identify current collision models and the necessary parameters for assessing the
  selected collision scenarios for vessels visiting offshore facilities.
- Compare and evaluate current collision models and the necessary parameters.
- Conclude and recommend best practice methodologies and data sources.
- Further develop the collision models.

There are three phases for all types of attendant vessels:

(a) Arrival on the field.
(b) Approach within the safety zone.
(c) Loading/off-loading/operation on the field.

A generic fault tree for attendant vessels collisions was developed in the project,
(see Fig. 9.5).

The project has also developed a generic barrier diagram for prevention of
collisions by attendant vessels (see Fig. 9.6), as well as a generic timeline for
collision scenarios (see Fig. 9.7).

The project has adapted the generic models to the different types of vessels and
collision scenarios (see [13]).

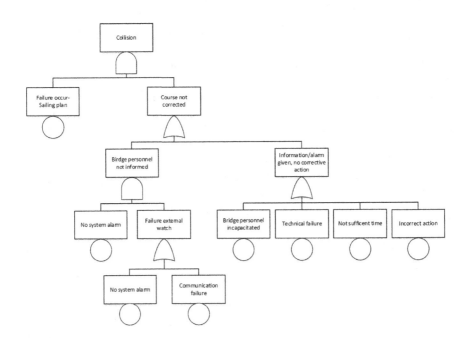

Fig. 9.5 Principal fault tree for attendant vessel collision [13]

Fig. 9.6 Overview of barrier functions [13]

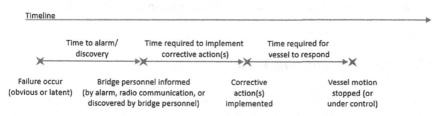

Fig. 9.7 Overview of generic timeline for collisions [13]

9.5 Collision Energy

9.5.1 Impact Energy and Platform Energy Absorption Capacity

The minimum collision design criteria for Norwegian offshore structures [7] correspond to an impact energy of 11 MJ for head-on collisions and 14 MJ for sideways collisions (5000 tons, 2 m/s, added mass of 10% and 40% respectively). The equation for calculating the kinetic energy is as follows:

$$E = \frac{1}{2}(m + a_m) \cdot v^2 \qquad (9.3)$$

where

m = mass of the vessel.
a_m = hydrodynamic added mass.
v = impact velocity.

It may be seen that a head-on collision with a 1000 ton vessel with added mass of 10% and a velocity of 10 knots would give an impact energy level just below the specified criterion. In other words, it is likely that a powered collision by a small merchant vessel may be inside the design criterion. Further, calculations of over-capacity in relation to collision and reserve strength implied by the design rules have shown that platforms may be able to tolerate considerably higher impact energies than the design criterion specifies, before global collapse occurs. For semi-submersibles, energy absorption capacities in the range 60 MJ have been found and for large concrete platforms the impact resistance is likely to be considerably higher. On the other hand, for smaller and less robust platforms e.g., small jackets, the capacity may be close to the design criterion.

Thus it may be argued that the smallest vessels do not represent any real hazard to a platform, even at full speed. This may be true, at least for large, integrated platforms on the Norwegian Continental Shelf. Small platforms which are less capable of withstanding collisions will on the other hand become more common in the future. Collision design criteria are also different in different countries.

9.5.2   Mass of Colliding Vessels

Vessels are often divided into different size groups, defined on the basis of the vessels deadweight (dwt). The mass of the vessel is expressed by the displacement. The problem with estimating the displacement is that it is varying, depending on whether the vessel is loaded or in ballast. In addition to the mass of the vessel itself, the hydrodynamic added mass must also be included.

9.5.3   Impact Velocity of Colliding Vessel

9.5.3.1   Powered Impacts

The velocity categories of the vessels should be calculated on the basis of typical service speeds for the different vessel categories. In addition to estimating an average speed, a certain distribution within each size category should be considered.

9.5.3.2   Drifting Speed of Disabled Vessels

When a vessel is disabled and lies idle in the sea, it will drift subject to the forces of the wind, waves and current. There will be a transient phase, during which the forces act on the vessel and turn it onto a different heading and the path that the vessel takes changes. This is then followed by a steady drift phase in which the vessel drifts with a set course and speed and a constant heading (unless wind and current headings are changing).

The drifting speeds in the transient phase are of limited interest since the distance the vessel travels before reaching its steady state condition is relatively short. The probability of a collision in this phase is therefore small. Drifting speeds are likely to be between 2 and 3 knots (less than 2 m/s, see Table 9.8).

Table 9.8 The mean value of the drifting speed (knots)

Vessel categories	Wind speed (Beaufort)						
	3	4	5	6	7	8	9
Merchant vessels	0.9	1.1	1.4	1.7	2.1	2.6	5.0[a]
0–1499 dwt	0.9	1.1	1.4	1.7	2.1	2.6	5.0[a]
1500–4999 dwt	1.1	1.3	1.7	2.1	2.5	2.3	5.0[a]
5000–14999 dwt	0.9	1.2	1.4	1.7	2.1	2.3	5.0[a]
15000–49999 dwt Over 40000 dwt	1.0	1.2	1.5	1.7	2.0	2.1	2.2
Fishing vessels	0.9	1.1	1.4	1.7	2.1	2.6	5.0[a]
Standby and supply vessels	1.1	1.3	1.6	1.9	2.3	3.1	5.0[a]
Shuttle tankers	1.0	1.2	1.5	1.7	2.0	2.1	2.2

[a]Conservatively calculated

9.5.4 Critical Collisions

There are many factors that may influence the collision consequences, depending upon the design of the structure, the actual detailed path followed by the vessel, and geometrical considerations. Among the most important are the following:

- Design/dimensioning principles applied for the structure of the platform
- Fendering/reinforcement
- Platform topology
- Mass and velocity of colliding vessel
- Distribution of collision energy between vessel and platform
- Strength of colliding vessel
- Relative orientation of platform and vessel
- Location of point of impact.

Some of these factors can be defined as clearly concept independent, (e.g. the mass and velocity of the colliding vessel), whereas others, like the design and dimensioning of the platform, are clearly dependent on the concept.

The most important factors in this list are probably the mass and velocity of the colliding vessel i.e., the kinetic energy. A simple and convenient assumption to apply is the principle of energy conservation i.e., that the kinetic energy arising from the vessel's movement prior to the impact is distributed between kinetic energy after the impact (may in principle be both vessel and installation) and deformation (elastic and plastic) energy. This principle should be used with care because the differences in design of ships are large. Thus some ships will absorb a larger proportion of the kinetic energy than other ships. It should be remembered that all the kinetic energy will only be absorbed in a head on impact which brings the ship to a standstill. For a glancing impact part of the impact energy will be retained as kinetic energy, implying that even higher initial impact energy is necessary to cause a critical collision.

During the collision all forces except collision forces and inertia forces are neglected. Assuming a central impact, the amount of energy to be dissipated as strain energy is given by:

$$E = \frac{1}{2}(m + a_m)\, v^2 \frac{\left(1 - \frac{v_i}{v_v}\right)}{1 + \frac{m + a_m}{M + A_m}} \qquad (9.4)$$

where

m	= vessel displacement
a_m	= vessel added mass
M	= installation displacement
A_m	= installation added mass
v_v	= vessel [impact] velocity
v_i	= installation velocity [after impact].

9.6 Collision Consequences

For a given impact energy the platform consequences are mainly dependant on the platform concept and the impact scenario. Some of the most important factors were listed in Sect. 9.5.4 above.

The critical impact velocity for different levels of damage is dependent on size of the vessel. Typical critical velocities for global structural failure are shown in Table 9.9.

Table 9.9 Critical impact velocities for global failure

	Colliding vessel	Vessel displacement (dwt)	Critical velocity (m/s)	Critical velocity (Knots)
Head-on	Fishing vessel	700	8.8	17.1
	Supply vessel	2,500	4.7	9.1
	Merchant vessel	10,000	2.3	4.5
	Semi-submersible	50,000	1.0	1.9
	Attendant vessel	2,200	5.0	9.7
Sideways	Attendant vessel	2,200	4.4	1.9
Stern	Attendant vessel	2,200	5.0	9.7

9.6.1 Failure Criteria

The failure criteria have to reflect the purpose of the study. Different criteria will be chosen if only the structure is being considered rather than the entire platform concept. The failure criteria most commonly selected for the structure, are:

1. Global failure: The impact leads to large deformations of the structure requiring personnel to be evacuated and operations to be shut down.
2. Local failure: The local stresses exceed the yield strength, and a plastic, irreversible deformation occurs. The platform's integrity is not threatened.

If consequences to the platform as a whole are considered, then there may be a requirement for additional failure criteria, for instance in relation to impact on risers or well conductors. These may be different from the global failure, especially for a steel jacket structure. The following sections address these scenarios.

9.6.2 Collision Geometry

The collision geometry may decide the distribution of energy between the vessel and the platform, reflecting whether hard spots are hit, what residual kinetic energy is retained in the vessel, etc. For a steel jacket structure, the following distinctions should be made:

• Hit of vertical column or bracing(s)	Hitting the bracings will result in larger plastic deformations, and thus higher energy absorption by the structure
• Central impact or 'glancing blow'	Considerable kinetic energy may be retained by the vessel, if the hit is a glancing blow, possibly resulting from last minute evasive actions
• Rotation of vessel	May transform energy to rotational, and thus limit the energy required to be absorbed by the platform structure
• Contact point on vessel	The contact point on the vessel may be important, especially for impact with concrete columns. If the contact point is a 'hard spot' on the vessel (e.g. heavily framed curvatures, such as bulb, stern, or edge) high puncture loads may be generated

9.6.3 Local Collision Damage

Assessment of damage to steel structures is considered extensively in the NORSOK Standard for Design of Steel Structures [14] Annex A, Design against Accidental Actions is devoted to accidental loads:

- Ship collisions
- Dropped objects
- Fires
- Explosions.

The following are aspects considered by the NORSOK Annex A in relation to consequences of ship collisions:

- Collision mechanics
- Dissipation of strain energy
- Collision forces
- Force–deformation relationships for denting of tubular members
- Force-deformation relationships for beams

    - Strength of connections
    - Strength of adjacent structures.

- Ductility limits
- Resistance of large diameter, stiffened columns
- Energy dissipation in floating production vessels
- Global integrity during impact.

9.6.4 Global Damage

For steel jacket platform s with four legs, global failure is often assumed to occur if one leg loses its load carrying capability. Analysis of the structure is normally needed on a case-by-case basis. Often other loads have determined the design of the leg and therefore there will be an in-built reserve capacity. Typically, global failure loads may occur above 20 MJ.

The collision energy causing global failure of a steel jacket platform with six or eight legs, is expected to be larger than for a four legged platform. In these cases it is often assumed that two legs have to fail to cause global failure.

As a simple illustration, it may be assumed that the collapse resistance of the second leg is only 50% of the original impact resistance. Typically, the total collision resistance will then be in the order of 30 MJ. To account for the fact that the 50% assumption is probably conservative and the ship has also to deform bracing in addition to the two legs, 30 MJ is often chosen as the criterion for global failure for six or eight legged jackets.

9.7 Risk Reducing Measures

9.7.1 Overview of Risk Reducing Measures

When risk-reducing measures are considered, account will have to be taken of which type of vessel represents the greatest risk to the installation. The effect of different risk reducing measures can be identified by considering the models that are used for quantification of collision risk for the different vessel groups. There are different models for passing vessels, nearby navigating vessels, drifting vessels and drifting floating units. The following discussion will be limited to passing vessel collision.

As mentioned in the previous section, it will be important to get a warning of a potential hazard as early as possible. This will apply to all vessel groups. A collision warning system on the platform will therefore be valuable. An ARPA type radar, with the antenna mounted high up on the platform, may start monitoring of vessels as much as 30–40 nm away from the platform. Installing an AIS station on board will improve the detection range because the AIS system transmits on the VHF band, and has a normal range of approximately 40 nm, with up to 60–80 nm during good conditions.

Warning of a vessel on a dangerous heading should not start around 40 nm distance, because there may be many vessels on a theoretical collision course at this distance, which plan to divert away from the platform at a later stage.

A study has been carried out by Safetec on behalf of the HSE Offshore Safety Division [15] to assess the benefits of a variety of collision control and avoidance systems under a number of different scenarios. Six different systems were reviewed to assess their benefits with respect to the reduction of collision probability. The systems investigated included:

- Standby vessel with standard marine radar
- Standby vessel with ARPA
- Installation-mounted radar early warning system
- Vessel traffic system (VTS).

The systems were assessed in varying environmental conditions and for different traffic pattern characteristics, e.g., high traffic density-high speed vessels, low traffic density–low speed vessels, etc. The probability of averting collision by having each of the systems in the different scenarios was determined. These probabilities may be used in order to calculate an overall risk reduction factor for other installations. The resulting level of risk reduction will be site-specific based on the characteristics of the location.

9.7.2 Passing Vessels

The different parameters in the model for collision by passing vessels are reviewed below with respect to possibilities for risk reduction.

• Traffic volume	Not practical to affect
• Probability of the platform being known	Can be increased by better distribution of information about the platform. Dedicated measures can be put into effect
• Probability of avoidance planning	Not likely that it can be affected significantly
• Probability of position-fixing planning	Not likely that it can be affected significantly
• Distance between lane centre line and platform	Can be affected by moving either the lane or the platform. Moving the lane will require international agreements (traffic regulations), and this is not feasible unless the risk is very high. Moving the platform is also often impossible, except in very occasional circumstances such as intermediate booster platforms and pipelines
• Lateral distribution of the lane traffic	Not likely that it can be affected significantly
• Probability of failure of ship-initiated recovery	Not likely that it can be affected significantly
• Probability of failure of platform-initiated recovery	Can be affected in a number of ways: • Calling vessel on VHF radio • Light and sound signals • Active use of helicopter • Active intervention by standby vessel

The possibilities to reduce the collision risk level are identified by reverting to the collision risk model, Eq. 10.2. Three parameters can essentially be affected:

- Probability of the platform being known.
- Distance between the shipping lane and the platform.
- Probability of failure of the platform initiated measures.

If the platform can be moved, this is probably the simplest measure, but this is not usually possible. Thus, the following measures are left:

- Improved efficiency of distribution of information about platform
- Procedures for warning incoming vessel of situation

    - Collision warning system
    - Calling vessel on VHF/radio
    - Use of light and sound signals on platform.

- Active use of standby vessel
- Active use of helicopter
- Availability of tugs.

9.7.3 Effect of Risk Reducing Measures

The different risk reducing measures, mainly relating to passing vessels, are briefly reviewed in the following sections, starting with a brief description of what each measure implies in terms of equipment, actions or procedures. A brief reference is made to the vessel groups that are affected, and parameters in the relevant collision model that are affected. A qualitative discussion of the measure is also included.

9.7.3.1 Collision Warning System

Collision warning systems have become relatively standard in the North Sea and surrounding areas. This is a radar based system which automatically gives a warning if a vessel is on a course which will take it close to the platform (i.e. maintains constant heading for the platform for a defined period). The system may be characterised as below:

• Vessel groups affected	To some extent all groups. Most important for passing vessels. System has limited use for visiting vessels
• Model parameters affected	No direct effect on the collision risk, but enables other measures to be put into effect earlier
• Discussion of effect	A dedicated collision warning system enables the platform crew to detect incoming vessels on a collision course far earlier and more reliably than would otherwise be the case. Such a system will follow all vessels which enter the radar coverage zone and can give warnings if any of the vessels have a heading which will take them too close to the platform. The warning limit can, for example, be set equal to the safety zone radius The increased warning time can be used to contact the vessel on VHF/ radio, to prepare a helicopter for take-off (if available), to position the stand-by vessel, etc. The possibilities of warning the vessel will be considerably increased. The longer time also implies that even if the watch-keeper is incapacitated, other members of the crew may be warned sufficiently early to perform the recovery In terms of the model for passing vessels, the effect is that the time to induce platform initiated recovery is increased from 20 to 30–40 min

The monitoring of vessel traffic around the majority of Norwegian installations has since 1998 been based on transmission of radar screen pictures from the installations to a shore based surveillance station. These traffic centres are continuously manned and perform the automatic and manual tracking for all the installations involved. It is believed that the reliability of such monitoring is very high.

The most recent development is that also AIS-data is integrated with the radar data, in order to achieve the most extensive identification and description of the vessels on possible collision course.

9.7.3.2 Distribution of Information About Platform

Improved distribution of information about the location and installation date of the platform is important to achieve the highest possible awareness. Various means can be used. This measure is applicable to production installations which are permanently in location over a long period, and mobile installations that have the AIS installed. The system may be characterised as below:

• Vessel groups affected	Passing vessels
• Model parameters affected	Probability of platform being known
• Discussion of effect	The probability of the platform being known generally increases sharply in the first few months after the platform has been installed. In some locations, there is therefore little to gain from improving the distribution of information. These measures will generally be effective to reduce the risk level only during the initial period after installation of the platform.

9.7.3.3 Warning the Vessel

This measure involves warning the vessel of its errant course either from the platform or from a standby vessel. Various means of warning an incoming vessel of the danger can be used. The primary means would be calling the vessel on VHF radio. Secondary means are foghorns, lights, etc. A standby vessel may also use its fire water monitors to create noise against the hull of the vessel, in order to notify personnel onboard.

• Vessel groups affected	Passing vessels and vessels navigating nearby
• Model parameters affected	Probability of failure of platform-initiated recovery
• Discussion of effect	With early warning of an incoming vessel, there will be much more time to put into effect various measures to warn the incoming vessel of the imminent danger

As a final option, to reduce the consequences of an imminent collision, shutdown of the platform and evacuation will ultimately have to be performed.

The following is a sketch of a procedure which could be used for collision warning and avoidance:

1. The automatic collision warning system will give a warning of a vessel on a collision course according to a predefined limit, often set to 45 min. This implies that the vessel may be tracked for some time (especially if monitored from a land-based station) from the time it enters the radar coverage zone. The

position, course, Closest Point of Approach (CPA), speed, and estimated time of arrival of the incoming vessel can be obtained from the radar-based warning system.

2. When the alarm is given, the standby vessel is immediately given the course and position of the vessel.
3. The first action from the platform and/or the standby vessel is to start calling the vessel on VHF radio. At the same time, the standby vessel starts to move into position between the platform and the vessel.
4. If there is no response from the vessel within the first few minutes of calling, the standby vessel starts steaming towards the incoming vessel. Before doing this, it must be assured that the standby vessel is able to travel at least at the same speed as the incoming vessel. At the same time, calling on VHF radio continues.
5. The standby vessel should be able to meet the incoming vessel around 20–30 min travelling time away from the platform. The standby vessel then takes up a position alongside the incoming vessel, as close as is regarded to be safe.
6. When in position, the standby vessel immediately starts using the various maritime sound and light signals which are available. This would include floodlights, white flares, foghorn etc.
7. If this gives no response after 5–10 min, further measures aimed at making noise against the hull of the vessel are put into effect. This may involve using fire monitors in order to make noise against the hull. This may in the extreme also involve firing small explosives or heavy objects against the hull, the aim being to raise the attention of not only the watch-keeper but any member of the crew.
8. Some 10–15 years ago, the prevailing thinking was that the standby vessel should prepare to try to deflect the incoming vessel, if there has been no response by the time the vessel is 15–20 min travelling time away from the platform. At this distance, even a deflection of less than 5° would be sufficient to avoid a collision. But unfortunate experience with actual contacts some years ago, has changed the thinking. It is currently unlikely that a standby vessel master will decide to make a physical contact.

The decision about contact will be the sole responsibility of the standby vessel's master, and his/her evaluation of the appropriateness of such an action. This action may involve serious structural damage to the standby vessel or the incoming vessel, depending on vessel sizes. The master's evaluation will have to involve the safety of both vessels and their crews.

With this procedure, there are several steps in the process which may cause the incoming vessel to discover that it is on a collision course. The effects of the different measures are therefore considered separately, anticipating that they are put into effect sequentially.

Recovery from a potential collision situation is divided into two effects in the collision model: Ship-initiated recovery and platform-initiated recovery. In the model, it is assumed that no warning system is present and that platform-initiated recovery only will take place during the last 20 min before the vessel reaches the platform. Before this, recovery is assumed to be initiated on the vessel itself.

When a collision warning system is in effect, this means that the platform-initiated recovery can start taking effect much earlier. In modelling terms, platform-initiated recovery will start having effect at the same time as the ship-initiated recovery is supposed to be affecting the possibility of collision.

9.7.4 Experience with Collision Avoidance

It is difficult to obtain data relating to experience with the use of field related resources to avoid threatening collisions. The data summarised below is therefore not a complete data set, but limited to what has been possible to establish from contacts with the companies involved. None of the sources are in the public domain. The experience discussed is not focused on fishing vessels, which for fishing purposes often navigate quite close to the installations, in fact often just outside the 500 m safety zone.

One of the important observations was that vessels that were contacted by radio, appeared to be less and less willing to respond with their identification and voyage particulars. This, however, has changed with the introduction of AIS.

In the Norwegian Sea (Haltenbanken) a German merchant vessel (about 2900 GRT) came very close to a collision on 24.6.1986, when the vessel was steaming for a long time directly towards a mobile drilling unit performing drilling operations. Both the drilling unit and the standby vessel tried repeatedly to contact the vessel on the radio, but no contact was established. The dangerous course was not altered until it was 200 m away from the installation, resulting in the safety zone being infringed. No other examples of collision avoidance actions are known.

One case is anecdotally known from the North Sea in 1982, where a standby vessel made physical contact outside the safety zone with a fishing vessel, in order to deflect its course away from the installation. The responsible oil company in charge ended up having to pay for the damage to the fishing vessel.

In one case where the standby vessel was mobilised, the standby vessel took up position immediately in line with an approaching large trawler, but the trawler kept coming towards the standby vessel without responding to radio contact attempts. Only when the standby vessel lit up its floodlights did the trawler alter its course.

A collision in the German sector occurred in 1995 (see Sect. 9.1.2.2), in which a small German freighter made contact with a pipeline riser platform (Norpipe H7 platform). There was virtually no damage to the installation and limited damage to the vessel. The following is a brief summary of the events and associated timing of the steps that occurred.

ca. 10:00   The vessel is detected by the standby vessel, whose master concluded that based upon manual observations the CPA would be 800–900 metres to the west side of the installation.

ca. 10:35   The standby vessel crew claim that the vessel altered course directly towards the installation at this time, however, this is not confirmed by the platform crew.

ca. 10:37   The standby vessel starts heading towards the approaching vessel using its sirens continuously (800 m distance).

ca. 10:42   General alarm and shutdown are initiated on the platform just seconds before the vessel hits the installation just scratching along one of the legs. The standby vessel is 50 m away from the merchant vessel when the contact occurs.

The avoidance actions were not successful in this instance and the procedures for warning and fending off threatening vessels have been changed by the operator of the platform based on this instance.

9.7.5   Illustration of Effect of Risk Reduction

The previous pages have illustrated that the effects of many of the risk-reducing measures are generally difficult to quantify, because there will be individual differences between platforms, locations and the traffic picture which will have to be taken into account.

For several reasons the potentially largest problem may be passing vessels. In many high risk locations, they are the most important contributor. Further, if other groups are more important they are usually all vessels which are related to the operation of the platform in some way, e.g. supply vessel s or offshore tankers. The risk associated with these vessels can usually be controlled more easily than the risk due to merchant vessels, mainly through operational restrictions or improved design. The same measures cannot be applied to merchant vessels.

Based on this, it can be concluded that the most important risk-reducing measure, at least to reduce the risk of merchant vessel collisions, will be the use of a collision warning system and procedures to give warning of an incoming vessel on a collision course.

To exemplify the effect, calculations have been performed for an actual platform in the North Sea, using the COLLIDE® model. This is a riser platform, and therefore the number of supply vessel visits is relatively limited. There are no other fields in the vicinity which would result in large activity of other offshore vessels near the platform. The fishing activity is significant, but fishing vessels are generally too small to give collision energies which represent any risk to the platform. We are thus left with passing merchant vessels as the main contributor.

In practice, there are two measures which can be effective in reducing the risk of collision with merchant vessels: Improved distribution of information about the platform and a collision warning system combined with procedures for warning the vessel.

Improved distribution of information is practically irrelevant as the platform has been on location for a number of years. The probability that the platform is known has therefore reached the 'plateau' level already (this is not the same example platform as 'XXXX' in the following section).

It may also be noted that even if the platform had been new, it would have been difficult to do very much with the distribution of information about the platform. Reaching the vessels directly before they leave port is difficult as the traffic is a combination of vessels from many different ports. The only foreseeable possibility would therefore be to send regular warning signals from the platform itself. The effect of this is difficult to assess.

The only measure left is thus a collision warning system in combination with procedures. An example calculation has been made for this installation, with and without the collision warning system. The total collision frequency is reduced by a little over 50% by the introduction of the collision warning system and procedures for warning the vessel. The initial collision frequency of 1.2×10^{-3} is reduced to $5.4 = 10^{-4}$ by these measures.

9.8 Collision Risk Case Study

9.8.1 Installation

Table 9.10 shows the input parameters for the installation 'XXXX'. The platform is considered as recently installed ('0 quarters' at the field). It is considered as an unmanned installation with no risk reducing measures implemented. The platform is a steel jacket with four legs, 18 m apart. In the input there is also the possibility to specify whether the platform is permanently manned and whether it is a fixed or a floating installation

Figure 9.8 shows a simplified section of the side view of the jacket structure. In addition to the trusses and columns, the conductors and risers are also shown.

Table 9.10 Input characteristics for platform XXXX

Installation number: 1								
Name: XXXX								
Position and dimension:								
Latitude			Longitude			Width (m)	Length (m)	Direction
59°	44'	5' N	2°	33'	26' E	18.00	18.00	45.00°
Season: All year								
Category: Permanent								
Quarters at location: 0								
Manned installation: No								
Risk reducing measures: No								
Floating unit: No								

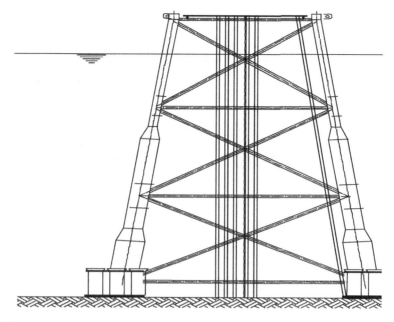

Fig. 9.8 Jacket structure side view

The case study is as noted above made with the previous version of the COLLIDE® software.

9.8.2 Routes

Using the COAST® database as input, COLLIDE® 2.60 identifies 13 routes, in the vicinity of the platform. Seven of these are routes of commercial shipping, and six are offshore related traffic i.e., supply vessel and shuttle tanker. The commercial routes (with annual volumes) are as given below (please note that this is not from the current version):

1. Lervick—Stavanger 35
2. Baltic—Seydhisfjordur 18
3. Kinnairds Head—Bergen 9
4. Scarborough—Bodø 9
5. Leith—Bergen 18
6. Stadt—Tyne/Tees 140
7. Lindesnes E—Iceland 120.

Table 9.11 shows data for the first route, Lervick—Stavanger. The table shows how the traffic is distributed over the year, shows terminal points for the route, distance to the installation, and a size distribution for the vessels. For each vessel size category there are three subgroups for each of the following parameters:

Table 9.11 Route characteristics for Lerwick–Stavanger

Route no.: 1
Name: Lerwick–Stavanger
Season: All year
Distribution: Gaussian
Direction: Both ways

Positions in path	Latitude			Longitude			Std.dev. (nm)
Lerwick	60	5'	0'N	1	5'	0'W	2.00
Stavanger	59	1'	0'N	5	17'	0'E	2.00

Minimum distance to XXXX (inst. no. 1): 12.21 (nm)

Merchant	<1500	1500–4999	5000–14999	15000–39999	≥40000
Vessels/year	26	9	0	0	0
Lgt	0.000000	1.000000	0.333333	0.333333	0.333333
Med	0.333333	0.000000	0.333333	0.333333	0.333333
Hvy	0.666667	0.000000	0.333333	0.333333	0.333333
Slow: 5.8 kn	0.100000	0.100000	0.100000	0.100000	0.100000
Ordn.: 9.7 kn	0.550000	0.550000	0.550000	0.550000	0.550000
Fast: 13.6 kn	0.350000	0.350000	0.350000	0.350000	0.350000
Flag cat. A	0.333333	1.000000	0.333333	0.333333	0.333333
Flag cat. B	0.333333	0.000000	0.333333	0.333333	0.333333
Flag cat. C	0.333333	0.000000	0.333333	0.333333	0.333333
Pass/year < 1	0.650000	0.650000	0.650000	0.650000	0.650000
1 < P./y. < 4	0.150000	0.150000	0.150000	0.150000	0.150000
4 < Pass/year	0.200000	0.200000	0.200000	0.200000	0.200000

- size within category
- speed
- flag
- number of passes per year.

The routes with offshore related traffic are the following:

- Frigg/Heimdal—Stavanger 104
- Odin—Kårstø 104
- Statfjord/Gullfaks—Rotterdam 280
- Statfjord/Gullfaks—LeHavre 80
- Statfjord/Gullfaks—Thames 100
- Brent—Rotterdam 100.

The first two are supply routes, while the other routes are shuttle tankers. Both heading and closest distance are important for the analysis. The closest of the commercial routes is Kinnairds Head—Bergen (0.74 nm), but the traffic volume on this route is low. The closest of the offshore routes is the supply traffic to/from Frigg/Heimdal (1.30 nm).

Figure 9.9 presents a sketch of the route headings. Please note that the main purpose of the map is to show the headings of the routes and thus distances between the routes and the installation should not be inferred from the sketch.

Two other installations that are close to XXXX are also shown as INST–1 and INST–2. These are manned installations with standby vessels present at all times, available for interception of incoming vessels. Please note that the collision risk of these nearby platforms is not considered in this example. The actual output from COAST® for the same area as in Fig. 9.9 is shown in Fig. 9.10.

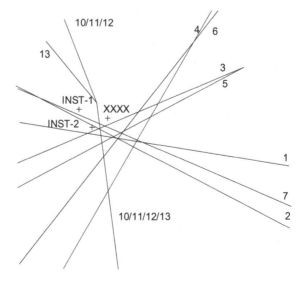

Fig. 9.9 Route sketch for traffic around installation XXXX

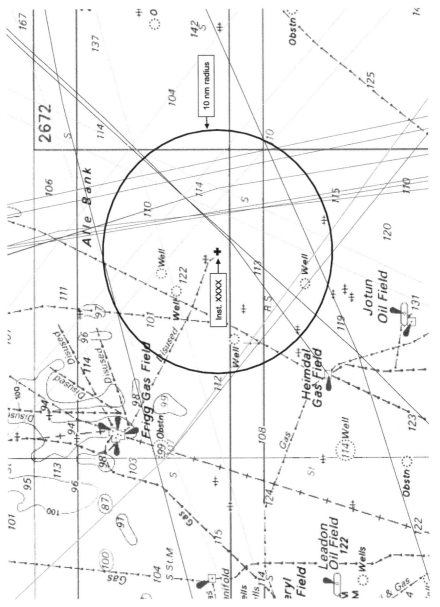

Fig. 9.10 COAST® traffic output for the same area around installation XXXX

Table 9.12 Collision results for platform XXXX

Collision mode	Annual frequency
Head-on	6.6×10^{-4}
Drifting	1.3×10^{-5}
Total, all modes	6.7×10^{-4}

9.8.3 Results

The overall results are shown in Table 9.12. It can be clearly seen that head-on collision by errant or blind vessels completely dominates the risk picture.

The following are the contributions to overall collision risk:

- Merchant vessels: 56.6%
- Supply vessels: 39.7%
- Shuttle tankers: 3.7%.

Only the head-on collisions are considered further in this example as they represent more than 98% of the overall risk.

9.8.4 Energy Distributions

The distribution of frequency against energy may be stated for all colliding vessels or for each vessel type separately. The distribution for all colliding vessels is shown in Fig. 2.14 in Chap. 2. It may actually be more illustrative however to consider the distributions for each type of vessel separately, because the distributions are rather different. Figures 10.7–10.9 present these distributions (Figures 9.11, 9.12, 9.13).

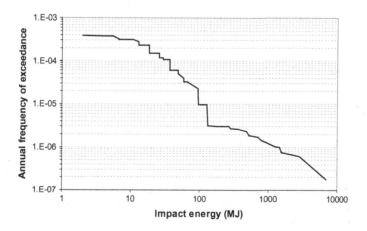

Fig. 9.11 Exceedance curve for merchant vessels

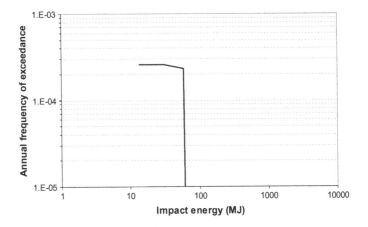

Fig. 9.12 Exceedance curve for supply vessels

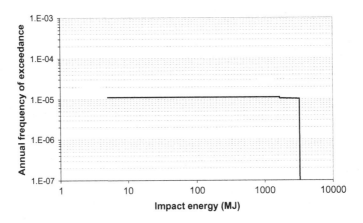

Fig. 9.13 Exceedance curve for shuttle tankers

9.8.5 *Intervention Options*

This subsection is concerned with the possibility of reducing the collision risk for the unmanned installation **XXXX**. Effective radar warning is assumed for this exercise, without considering how such warning could be achieved technically.

The following is a brief discussion of possibilities of how the standby vessels from INST–1 and INST–2 could be used to intercept (not necessarily physically) errant or blind vessels. This assumes that warning is possible at a distance of approximately 20 nm, (i.e. around 90 min prior to potential collision in the case of the fastest moving vessels).

Route 2	Southeasterly heading	INST–1 (also INST–2) standby is available to intercept vessels.
	Northwesterly heading	INST–2 standby should be able to intercept the approaching vessel in time.
Route 3	Southwesterly heading	None of the standbys are favourably positioned to intercept vessels.
	Northeasterly heading	INST–2 standby is available to intercept the approaching vessel in time.
Route 6	Southwesterly heading	None of the standbys are favourably positioned to intercept vessels.
	Northwesterly heading	INST–2 standby is available to intercept the approaching vessel in time.
Routes 10, 11, 12, 13	Southeasterly heading	INST–1 standby is available to intercept vessels.
	Northeasterly heading	INST–2 standby is available to intercept vessels.

The consideration above does not address whether a standby vessel is physically able to intercept a shuttle tanker. For a shuttle tanker, however, radio contact should be sufficient, because they must be assumed to have crew on the bridge at all times.

For the other routes, the possibilities to intercept (such as by blocking the course line) are good, considering the following vessel size information:

Route   2 All vessels less than 5000 dwt
        3 All vessels less than 1500 dwt
        6 98% of traffic less than 5000 dwt, 2% of traffic above 15000 dwt.

We have assumed equal number of passes in each direction (COAST® actually has route data on this aspect). Thus, half of the volume in routes 3 and 6 could not be intercepted effectively, due to long steaming time for standby vessels.

If it is assumed simplistically that the possibility to intercept (not physical, but for instance by blocking) eliminates the risk, then the scenarios which still represent a risk are the following:

Route 3. Southwesterly heading
Route 6. Southwesterly heading.

Using this approach the risk from both directions in Routes 2, 10, 11, 12, 13 and in the northeasterly direction in Routes 3 and 6 is eliminated.

9.8.6   Collision Geometry

Fig. 9.14 shows the orientation of the main support structure in relation to Route 6, the dominating route. It is noted first of all that the route is almost parallel to one of the sides. The marginal angle is disregarded.

The waterline section is 18 × 18 m. The two route centre lines show the middle line and the extreme position which will result in a head-on collision with the platform. This distance 'b' is calculated as:

half platform width + quarter of vessel breadth = 9 + 3 = 12 m

Typical vessel breadth = 12 m, for 1500–2500 dwt. For the half width shown, the central zone causing impact against braces only, is 3 m. The total hit zone is 18 + 12 = 30 m. This can now be split into the following partial zones (breadth is measured across the route centre line, negative values on the starboard side, positive values on the port side of the centre line):

- −15 to −12: Glancing collision
- −12 to −3: Corner column collision
- −3 to +3: Bracing collision
- +3 to +12: Corner column collision
- +12 to +15: Glancing collision.

The probability is assumed constant across the total zone. Thus given that a collision occurs, the conditional probabilities of the collision scenarios are as follows:

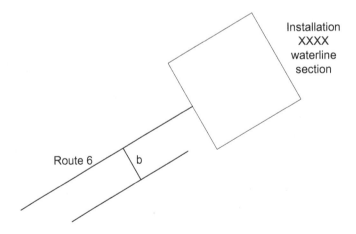

Fig. 9.14 Route 6 in relation to orientation of structure

- Glancing collision: 20%
- Column collision: 60%
- Bracing collision: 20%.

Based on these scenarios, an event tree for collision scenarios and consequences may now be established, as shown in Fig. 9.15. For each of the scenarios discussed above, a distinction must be made between local failure, progressive failure, and complete push-over (foundation failure) of the jacket structure. The terminal events in the event tree may be characterised as follows:

Nodes 1, 4, 8: Local failure
Nodes 2, 6, 10: Uncontrolled fire due to riser or well conductor rupture
Nodes 3, 7, 11: Critical structural failure, load bearing ability not impaired
Nodes 5,9: Jacket topples over.

The conditional probabilities for the collision scenarios (first branching point in the event tree) have already been determined above. The probabilities of different levels of damage are now determined for Route 6, as an illustration of the applicable approach. Figure 9.16 shows the conditional probability distribution of impact energy in Route 6.

For the jacket in question the following energy levels are found to be critical:

- 3 MJ Plastic failure of one bracing.
- 8 MJ Plastic failure of several bracings.
- 14 MJ: Plastic deformation of corner leg, able to withstand 100 year storm.
- 40 MJ: Substantial plastic deformation of leg, damaged in 100 year storm.
- 50 MJ Rupture of riser. Conductors also assumed to rupture.
- 60 MJ Jacket push-over.

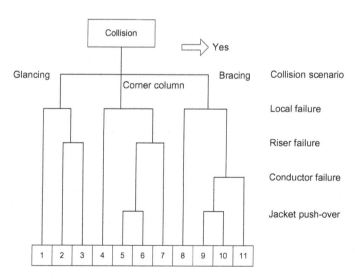

Fig. 9.15 Event tree for collision scenarios and consequences

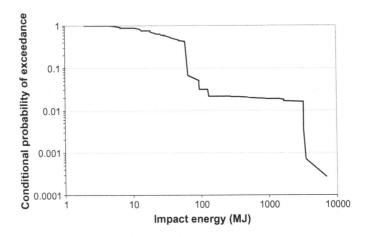

Fig. 9.16 Conditional probability of exceedance for Route 6

These energy levels apply to energy absorbed by the structure and it must be remembered that in many impacts some energy will be retained by vessel. For this purpose, the following assumptions are used:

Corner column impact: 25% of energy retained in vessel
Glancing impact: 75% of energy retained in vessel
Bracing impact: 40% of energy retained in vessel.

The conditional probabilities of impact energies equal to those different levels of damage to the platform may then be determined from Fig. 9.16. The values in Table 9.13 were found from interpolation of the values produced by the software, in order to get exact values.

Table 9.13 Energy levels and conditional exceedance probabilities for Route 6

Failure cases	Scenario						
	Basic failure level (MJ)	Corner column		Bracing impact		Glancing blow	
		MJ	Prob	MJ	Prob	MJ	Prob
Several bracings	8	Not applicable		13.3	0.528	Not applicable	
Corner column	14	18.7	0.365	Not applicable		56	0.102
Riser/conductor failure	50	67.7	0.0803	83.3	0.06	200	0.01
Push-over	60	80	0.0667	100	0.03	240	0.01

Table 9.14 Event tree results for Route 6

Consequence	Conditional probability given collision	Annual frequency
Local failure (Nodes 1, 4, 8)	0.655	2.03×10^{-4}
Uncontrolled fire, riser or well conductor rupture (Nodes 2, 6, 10)	0.0165	5.13×10^{-6}
Critical structural failure, load bearing ability not impaired (Nodes 3, 7, 11)	0.282	8.75×10^{-5}
Jacket topples over (Nodes 5, 9)	0.0462	1.43×10^{-5}

The results for Route 6 are then summarised in Table 9.14. The total frequency of failures causing total loss (that is Rows 2 and 4 in Table 10.14) is $1.94 = 10^{-5}$ per year.

References

1. Dnv GL (1998) WOAD. Høvik; DNV, Worldwide Offshore Accident Database
2. Kenny JP (1988) Protection of offshore installations against impact, background report. Report No.: OTI 88 535; Prepared for Department of Energy
3. HSE (2003) Ship/platform collision incident database (2001). HMSO, London
4. PSA (2017) Regulations relating to design and outfitting of facilities, etc. in the petroleum activities (the facilities regulations). Last amended 18 December 2017
5. AIBN (2018) Preliminary marine accident report—collision between the frigate 'KNM Helge Ingstad' and the Oil Tanker 'Sola TS' on 8 November 2018, Report 29.11.2018. https://www.aibn.no/Marine/Investigations/18-968?iid=25573&pid=SHT-Report-Attachments.Native-InnerFile-File&attach=1
6. Haugen S (1991) Probabilistic evaluation of frequency of collision between ships and offshore platforms. Trondheim; NTNU; Dr.ing. Thesis. Report No.: MTA 1991:80
7. DNV GL (1981) Technical note—fixed offshore installations—Impact Loads from Boats. Høvik; DNV. Report No.: TN–202
8. Safetec Nordic AS (1994) Collide II, Reference Manual, Rev 2. Trondheim; Safetec; 1994 Mar. Report No.: ST–91–RF–032–02
9. Hassel M (2017) Risk analysis and modelling of allisions between passing vessels and offshore installations. Thesis for the Degree of Doctor Philosophy, NTNU, 2017/305
10. Safetec Nordic AS (2002) COAST Norway, Development of COAST for Norwegian waters, Rev 00. Sandvika; Safetec; 2002 Oct. Document No. ST–20135–TS–1–Rev 00. http://www.safetec.no/article.php?id=105
11. Vinnem JE (2011) Evaluation of offshore emergency preparedness in view of rare accidents. Safety Sci 49(2):178–191
12. PSA (2012) Trends in risk level on the Norwegian Continental Shelf, main report, (in Norwegian only, English summery report). Petroleum Safety Authority, Stavanger 25(4):2012
13. LRC (2017) JIP visiting vessel collision risk assessment methodology, Report no: 106074/R1, 20 October 2017
14. NORSOK (2004) Design of steel structures, N–004, Rev. 2. Standards Norway: Oslo
15. HSE (1997) The effectiveness of collision control and avoidance systems. London: HMSO; 1997 Jan. Report No.: DST–96–CR–052–01

Chapter 10
Marine Systems Risk Modelling

10.1 Ballast System Failure

10.1.1 Background

It is more than 30 years since serious accidents or incidents associated with loss of buoyancy or stability occurred in the Norwegian sector. But there have been incidents and accidents in other areas, which remind us that this hazard has not been eliminated. The most serious accident ever in this category was the capsize and total loss of semi-submersible drilling unit Ocean Ranger in 1982 off Newfoundland (see Sect. 4.21).

Two incidents in the Norwegian sector during the autumn of 2012 may be seen as indications that the lessons from the 1980-ties are forgotten and this hazard is being neglected by operating personnel. In one incident the operator put personnel without adequate competence in charge of ballasting operations, this resulted in faults during ballast operations, significant listing and mustering of personnel in their lifeboats. In the second case an anchor had not been properly secured before bad weather conditions, causing the anchor to puncture a ballast tank, which resulted in a significant listing and precautionary evacuation of non-essential personnel by helicopter. Both incidents are being investigated by PSA, in addition to the owners' investigations.

10.1.2 Regulatory Requirements

The regulatory requirements discussed here are those applicable to analysis of reliability, vulnerability and risk associated with loss of buoyancy and stability.

There are requirements for the design of ballast systems and for the stability of floating units. The requirements for probabilistic/risk analysis of these systems are

© Springer-Verlag London Ltd., part of Springer Nature 2020
J.-E. Vinnem and W. Røed, *Offshore Risk Assessment Vol. 1*,
Springer Series in Reliability Engineering,
https://doi.org/10.1007/978-1-4471-7444-8_10

somewhat indirect. The survivability of the units is included in the phrase 'main support structure', which in the facilities regulations [1] Sect. 7 is defined as a Main Safety Function. The facilities regulations Sect. 11 specifies limits for the frequency of loads that may impair the Main Safety Functions. The HES management regulations [2] require that QRA studies are conducted for the Main Safety Functions.

The facilities regulations refer to the detailed requirements of the Norwegian Maritime Directorate (NMD) regulations for ballast systems on mobile units and the regulations for stability and watertight divisions. These again refer to NMD's risk analysis regulations. These regulations do not have explicit requirements for probabilistic/risk analysis of ballast systems, but there are requirements for demonstration of accordance with regulations, which may be satisfied through risk or reliability analysis.

PSA has for almost 10 years focused attention on hazards associated with floating installations, and have requested the industry to put more focus on hazards associated with buoyancy, stability and station-keeping. Several reports were issued in order to five the industry relevant documentation [3, 4]. Over time there has been some effect, but the comment was reiterated in 2014 [5] that structural and marine systems incidents appeared to still have only limited attention.

One of the authors advised an oil company some years ago in a development project on risk management issues, including the execution of QRA studies in the FEED phase. The QRA contractor had initially not intended to include stability as one of the applicable hazards, but was required to include also this hazard. At the end, it turned out to be the hazards with the highest contribution to the annual PLL value for the concept.

10.1.3   Relevant Hazards

Loss of stability may be caused by a single failure or perhaps more likely by a combination of different causes for mobile units and floating production installations including:

- Ballast system failure, including pumps, valves and control systems.
- Operational failure of ballast systems.
- Filling of buoyancy volumes or water filling of volumes on the deck from errors or maloperation of internal water sources, such as fire water or water tanks.
- Filling of buoyancy volumes due water ingress caused by collision impact.
- Filling of buoyancy volumes due to design or construction errors.
- Filling of buoyancy volumes or water filling of volumes on the deck due to fire or explosion, including fire water.
- Filling of pump rooms.
- Displacement of large weights on deck (SS).
- Loss of weights due to anchor line failure or failures in the anchor line brakes (SS).

- Ballast system failure or maloperation during transition of mobile units (JU).
- Loading system failure which leads to abnormal weight condition (FPSO).
- Failure during operation of loading system which leads to abnormal weight condition (FPSO).

The first seven items in the list above are general with applicability for all floating concepts, the last five are special for the following concepts; SS— semi-submersible units; JU—jack-up units; FPSO—floating production, storage and offloading tankers.

10.1.4   Previous Studies

The R&D programme Risk Assessment of Buoyancy Loss (RABL) was conducted in the middle 1980s [6]. This programme developed an approach for analysis of ballast system failures, based on event trees and fault trees. It was actually found that in the almost 20 years period after completion of the RABL programme, no other studies had performed similar detailed studies of loss of buoyancy or stability [4].

The approach normally adopted in QRA studies is discussed in Sect. 10.1.7 below. An alternative approach to QRA studies has been adopted by Lotsberg et al. [7]. The approach which was adopted for the Kristin field in the Norwegian sea, is presented in Fig. 10.1.

This approach is an improvement compared to traditional QRA approaches currently being used. One aspect where this approach falls somewhat short is the lack of ability to identify what could be risk reducing measures and their effects.

10.1.5   Stability Incidents and Accidents

10.1.5.1   North Sea Events

One total loss has occurred in the North Sea and North Atlantic and Norwegian Sea areas during the last 20 years, the water filling and sinking of jack-up West Gamma in 1990 (see Sect. 4.24). Many minor incidents have been recorded by HSE [4, 8], where equipment malfunction or maloperation have been corrected before severe consequences resulted. The most serious occurrences are:

- Loss of two anchor lines due to winch failure caused 160 m drift-off and a transient tilting of some 10°.
- Malfunction of the ballast control system caused a 9° list, lasting for 90 min before the rig was uprighted.

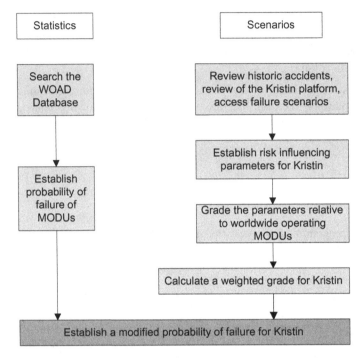

Fig. 10.1 Approach to failure frequency assessment based on gross errors [7] (Reprinted from Marine Structures, Vol 17/7, Lotsberg et al., Risk assessment of loss of structural integrity of a floating production platform due to gross errors, pp. 551–573, 2004, with permission from Elsevier)

- 6–8° inclination due to activation of deluge system, caused by loss of main power.
- Unknown inclination due to opening of ballast valves caused by failure of control desk.
- Several cases where water has leaked into ballast tanks due to cracks and puncture of tanks.
- Inclination due to maloperation of the ballast system by person without relevant competence.

10.1.5.2   Worldwide Occurrences

There are some occurrences in the worldwide operations that are well known, and which form an important basis for the evaluations, brief summaries are provided below:

- Ocean Ranger, 15th February 1982 (see Sect. 4.21)
- West Gamma, 21st August 1989 (see Sect. 4.24)

- Ocean Developer, 14th August 1995
- P–36, 15th March 2001 (see Sect. 4.26)
- P–34, 13th October 2002 (see Sect. 4.27)
- Thunder Horse, 11th July 2005

The semi-submersible mobile drilling unit Ocean Ranger capsized on 15th February 1982 in Canadian waters (see Sect. 4.21.1) for a description of the sequence of events.

Ocean Developer was under tow between two African ports on 14th August 1995 when it capsized and sunk without loss of life. The investigation report indicates that ballast operation by inexperienced personnel may be one of the causes.

The floating production unit P–36 (see Sect. 4.26) capsized and sunk on the Roncador field in Brazil. A ruptured drain tank in a column caused an explosion that destroyed a fire water pipe, killed 11 persons, and caused subsequent water ingress into watertight compartments, pump room s and thruster rooms.

The FPSO P–34 (see Sect. 4.27) developed a serious list due to malfunction of ballast and loading systems, caused by electrical faults. The vessel was close to capsizing before control was re-established. No fatalities occurred.

Following the passage of Hurricane Dennis in the Gulf of Mexico in 2005, personnel returned to the Thunder Horse facility to find it listing at approximately 20° with the top deck in the water on the port side [9]. The exact source/cause of the water influx/listing has not been determined; however, preliminary findings from the investigation indicate that water movement among the access spaces occurred through failed multiple cable transits (MCTs). MCTs are the points in the watertight bulkheads where cables that carry electrical power and instrument signals pass through the watertight bulkheads. Essentially, MCTs are molded blocks of plastic that seal around each cable. Failure occurred in the spaces filled with blank blocks. Specifically, the findings indicate that either the MCTs may not have been installed properly, may have been installed using the wrong procedures, or may not have been properly pressure rated for the configurations being used.

10.1.6 Observations from Incidents and Accidents

Figure 10.2 presents a summary of causes for stability failures, based on worldwide accidents and incidents discussed in Vinnem et al. [4].

Minor problems are not included. The diagram pinpoints clearly that valve failures are the main cause category for accidents and incidents. It may further be observed that the two total loss accidents were caused by operational failures. It may further be observed that 58% of all accidents, incidents and minor problems are associated with technical problems. This is unusually high.

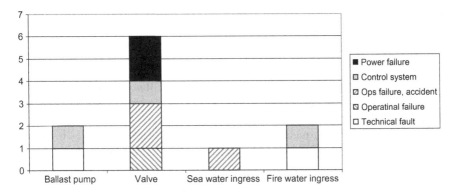

Fig. 10.2 Distribution of causes of stability failures

10.1.7   Evaluation of Typical QRA Studies

Current practice in Norwegian QRA studies related to stability of mobile units and floating production units was surveyed by Nilsen [10]. The conclusion is that current practice in QRA studies are not suitable for identification of possible risk reducing measures, nor are they suitable for quantification of the effect of such measures for the risk levels. Deficiencies in a majority of the studies have been demonstrated, including:

- Accident scenarios are not modelled. The possible failure categories are considered on a superficial level, without the possibility to identify how the scenarios could develop.
- Several failure mechanisms are not considered at all, as rupture of fire ring mains, major displacement of heavy loads on deck, operator error during ballasting or loading operations, and water ingress due to collision impact.
- Experience data are not considered. Some data were mention above, and are further documented in Ref. [4]. None of these are usually considered in QRA studies.
- Assumptions, premises and simplifications are not addressed. The PSA regulations require the assumptions and premises to be documented and to be traceable. The studies do not comply with this requirement.
- Presentation of results is without traceability. Some of the studies do not present quantitative results at all, but are limited to conclude that 'the design is considered to be safe'. This is virtually worthless when it comes to transparency, as it fails completely to document how this was reached, and what are the limitations and underlying assumptions.

10.1.8 Proposed Approach to Analysis of Stability Hazards

10.1.8.1 Life Cycle Phases

The main analytical efforts should be made during design and engineering, in order to give good opportunities for implementation of risk reducing measures. Updating of the analysis may be done after completion of construction, and sometimes during the operations. Detailed studies will be particularly important when untraditional concepts and solutions are adopted, including solutions that are not addressed in the regulations and standards.

The studies should also address special conditions that may occur, such as during displacement of heavy loads, rupture of fire water ring main, as well as special conditions during inspection and maintenance when doors and manholes may be opened, or systems deactivated.

10.1.8.2 Analytical Approach

The proposed analytical approach is presented in Ref. [4], adapted from Haugen [11].

Collection of experience data should be the starting point for the analysis, which should continue with hazard identification (HAZID), in order to identify those scenarios that may result in critical consequences, particularly with respect to combinations of failure cases and effect of operational error.

A detailed analysis should be performed for the critical scenarios, limited in this context to marine systems or systems that may influence marine systems. If a FMECA and/or task analyses have been carried out, then these may serve as the starting point for the detailed analysis, including fault trees and event trees.

Fault trees and event trees may be used in order to calculate risk values, as well as to identify where the most effective modifications (risk reducing measures) in order to improve the situation. During this part of the analysis, efforts should be made in order to document assumptions and premises, relating to:

- Technical conditions
- Conditions associated with operations and maintenance
- Assumptions related to analysis methodology and modelling.

The management regulations [2] have a general requirement for consideration of uncertainty to be addressed for all risk elements, not only marine systems. This is usually most effectively implemented through sensitivity studies, in relation to data and variations in assumptions and premises.

10.1.8.3 Detailed Analysis of Ballast System Failures

The approach outlined in Fig. 10.3 should be used in order to analyse risk due to
failures in ballast system components. This implies adoption of the same approach
as developed in the RABL project. The main elements of this approach are fault
trees and event trees; see Figs. 10.4 and 10.5. The importance of choosing the
approach is that it enables a detailed identification of system modifications and
operational changes that may be most effective in order to reduce the risk level, and
the likely effect of such actions.

This is one of the important requirements in the management regulations [2].
Identification of possible risk reducing measures is also an important element in the
ALARP demonstration, which is essential in the Norwegian as well as UK regulations. The following additions to the approach described in the RABL project
should be implemented:

- Fault tree analysis should also be performed for the most critical nodes in the
  event tree (see example in Fig. 10.5).
- Human and organisational errors
- Common mode failures and dependencies.

As input to the analysis, a detailed analysis of collision risk may be required,
depending on the circumstances. Also fatigue failures may be required as input, in
addition to failures during loading of FPSOs.

10.1.8.4 Other Failures

Other failures that according to the approach in Fig. 10.3 are not considered critical
may be analysed using the approach suggested by Lotsberg et al. [7] (see Fig. 10.3).
The disadvantage of this approach is the inability to identify risk reduction
proposals.

10.1.8.5 Analysis of Human and Organisational Aspects

Human and organisational errors should be included in the fault tree analysis where
relevant, using, e.g. the BORA approach [12, 13].

10.1.8.6 Analysis of Dependencies in Barriers

Common mode failures and dependencies should be analysed as appropriate.
Standard approaches in fault tree analysis for both of these aspects exist, normally
used in the analysis of failures of nuclear power plants. For offshore installations
however, it has not been common practice to include these aspects in the analysis.

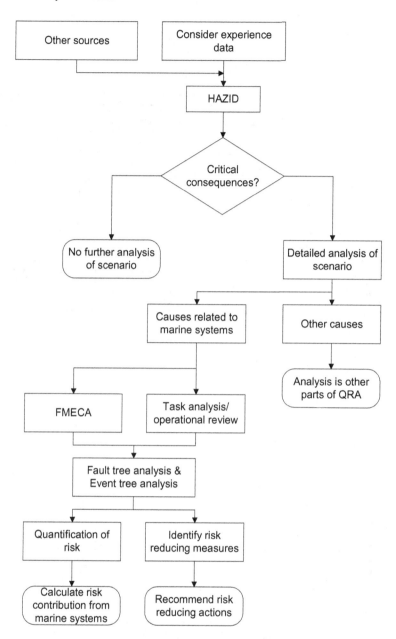

Fig. 10.3 Proposed analytical process for marine systems

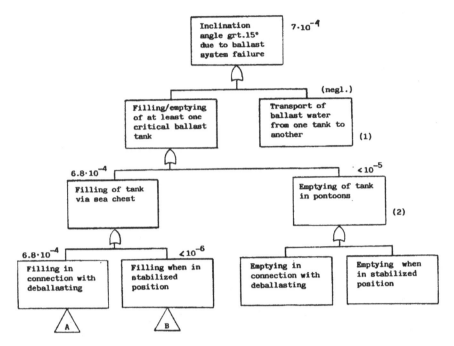

Fig. 10.4 Example of top levels in fault tree for analysis of ballast systems [6]

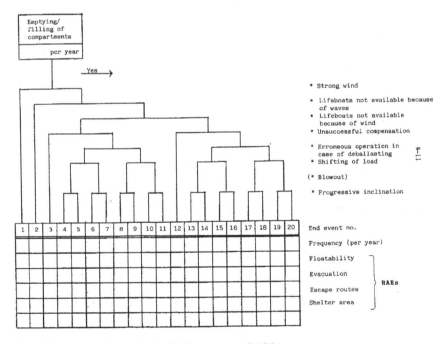

Fig. 10.5 Event tree for analysis of ballast systems (RABL)

10.1.8.7   Analysis of Barriers

In addition to the initiating events, several conditions might contribute to escalation of the events—improper sectioning of the hull, lack of draining capacity on the deck, lack of or failure in the leak detection systems, lack of pumping capacity, lack of training of personnel in emergency situations or open doors and manholes.

The Norwegian management regulations Section 15 require that QRAs shall model accident sequences and consequences so that possible dependencies between physical barriers can be revealed, and so that the requirements that must be set in respect of the performance of the barriers, can be calculated. A method to analyse barriers related to stability is demonstrated by Ersdal and Friis-Hansen [14].

10.1.8.8   Discussion of Approach

The capsize of the flotel 'Alexander Kielland' in 1980 was the last serious accident involving loss of buoyancy or stability on the Norwegian Continental Shelf. The jack-up 'West Gamma' capsized and sank in 1989 during tow from the Ekofisk field, the accident occurred in the Danish sector and the platform finally sank in the German sector, without fatalities. But there have been near-misses also during recent years, which could have developed into serious accidents.

It has been claimed by most experts that gross errors during platform design and construction cannot be analysed by traditional risk analysis methods. The full analysis of this is outside the scope of this discussion, but it appears that this view has been adopted also for hazards relating to marine systems. This is considered to be a misunderstanding.

There is little similarity between analysis of design and construction defaults and failures of marine systems. Gross errors in design and construction are events that may jeopardise the integrity of the structure, and may prevent the normal redis-tribution of forces to compensate for local failures. This is difficult to analyse by normal risk analysis methods. Gross errors may be caused by single failures.

Marine systems like ballast systems are quite different, there is redundancy and possible dependencies, to the extent that it is important to analyse failure event combinations. Traditional risk analysis methods, like fault trees and event trees or similar, may be used in order to analyse such scenarios. It is recommended that marine systems are analysed according to the approach outlined in Fig. 10.3, including HAZID, FMECA and task analysis (or similar), fault trees and event trees.

It should be noted that also human errors and organisational failures need to be considered, and that risk influencing diagrams or Bayesian belief networks may be appropriate tools.

One alternative could be to employ the approach used by Lotsberg et al. [7] for ballast systems and failure of stability scenarios.

A second alternative could be to use event trees for the most critical scenarios, but omit quantification of accident probabilities through fault trees and similar.

The disadvantage of these two approaches is that the basis for identification of possible improvement measures will be significantly weaker. Identification of possible improvements is one of the main objectives of risk analysis. This is the main reason why the proposed approach should be selected.

10.1.9   Comparison of QRA Results with Experienced Events

If we restrict the consideration to semi-submersible installations, it is more than 30 years since the last serious Norwegian accident, as noted above. Therefore, there is no basis for making a comparison of QRA results with accident statistics.

If we take a 35-year perspective in the Norwegian sector, we have one total loss, the capsize of 'Alexander Kielland'. The regulatory requirements were changed in the early 1980s, to the extent that this accident might be unrepresentative for risk level s implied by current standards. The current standards specify weather conditions where failure of one brace should not escalate, if the weather exceeds this level (one year environmental conditions plus safety factors), similar situations might occur. Nevertheless, it may be interesting to consider what the accident statistics implies.

Based on data in the RNNP project [15] the number of mobile installations per year may be calculated. The sum for the period 1990–2011 is 402 unit years. A rough calculation for the period 1977–1989 is 106 unit years. The total value for the 35 years is 408 unit years.

If we assume a Poisson distribution, the expected value is 1/408 per unit years, 2.5×10^{-3}, as the frequency of total loss per unit years. If we consider a prediction interval, the upper 95% limit would be 9.8×10^{-3} per unit year. Assuming a normal manning level during a period, the FAR value is almost 40!

As noted above, it would be expected that the frequency of total loss is lower with today's standards, due to the stricter requirements for damage stability. How much lower it would be, is impossible to know, based on present knowledge. On the other hand, there are no reasons to assume that the frequency would be much more than one order of magnitude lower for modern installations. This implies that it is unlikely that the failure frequency historically is lower than 1.0×10^{-4} per unit year.

Table 10.1 presents illustrative results from a concept study for a floating gas production installation considered for NCS recently, where analysis was carried out of the ballast system as well as the anchoring system. Table 10.1 shows impairment frequencies for the loss of structure main safety function, interpreted also to include loss of buoyancy and stability. The concept in question was considered to be a robust and safe installation. Nevertheless, ballast system failure and anchor system failure turned out to have the highest frequency of impairment of the structure.

Table 10.1 Impairment frequency results for floating production installation concept

Accident type	Loss of structure as MSF (buoyancy/stability) per 10.000 years
Fire	0.01
Explosion	Negligible
Ship collision	0.29
Anchor system failure	0.40
Ballast system failure	0.50
Gross error	0.30
Extreme weather	0.30
Dropped object	0.30

10.1.10   Observations

The QRA studies normally conducted for floating installations are inappropriate in several ways. It has been implied that when the studies claim that there is an insignificant risk level, there is no basis in available data to draw these conclusions. The study referenced in Table 10.1 also confirms that the risk associated with ballast system failures is far from insignificant.

What is more disturbing, is that the studies do not give any basis for identification of risk reducing measures, which is one of the main objectives of risk analysis. An approach based on fault trees and events trees should be implemented for scenarios identified as critical.

10.2   Anchoring System Failure

The safety of anchoring systems for use in the petroleum activity on NCS is regulated through the facility regulation Section 62 stating that the anchoring system for mobile offshore units shall be in accordance with the Norwegian Maritime Directorate's (NMD) regulations of 4 September 1987 No. 857 concerning anchoring/positioning systems on mobile offshore units. In addition, the anchoring system for facilities with production plants and facilities located adjacent to another facility, shall also be in accordance with the Norwegian Maritime Directorate's regulations of 10 February 1994 No. 123 for mobile offshore units with production plants and equipment. The calculations shall not include the advantage of active operation of anchoring winches.

The NMD regulations state in general, that the environmental actions shall be stipulated with an annual probability of 10^{-2} and a set of safety factors are stipulated. The main differences between the two NMD regulations are different safety factors and the requirements regarding loss of anchor lines. While the NMD 857 requires that the unit shall maintain position during loss of one anchor line, the 123 regulations state that the position shall be maintained in case of two simultaneous line failures if the unit is adjacent to another facility.

According to NMD, drilling rigs will normally not have to comply with the NMD 123 regulation. But through the PSA regulation all units, whether they are mobile or not, are linked to the NMD123 regulation if the unit operates adjacent to another facility. Hence, the main difference between the NMD and PSA regulations is the requirement regarding adjacent platforms—as flotels. In the PSA regulation they should be analysed as the anchoring systems on production platforms. From PSA point of view, accidents have demonstrated the need to have this requirement. A flotel is typically 150 m away from the adjacent platform in storm conditions. Drifting of more than 150 m has occurred after multiple anchor failures.

The regulations should give reasonable protection against accidents, but incidents have occurred frequently. Our main conclusion is that the requirements in the regulations are reasonable, but improvements have to be made in industry to comply with the requirements.

10.2.1 Incidents Involving More Than One Anchor Line

The requirement that loss of two anchor lines should be analysed, has been frequently discussed in Norway, in connection with the introduction of ISO–DIS–19901-7. The experience nevertheless demonstrates that loss of two lines is a realistic case, and the regulations should continue to have this requirement. The incidents discussed in the following are extracted from Næss et al. [16].

Bideford Dolphin experienced three anchor line failures close to Snorre A in the North Sea, in a summer storm on 13th June 2000. The failures occurred in shackles (CR-links). The CR-link was used as connecting links between chain and wire in the mooring system. The shackles failed because of fatigue and tear-off fractures. The shackles were only two years old. The tension was about one third of its proof capacity. The platform got a drift-off of about 250–300 m from its target position. The well was secured. The anchor lines crossed several export pipelines, but they were not damaged. The ten minutes average wind velocity was about 20 m/s, and the significant wave height was about 8.5 m.

Transocean Prospect at the Heidrun field in the Norwegian Sea experienced dragging of two anchors about 50 m on 11th November 2001. They used eight 12 ton anchors. The 10 min average wind velocity was about 21 m/s, and the significant wave height was measured at Heidrun to be about 13–14 m [17].

Scarabeo 6 drilling at the Grane field in the North Sea experienced anchor dragging on 24th December 2002. They used eight 15 ton anchors. The tension experienced was about 50% higher than the test tension. The event escalated when a chain fractured in the fairlead. According to calculations the line broke at about 80% of its holding capacity. The fairlead had only five pockets, and some bending and a reduction in the breaking load in the link was anticipated. The 10 min average wind velocity was about 22 m/s and the significant wave heights were 9–9.5 m. The well was secured, with the drilling riser hanging in the sea.

On 14th December 2004 an accident occurred on the Ocean Vanguard drilling rig at Haltenbanken in the Norwegian Sea, as discussed in Sect. 4.28.

Of the four cases involving more than one anchor line, one was connected to the brakes, one in the lines, one caused by failures in the soil and one as a combination of soil and chain. Failures in the brakes, failure of the lines and dragging are the three fundamental failure modes in anchoring systems in storms. Statistics of incidents will be presented individually for the three causes.

Recently there have been some cases of multiple anchor line failure in severe storms in the UK sector of the North Sea. Gryphon Alpha FPSO experienced such an incident on 4th February 2011 (see Sect. 4.34). Later the same year, 8th December 2011, the Petrojarl Banff FPSO experienced five of 10 anchor lines failing during severe storm, also in the UK sector in the North Sea. Petrojarl Banff FPSO also had a similar incident in 2016.

The severity of such incidents is seen clearly from the Gryphon Alpha incident, where several risers failed, leading to release of gas cloud, and substantial damage to subsea installations from failing risers. The installation also spent over two years in the yard for repairs and refurbishment before returning to service.

10.2.2 Release of Chains in Winches

A comparison of the incidents, demonstrate that malfunction of the band brake is the most common cause of failure. The most common cause has been errors in adjusting the band brakes and/or corrosion and wear. The brakes did not take more that 16% of its documented holding capacity in one case. None of the incidents involving uncontrolled release of anchor chain have involved tension in the chain exceeding the theoretical holding capacity, Ref. [16].

Figure 10.6 specifies the operational modes when the failures occurred in the winches. As expected, most of the incidents occurred during anchoring operations. During anchoring operations the winches are active and the brakes are deactivated. Consequently, the system is vulnerable to both technical and operational errors.

The most common root causes in use are failure in the band brake and lack of maintenance. Other causes are maloperation by personnel, instructions from the supplier are not followed and errors in the procedures.

10.2.3 Failures in Anchor Lines

The number of anchor line failures reported to PSA on NCS in the period 1996–2010 is 20 cases. For the period 1996–2005 there are five cases with chain failure, two cases of loss of fibre ropes, and three cases with failures in shackles, Ref. [16].

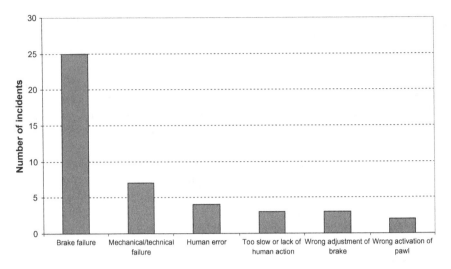

Fig. 10.6 Number of loss of lines related to winches according to type of error

Shackles are used to connect line segments. Experience from UK demonstrates that the failure frequencies in shackles are significantly higher than for elements in the chain itself [18]. The Norwegian experience is similar. Two of the cases (CR-links) were fatigue failures. Local stresses in the connection initiated fatigue cracks, due to the material impact toughness that did not meet the requirements. It was impossible to get the shackle up from the seabed in one case, to determine the cause.

No wire has been reported failing in use in an anchoring system, even when the data have been reviewed back to 1990. Wires are not considered any further.

Failures in the chains are most frequently caused by poor quality of the chain at the time of failure. The causes are split evenly between brittle fracture and fatigue. Some chains had been exposed to bending loads, with corrosion and loose studs contributing to failure. The bending probably occurred in five-inch fairleads.

Three failures in fibre rope have been experienced over a short period of three years. The fibre ropes are not as robust to mechanical actions as chains. The ropes are normally used in a combination of chains and ropes on NCS, to protect pipelines from damage. In recent years about 20 fibre ropes have been used on the NCS. The failures were caused by wire from a fishing vessel to a trawl board, by a wire to a ROV and by a wire connected to a hook.

Buoys are used to connect anchor lines to lift the anchor lines above obstacles on the sea floor. Only two cases of loss of buoys have been reported to PSA in the period from 1996. The low number might be caused by underreporting—since the rules are not clear about whether reporting is necessary in such cases.

10.2.4   RNNP 2013 Study

In 2013, the Petroleum Safety Authority Norway commissioned a study [19] on the causes of structural and maritime incidents on the NCS. The background for the study was the negative trend in reported structural and maritime systems incidents on the NCS in the last three years, as well as the serious incidents on the installations Floatel Superior and Scarabeo 8 in 2012. The study is focused on incidents with major accident potential, with the following objectives:

- Collect data from literature, investigations, interviews and questionnaires concerning causes and measures associated with structural and maritime incidents.
- Perform a complete assessment and analysis of human, technical and organizational causes and underlying factors.
- Suggest areas for improvement and concrete measures which the industry should address on the basis of identified causes.

Technical experts from oil companies, engineering companies, shipowners, other suppliers and research institutions provided information for the study. Viewed in the context of the major accident potential, the study showed that the focus on structural and maritime incidents and the disciplines involved is inadequate.

A brief overview of initiating and root causes of the incidents considered is shown in Fig. 10.7. The investigations of maritime incidents are of variable, sometimes poor, quality, while few structural incidents are investigated at all. Overall, the investigations contribute less than is desirable to an improved understanding of underlying causes and to a basis for sound risk-reducing measures.

Furthermore, the industry's own expert technical personnel find that the status of the structural and marine systems engineering profession has been diminished and that more attention needs to be focused on maritime systems and operations. Based on the results of this study, four main challenges were identified:

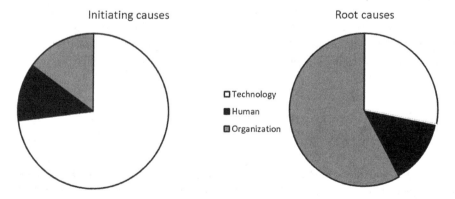

Fig. 10.7 Overview of causes in RNNP study [19]

i. Increase the quality and quantity of investigations of structural and maritime incidents
ii. Improve information exchange between participants and between different phases
iii. Improve knowledge and practice associated with marine systems
iv. Need for more systematic safety work and prevention of major accidents linked to both structural and maritime incidents.

10.2.5   Dragging of Anchors

Four cases of loss of anchor holding capacity are reported in Norway in the period. All the events occurred in bad weather. In the period 1996–2003 no storm dragging event was reported in the UK [8, 16].

Dragging of anchors occurs when they are not properly fixed to the seabed. Soil investigations are normally necessary to calculate the capacity with high accuracy. For exploration drilling, it has been accepted to use general information of the soil conditions in the area. Typically a test tension of 150–200 tons is performed. The test tension restricts when the platform can operate connected to a well, or in the vicinity of other platforms. A major part of the test tension is used to overcome the chain friction on the sea bed. The dragging incidents demonstrate that testing to 150–200 tons tension has been insufficient to avoid dragging.

Dragging is not an accident, but it causes an unintended redistribution of tension in the lines, causing other lines to fail, and unintended loss of position.

10.2.6   Other Risks with Anchoring Systems

Anchor handling is an operation with high risk for personnel. Fatal accidents occurred on the anchor handling vessels Maersk Terrier and Far Minara in 1996, Maersk Seeker in 2000 and Viking Queen in 2001. In addition incidents have been reported where the anchoring system has damaged equipment on platforms and vessels, and on seabed pipelines. These risks are not discussed further.

10.2.7   Risk Analysis of Anchoring Systems on MODUs on the NCS

The NCS and the UK shelf [8] had for the period 1998–2003 about the same number of failures in anchoring systems, indicating that the reported Norwegian frequencies are higher because of the higher number of MODUs in the UK sector.

In particular, the number of reported dragging events is higher in the Norwegian data.

Ten QRAs for six mobile offshore drilling units, one flotel and three production platforms have been reviewed with respect to an analysis of anchoring system failures [10]. The incident 'loss of position' is analysed very coarsely in most of the QRAs. Several of the hazards that can cause loss of position are not identified. Only one of the QRAs identified winch failure hazard. Other hazards missing in the majority of the QRAs are loss of buoys, fatigue, fishing vessels in contact with ropes and anchor dragging. The analysis methods and the applied data are not well documented in several studies. Only three of the analyses specified the assumptions. The use of different data sources gives large variations in the calculated risk level s for similar systems. Fault trees are not used in the analyses.

10.2.8 Use of Fault Trees in QRA of Anchoring Systems

Fault trees have been produced for the cases of failures for active operation and in a storm situation, failures in the lines (chain or rope) and dragging of the anchor. Generic fault trees for the failure in the brakes during storm, failure in the chain and dragging of the anchor are presented in Ref. [3]. A fault tree for paying out the chain during marine operation can be found in Ref. [16]. The fault trees are made through a process of hazard identification and a review of the causes of incidents in Norway and in available publications. Frequencies are calculated based on Norwegian incident data. The fault trees give the causes leading to the top events.

10.2.9 Summary

There is a high number of anchoring system incidents on MODUs on NCS. It is proposed that training and organisational factors should receive more attention. Several of the incidents would probably not have occurred if the industry had a good system for exchanging experience—and if the crew had adequate competence on anchor systems and their function. Maintenance should also receive more attention.

Many of the incidents occurred in connection with critical operations. The facility has been connected to the well—or has been alongside another facility. Even though the anchoring system is designed to withstand a line failure, this is still an undesirable incident.

According to the regulations, two independent brake systems shall be in use at any time. To get an incident both brakes have to fail. None of the cases would have occurred if the winches had been according to the regulations. It has not been possible to determine how often each individual brake system fails. However, the high number of incidents involving failure of both brake systems with resulting chain deployment indicates that the failure rate is high.

Failures in the anchor line itself are the most frequent cause of failures in the anchoring system. The quality and frequency of inspections and repairs performed in connection with recertification of the chains are of major importance. Chains that are more than 20 years old are still in use. Therefore, the inspections and repairs conducted in connection with recertification are essential in ensuring that the chain meets the applicable quality requirements for the anchoring line. The chain owners must know the history of each individual line (traceability) in order to ensure a successful recertification. Several fatigue failures have occurred on anchor chains, caused by bending stress. It is reasonable to assume that the bending stress has occurred at the fairleads.

The number of failures in shackles is about the same as in chains, and both types of failures have the same consequence. Since the number of shackles is small compared to the number of chain links, each individual shackle has a significantly higher failure frequency than the chain links. Special attention should be given to the selection of shackles, as well as in connection with assessments of the condition of the shackles.

Use of fibre rope in the anchor lines may be advantageous in some cases, with respect to safety as well as operational aspects. Fibre ropes have on the other hand been proven to be vulnerable to mechanical exposure, e.g. when in contact with wire. Activities carried out within the anchor pattern must be better supervised because of the vulnerability of the fibres.

The number of dragging events demonstrates a need for increased pretension capacity or use of other anchor solutions. With the present test tensioning capacity, it might be impossible to get a safe anchoring with traditional fluke anchors. The anchor holding capacity on mobile units must be stipulated more precisely, than is the practice so far. Limited dragging of the anchor will not necessarily cause major consequences for a mobile unit drilling an exploration well. On the other hand, experience has shown that dragging an anchor can cause failures of neighbouring lines. Frequently, drilling units are anchored in areas with many subsea facilities. An anchor dragging may damage these installations. There is a need to increase the anchors' test tension on mobile units. Even with good knowledge of the soil conditions, it may be difficult to achieve a good foundation solution based on conventional anchors (drag anchors). Alternative types of anchoring should then be evaluated. Dragging anchors can only be accepted after a consequence evaluation for the tension in the other lines as well as the possibility of damage to subsea facilities and neighbouring platforms.

One should expect compliance between the results from the anchoring analyses and the practical anchoring work on the facility. Good anchoring analyses will not enhance safety if they are merely an academic exercise.

Quantitative risk analysis is a common approach to find and quantify risk reducing measures. The QRAs that have been reviewed have not addressed anchor systems and operations in detail. Several failure modes have not been identified or analysed, nor have the analyses been used as a basis for reducing risk. The QRAs are also, in general, too coarse for this purpose.

10.3 Failure of Drilling DP Systems

The barrier approach had been used for analysing the safety of DP (dynamic positioning) drilling operations on the Norwegian Continental Shelf [20]. An illustration of such an operation is presented in Fig. 10.8.

There are three barrier functions modelled in this study based on the critical failure events in DP drilling operations:

- Barrier function 1: to prevent loss of position
- Barrier function 2: to arrest vessel movement
- Barrier function 3: to prevent loss of well integrity.

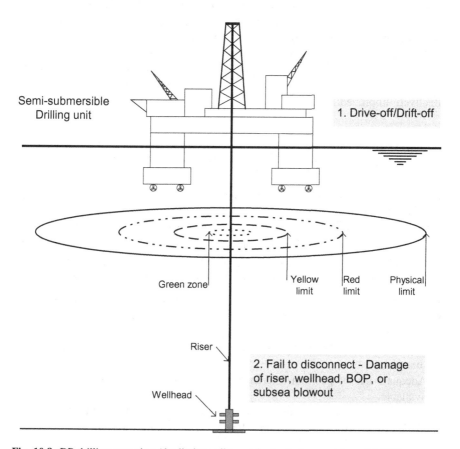

Fig. 10.8 DP drilling operation (the limits: yellow, red, physical are not to scale) [22]

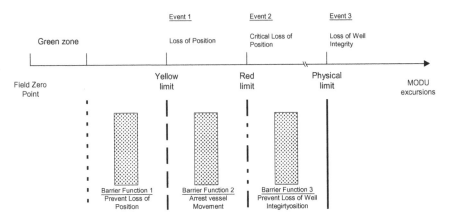

Fig. 10.9 Three barrier functions to safeguard DP drilling operation [20]

An illustration of the barrier functions in relation to the horizontal excursion limits and the critical failure events is given in Fig. 10.9. This is the overall safety model developed in this study for DP drilling operation.

By analyzing the three barrier functions (via systematic identification and analysis of the involved barrier elements and the associated technical, human and organisational factors), safety of DP drilling operation can be diagnosed and effectively improved.

10.3.1 Barrier Function 1—Prevent Loss of Position

The drive-off is initiated by the DP control system due to erroneous position data from two Differential Global Positioning System (DGPS) units [21].

Barrier elements to prevent DGPSs generating erroneous position data, as well as to prevent erroneous DGPS position data being used by DP software are identified, and their deficiencies are revealed based on the incidents and operational experiences on NCS. Recommended risk reductions include the following actions:

- Correction of the current lack of independence between the two DGPSs.
- Assessment and verification of GPS and differential link antenna locations on the vessel.
- Configuration of the DGPS software with respect to the DGPS quality control function.
- Adequate use of position reference systems with three different principles if available.
- Design, testing and configuration of DP software with respect to the DGPS input validation function and position reference error testing function.

These recommendations are to be incorporated in the ongoing revision work for DP requirement documents from several oil companies on the NCS, including Statoil (now Equinor). In addition, the research work has been summarised in two papers, Refs. [23, 24].

10.3.2 Barrier Function 2—Arrest Vessel Movement

Effective arrest of vessel movement must be performed in order to prevent a loss of position from escalating into a critical loss of position. The DP operator is identified as the only barrier element associated with this barrier function. A number of deficiencies that could significantly affect the DP operator's reactions in a time-critical drive-off scenario are identified. They are considered to be associated with the following four influencing factor categories i.e., bridge ergonomics, alarm system, procedures, and training. Risk reduction can be achieved by:

- Design of bridge layout, DP operator workstation, and information presentation on visual display units.
- Design of alarm system with respect to alarm generation, alarm perception, alarm comprehension, alarm handling and alarm philosophy.
- Procedures for DP operator to detect deviations, handle failures of position reference systems, and perform various recovery tasks under possible emergency scenarios.
- Training programme for DP operators on DP drilling units in general, and DP simulator training based on real and/or assumed worst case DP incidents in particular.

These recommendations are to be implemented by the operational management from each vessel owner. In two areas i.e., design of bridge and assessment of DP alarm system on DP drilling unit, further joint industry efforts involving oil companies, vessel owners, classification societies, equipment vendors, training institutions are needed in order to systematically strengthen the human barrier element.

10.3.3 Barrier Function 3—Prevent Loss of Well Integrity

Given a critical loss of position, failure of emergency disconnection could lead to loss of well integrity i.e., an open-hole situation in the well. Three barrier elements are identified in order to prevent loss of well integrity. They are Emergency Quick Disconnection system (EQD), Safe Disconnection System (SDS) and well shut-in function. Recommendations are proposed in the following areas:

- Technical integrity programme for the systems related to the emergency disconnection, in order to minimise the failure on demand.

- Use of auto EQD and SDS for drilling on shallow waters if these systems are available.
- Competence of DP operator to evaluate the situation and activate the red status in time when needed, and support from the operational management to the DP operator's decision.
- Drills onboard involving DP operator and driller for manually activating EQD.
- Special operational precautions when there are non-shearable items through the BOP.

10.4 Shuttle Tanker Collision Risk

The assessment of collision risk associated with shuttle tanker impact during off-loading, is a relatively coarse assessment, based on incidents and accidents in the period 1995–2011. A general discussion and assessment of shuttle tanker collision risk is presented. This assessment is limited to DP operated shuttle tankers, and disregards tankers operated with taut hawser. Collision frequencies are presented for DP1 and DP2 tankers, where also the trends in the occurrence frequencies have been taken into account.

10.4.1 Background

Turret moored FPSOs and FSUs of the mono-hull type have been used in the North-west European waters since 1986 (Petrojarl I), so far without serious accidents to personnel or the environment. The use of such vessels for field development has increased during the last 10–15 years, in UK and Norwegian waters, as well as in many other offshore areas worldwide.

FPSOs are not new as petroleum production units, they have indeed been employed in other parts of the world already for some time, and in quite significant numbers compared to the current North Sea fleet. Where such vessels are installed in benign waters, they have usually been converted cargo tankers with mooring and fluid transfer in the bow of the vessel, or sometimes transferred from a loading buoy. The last 10 years has seen some massive FPSOs installed in fields west of Africa, purpose built, with extensive processing capacities.

The vessels installed in the North Sea, North Atlantic and Norwegian Sea fields have traditionally been designed for considerably higher environmental loads and often also higher throughput compared to vessels in more benign waters. Without exception, the ones so far installed or under construction for these areas have what is termed 'internal' turret, in the bow or well forward of midships, with transfer of pressurised production and injection streams through piping systems in the turret.

Although FPSOs are becoming common, operational safety performance may still be considered somewhat unproven, especially when compared to fixed installations. Floating installations are more dependent on manual control of some of the marine systems, during normal operations as well as during critical situations. There is accordingly a need to understand the aspects of operational safety for FPSOs, in order to enable a proactive approach to safety, particularly in the following areas:

- Turret operations and flexible risers
- Simultaneous marine and production activities
- Vessel movement/weather exposure
- Production, ballasting and off-loading.

Accidents are often initiated by errors induced by human and organisational factors, technical (design) failures or a combination of these factors. Effective means to prevent or mitigate the effects of potential operational accidents are thus important. The scope of the assessment is limited to the collision risk between shuttle tanker s and FPSO/FSU during off-loading of cargo from the stationary vessels, including the approach, connection and disconnection phases.

Five accidents occurred in a 4.5 year period from 1996 to 2000, and a lot of attention was paid to this hazard in some of the projects where FPSO concepts were chosen.

An R&D project on 'FPSO Operational Safety' was conducted in the period 1996–2003 [25–28]. The work was organised as a Joint Industry Project (JIP), with funding from Statoil (now Equinor), Exxon (now Vår Energi) and HSE, and with Navion (now Teekay) as technology sponsor.

A detailed assessment of the collision risk between shuttle tanker and FPSO/FSU during off-loading in tandem configuration was performed during the period 2000–2003. The general experience from the work is documented in an HSE report 'Operational Safety of FPSOs, Shuttle Tanker Collision Risk, Summary Report' [23]. The assessment builds strongly on the work in this report.

A PhD study was conducted in parallel with the JIP project, with the same topic [33]. The work of Dr. Haibo Chen is also part of the basis for the current assessment. The assessment considers the total off-loading system (see Fig. 10.10), consisting of:

- FPSO during all phases of off-loading
- Off-loading arrangements
- Shuttle tanker during all phases of off-loading.

The operational aspects (human and organisational factors) that are addressed in the assessment are applicable to organisations within the total analysis envelope. This implies that the operating organisations of both the FPSO and the shuttle tanker during all phases of off-loading are within the scope of the analysis. The assessment is limited as follows:

Analysis envelope in study

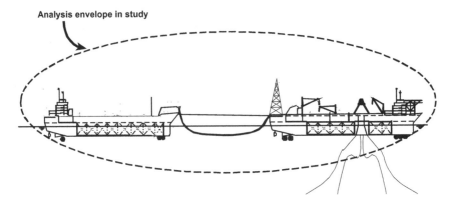

Fig. 10.10 Analysis envelope

- It applies to DP operated shuttle tanker s, and not field configurations where control of distance between the two vessels is by means of 'taut hawser' (see also Sects. 10.4.3 and 10.4.9).
- Only tandem off-loading configuration is considered.
- Operations before 1995 are not considered.
- The assessment is intended to be used in a coarse risk assessment, thus a detailed assessment approach involving field specific configurations and characteristics is not attempted.
- The off-loading phase is demonstrated to be the main source of risk, thus the other phases approach, connection and disconnection are not considered explicitly.

10.4.2   Tandem Off-Loading Configurations

The discussion here is limited to off-loading from FPSO and FSU systems. It is taken as a presumption that the off-loading system will be configured such that the FPSO/FSU and shuttle tanker will be at relatively close distance, say in the range (theoretically) from 40 to 150 m. (One new FPSO has recently been installed with distance of 150 m between vessels.)

The focus in this section is on aspects which have importance for the collision hazard, some of these aspects are not considered in the coarse assessment, but will have to be included in a more detailed assessment.

One of the particular aspects of tandem off-loading systems is that purpose built and commercially available systems are combined. Hence there are some quite wide differences between configurations applied to comparable situations.

Table 10.2 Variations in FPSO/FSU field configurations

Characteristic	Variations
FPSO station keeping capabilities	Internal turret with 8–12 point mooring system
FPSO heading keeping capabilities	Without heading control With heading control
ST heading and station keeping capabilities	No propulsion Main propulsion (single or twin screw) No DP system DP1, DP2 or DP3 systems
Off-loading mode	DP operated Taut hawser operated
Interface systems	With hawser connection Without hawser connection
Distance FPSO– ST	50–100 m 80 m 150 m (proposed by some, but not used at present)

The FPSO may be termed 'purpose built'. When an FPSO is a new build for a specific field, then it may be perfectly tailored to the needs and requirements. Conversion of commercial tankers to FPSO may often be the main option in some areas where the environmental conditions are quite benign, and where the challenges in off-loading are more limited.

However, conversion is also adopted in the North Sea and other areas where the environmental conditions may be severe, and where the challenging tandem mode of off-loading has to be adopted. This implies that there are quite considerable variations between system configurations.

Shuttle tankers are used for off-loading purposes from FPSO/FSU units, in largely the same manner as from fixed installations i.e., from fixed, floating or subsea buoy systems. These tankers are usually not built only for one type of service, but the capabilities of the tanker may imply the type of services that it is suitable for. Table 10.2 illustrates some of the variations that may exist, in relation to some of the aspects that are important for avoiding collisions between the shuttle tanker and the FPSO.

10.4.3 Overview of Current Field Configurations

If aspects of vital importance for the collision frequency are considered, quite extensive variations between the different field configurations may be found. Some of these are briefly outlined below.

The distribution of DP-based off-loading, taut hawser and other off-loading modes are shown in Fig. 10.11, for UK and Norwegian sectors.

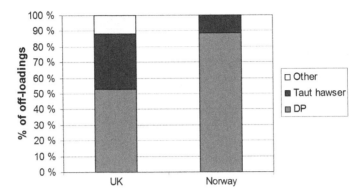

Fig. 10.11 Overview of off-loading modes in UK and Norway

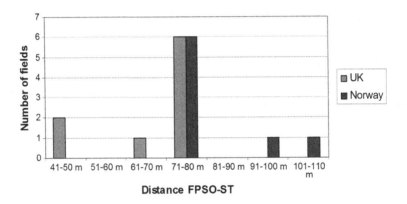

Fig. 10.12 Overview of off-loading distances between FPSO and ST

It is shown that DP-based off-loading dominates (about 90%) in the Norwegian sector, whereas the fraction is just above 50% in the UK.

The distributions for UK and Norwegian sectors of the off-loading distances between the vessels are shown in Fig. 10.12, which is limited to fields with off-loading based on DP operation, thus excluding fields with taut hawser and pipeline or buoy based off-loading.

It is worthwhile noting that the distances range from 50 m up to 75–80 m in the UK fields, whereas they range from 75–80 m up to 100–110 m in the fields of the Norwegian sector. One field with 150 m distance is not included.

An assessment of the available extent of thruster capacity on the FPSOs and FSUs has also been performed. There is a tendency that the Norwegian fields have more thruster power installed, but the difference is not as extensive as that seen for distances in Fig. 10.12.

10.4.4   Characterisation of Shuttle Tanker Collision Hazard

It is common to view the FPSO–shuttle tanker collision hazard as almost entirely a function of the shuttle tanker's technical and operational capabilities. This has been proven to be too simplified a view in Ref. [32]. It is shown that the collision failure model in tandem off-loading may practically be structured into the following two phases:

1. *Initiation* phase: shuttle tanker (ST) in drive off position forward.
2. *Recovery* phase: recovery action fails to avoid collision.

This model applies to collision caused by drive-off of the shuttle tanker, which is shown in Ref. [32] to be by far the most important collision mechanism. The shuttle tanker drift-off forward scenario is considered to have low probability and low consequence in tandem off-loading. Two parameters are defined to characterise these phases:

• Resistance to drive-off—in the *Initiation* phase.
• Robustness of recovery—in the *Recovery* phase.

These two parameters are used in order to identify necessary requirements for FPSO– ST field configuration, which are as favourable as possible in order to minimise or reduce the contributions from HOF aspects to collision probability.

Resistance to drive-off implies various factors relating to control of vessel movements on both FPSO and ST, as listed in the following.

• Shuttle tanker

  – Station-keeping system (including DP, PRS(s), vessel sensors, main CPP(s), thruster(s), and associated propulsion systems)
  – DP operator

• FPSO

  – Station-keeping system (including possible DP, thruster(s), main propulsion, rudder, in addition to turret mooring)
  – Station-keeping operator.

Robustness of recovery implies the capability of the shuttle tanker to initiate successful recovery in drive-off situation, so that collision is avoided. We have to notice further that in the drive-off situation, it is the ST DP operator who initiates and performs recovery actions.

Significant time pressure exists during recovery, since it may only take 120 s for ST in a full ahead drive-off to collide with FPSO stern. Subsequently, recovery actions have to be initiated early enough in order that collision is successfully avoided.

The robustness of recovery is therefore addressed from human action-time perspective i.e., to clarify what factors influence the available time (time window) and what factors influence the time needed for action initiation, respectively. It is considered that the robustness of recovery (made by ST DP operator) is higher if the time window is longer and/or the time needed for recovery action initiation is shorter.

- Factors influencing the available time (time window)

    - Separation distance between FPSO and ST
    - DP class and main propulsion capacity of ST
    - Operational phase

- Factors influencing the time needed for action initiation

    - Alarm design and setting
    - Job attitude and attention level
    - Operator competence via training and operational experience.

A simplified version of this approach considers the following aspects:

• Resistance to drive-off:	FPSO—ST station/heading keeping capabilities • Extensive thruster systems (FPSO) • DP2 class tanker
• Robustness of recovery:	FPSO—ST distance

Figure 10.13 presents a simplified illustration of the robustness-recovery principles, and indicates how current concepts in the North Sea fall in the diagram.

10.4.5 Barrier Modelling

Figure 10.14 outlines a barrier model for the drive-off collision hazard. The model underlines that once an abnormal forward movement has occurred, the DP operator is the only barrier element who may provide the required barrier function of

Fig. 10.13 Simplified illustration of resistance to drive-off and robustness of recovery

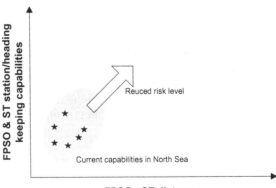

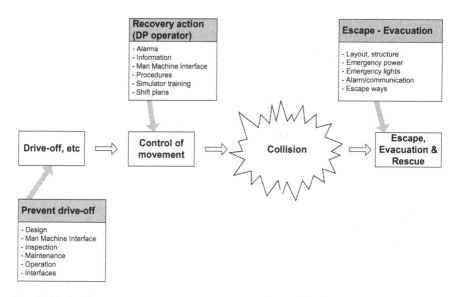

Fig. 10.14 Barrier modelling related to prevention of collision risk consequences

recovery. His or her chance of success depends on a number of technical and operational and personal factors, but the correct and timely action must be performed by the DP operator.

The RIFs that determine the likelihood of success of the collision avoidance performed by the DP operator were identified and structured in the FPSO Operational Safety R&D project (see [32]).

10.4.6 Analysis of Risk Aspects

10.4.6.1 Experience from Incidents, Near-Misses and Questionnaires

Table 10.3 presents an overview of 19 reported incidents e.g., collisions, near misses and 'other' events, for tandem loading with DP tankers in the North Sea in the time period 1995 up to the end of 2002. The field names of the near-misses are not publicly available.

Five accidents occurred in the period 1996–2000. No further accidents then occurred for almost seven years. Then on 13th November 2006, a minor collision occurred on the Njord field, in the Norwegian sector, due to main engine black-out. No other incidents or accidents are known on NCS after 2002, but this has not been

---

Table 10.3 FPSO/shuttle tanker collision incidents and near-misses 1995–2002 (Copyright 2003, Offshore Technology Conference Copyright 2003, Offshore Technology Conference. Reproduced with permission of OTC. Further reproduction prohibited without permission.)

| Year | Sector | Phase | Cause | Type of incident | | | DP class |
				Near-miss	Collision	Other	
1996	UK	Loading	DP failure		X		1
1997	UK	Loading	PRS failure		X		1
1997	UK	Loading	Operator error		X		1
1997	UK	Loading	PRS failure			X	1
1998	UK	Loading	Operator error		X		1
1998	UK	Loading	CPP failure			X	1
1999	Norway	Loading	DP failure	X			2
1999	Norway	Loading	DP failure	X			2
1999	UK	Disconnection	FPSO thrusters tripped	X			1
1999	UK	Approach	DP failure	X			1
2000	Norway	Loading	Operator			X	2
2000	Norway	Disconnection	Manually initiated drive off		X		2
2000	Norway	Approach	DP failure	X			2
2000	Norway	Connection	Technically initiated drive off	X			2
2000	UK	Connection	Operator error	X			1
2001	Norway	Loading	PRS/DP failure	X			2
2001	UK	Loading	Technically initiated drive off	X			1
2002	UK	Loading	Rapid wind change	X			1
2002	UK	Loading	Engine failure			X	1
2003	Norway	Loading	PRS/DP failure	X			1

researched as thoroughly as for the period 1996–2002, and therefore not included in Table 10.3. The following are the accidents that have occurred [33]:

- 1996: Emerald FSU, 28th February 1996
- 1997: Gryphon FPSO, 26th July 1997
        Captain, 12th August 1997
- 1998: Schiehallion FPSO, 25th September 1998
- 2000: Norne FPSO, 5th March 2000
- 2006: Njord B FSU, 13th November 2006
- 2009: Schiehallion FPSO, 8th October 2009
- 2012: Offshore Brazil, 31st May 2012.

There is no publicly available investigation report for the 2009 Schiehallion collision by Loch Rannoch tanker into the stern of the Schiehallion FPSO.

Table 10.4 Classification of collision incidents and near-misses, 1996–2002

	Off-loading phase			
	Approach	Connection	Loading	Disconnection
Collisions				
DP1 tanker	0	0	4	0
DP2 tanker	0	0	0	1
Near-misses				
DP1 tanker	1	1	3	1
DP2 tanker	1	1	4	0

Available information suggests it occurred during the connection phase between the two vessels, without significant oil spill nor injury to personnel. Reference systems errors are mentioned. Details about the collision offshore Brazil in May 2012 are also unknown, the incident may have been a DP1 drive-off scenario.

Table 10.4 presents a summary of the incidents and near-misses in the period 1995–2002 (excluding the category 'other' in Table 10.3), focusing on which off-loading phase and the DP class of the shuttle tanker.

For the collision incidents, the details are known from the period 1996–2002. Four of the incidents occurred with DP1 tankers, only one incident with DP2 tankers. The four incidents with DP1 tankers occurred during the loading phase, the one incident with a DP2 tanker occurred during disconnection.

For the 12 near-misses, just over half of them occurred during loading, two during approach and connection, with one near-miss during disconnection. The near-misses are evenly distributed between DP1 tankers and DP2 tankers. Another way to express this is through the ratio between near-misses and incidents, which is quite different for DP1 and DP2 tankers:

- DP1 tankers: 6:4
- DP2 tankers: 6:1.

Although the data basis is limited, these can be interpreted as indications that the chance to control abnormal occurrences is better for DP2 tankers than for DP1 tankers, which is not a big surprise.

What is the significance of the differences between DP1 and DP2 shuttle tankers, the latter with redundancy in all active components of the DP system? This is important both for the occurrence frequency of abnormal behaviour (such as forward movement during off-loading), and the ability to recover the situation.

10.4.7   Trends in Occurrence Frequencies

Figure 10.15 presents a summary of the near-misses ('incidents') and accidents in UK and Norwegian sectors. It may be observed that the accidents occurred in the

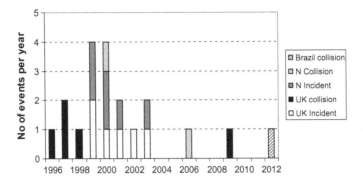

Fig. 10.15 Overview of incidents and collisions in UK and Norwegian sectors

UK sector in the period 1996–1998, and that a similar number of incidents have occurred since then. Detailed information is unavailable for UK after 2003.

In the Norwegian sector, there were incidents in the period 1999–2003, and accidents in 2000 and 2006. The overall picture is a falling trend, for accidents as well as incidents. There is on the other hand insufficient data in order to determine conclusive trends.

10.4.8 Collision Energy and Consequences

Most of the collision impacts that have occurred have been at very low speed. Of the five accidents listed in Table 10.3, the collision in 1996 is completely unknown with respect to the circumstances of the impact, although the contact in general is known to have been very light. For the contacts in the UK sector, the following contact speeds have been calculated from available information:

- 0.10 knots
- 0.11 knots
- 0.16 knots.

Thus the contact energy in these incidents has been very low, in the range 0.13–0.32 MJ. The impact speed of the Norne accident was stated as 0.6 m/s (1.2 knots), and the impact energy was 31 MJ, which is one of the most powerful collisions ever in north-west European waters. The impact speed of the Njord accident was stated as 1.2 m/s (2.4 knots, see Sect. 9.1.2.5), and the impact energy was 55 MJ This emphasises the potential for high collision energies in these scenarios.

The consequences in the UK accidents have been negligible. Only minimal structural damage was caused to FPSO and shuttle tanker in the Norne and Njord impacts.

It should be noted that quite high collision energy would be needed in order to damage a cargo tank. Most shuttle tanker impacts will occur in the stern of the FPSO/FSU, which is quite some way from the cargo holding section, usually shielded by engine room(s) and similar.

10.4.9   Accidents and Incidents for Taut Hawser Configurations

It was stated in Sect. 10.4 that taut hawser is not considered in the risk assessment. It may nevertheless be worthwhile to comment on the occurrences of accidents and incidents.

No report on collisions or near misses during loading with taut hawser has been found, but one incident of hawser breakage is known. It is known that Petrojarl 1 (taut hawser) had one near-miss during its first months of operation, in late 1986, but the actual number of incidents with taut hawser prior to 1995 is unknown.

It should also be noted that the extent of experience data for taut hawser off-loading is substantially lower than for DP operated off-loading, probably 20% or less. Even if the probability of collision was the same with taut hawser as for DP tankers, it would not be unlikely that no accidents had occurred.

10.4.10   Main Contributors to Collision Frequency, in Drive-Off

An assessment of the main contributions to collision frequency has been performed in Ref. [26], based on incident experience as well as various expert evaluations (see Table 10.5). The contributions to collision frequency in the table, which also are presented in the diagram, should be considered order-of-magnitude values, rather than exact predictions.

The combination of technical and human/operational dependability is judged to be the most significant contributor; assessed by expert judgement to cause about 40% of the collisions.

Human/operational factors contribute alone as well as in combination with other factors. Actually it is assessed that human/operational factors, possibly in combination with other factors, may contribute to 80% of all collisions. This results when all sectors of the 'human/operational' circle in the diagram are considered, including the sectors that overlap with other factors.

Similarly, technical dependability (possible in combination with other factors) is judged to contribute to about 70%, and external conditions is in total judged to contribute to about 35% of the collision incidents. The percentages add up to more than 100%, due to the overlaps being counted twice.

Table 10.5 Ranking of RIF group combinations (expert judgments)

RIF group/RIF group combination	Ranking	Contribution (%)
1. Technical dependency alone	4	10
2. Human/operational dependency alone	2	15
3. External conditions alone	7	2
4. Technical *and* Human/operational (in combination)	1	40
5. Technical *and* External (in combination)	6	8
6. Human/Operational *and* External (in combination)	5	10
7. Technical *and* Human/operational *and* External (in combination)	3	15

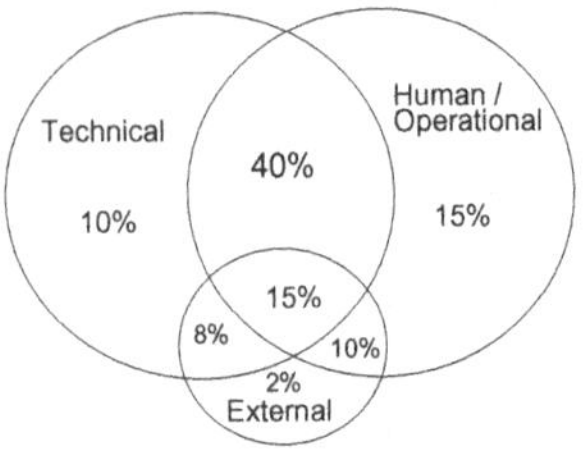

Venn diagram labels: Technical 10%, Human/Operational 15%, 40%, 15%, 8%, 10%, 2% External.

10.4.11 Experience Data

10.4.11.1 Norwegian Sector

The assessment of experience data has been based on input from the following sources:

- Statoil (now Equinor) [29]
- COAST [30].

The main statistics in this subsection is based on work done in the research project until 2003. Values are also available for NCS for the recent years. The total number of shuttle tanker loadings in the period 1 January 1995 to 31 December 2011 is calculated to over 3000 operations, with a build-up as shown in Fig. 10.16. From a modest start in 1996, the volume of visits has been quite stable since year 2000, somewhat above 300 loadings per year in the early start of the decennium. Many of the fields are in their decline phases at the time of writing. The additional data is based on RNNP [15].

In Norway, there is only one field with taut hawser operated off-loading, the Glitne field, with Petrolarl 1 FPSO, which is not included in the diagram.

10.4.11.2 UK Sector

The assessment of experience data has been based on input from the following sources:

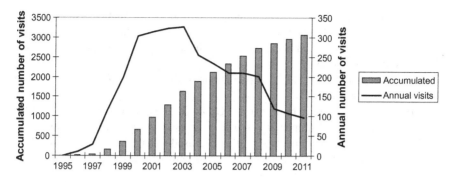

Fig. 10.16 Shuttle tanker loadings from FPSOs/FSUs, DP operated, Norway

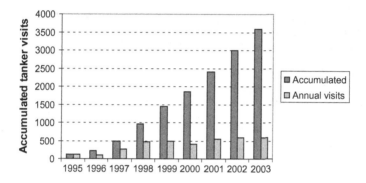

Fig. 10.17 Shuttle tanker loadings from FPSOs/FSUs, DP operated, UK

- COAST UK [12]
- Input from Navion [31].

The total number of shuttle tanker loadings in the period 1 January 1995 to 31 December 2003 is calculated to 3600 operations, with a build-up as shown in Fig. 10.17.

These operations started earlier in the UK than in Norway, but still had a modest volume in 1995. The annual volume of visits was in the 1999–2003 period around 500 operations per year, with an increasing trend.

In the UK, there is a significant volume of operations on fields with taut hawser based off-loadings. This is not included in Fig. 10.17.

10.4.11.3   DP1 Versus DP2 Tankers

In Norway, there has been an industry practice to require DP2 tankers for all fields North of the 62nd parallel since the first FPSO was installed in these waters. Since

1 January 2002, there has been a regulatory requirement [32] that DP2 tankers shall be used on all fields. These regulations do not apply to installations having consent dating to before 2002, and there are some fields in the North Sea where this requirement does not apply.

It has been assumed that 90% of operations in the Norwegian sector have been done with DP2 tankers.

In the UK there is a requirement in the Oil & Gas UK FPSO Guidelines [33] for DP2 to be used in environmentally sensitive areas or Atlantic frontier, otherwise there are no specific requirements for DP2 to be used. It has been assumed that 20% of operations in the UK sector have been done with DP2 tankers.

10.4.11.4 Installation Years

An alternative way to express exposure data is by means of installation years for the FPSO/FSU installations that are being served by the shuttle tanker s. For the period 1995–2003, the following values have been calculated:

- Norway: 39.5 installation years
- UK: 70.1 installation years
- Total: 109.5 installation years.

The total number of FPSO/FSU (DP operated) installation years in the period 1996–2011 is 92.

10.4.12 Accident Frequencies—1996–2003

The principles adopted in the R&D project on 'FPSO Operational Safety' [34] and in the work by Chen [35] suggest a detailed modelling of frequency of FPSO–shuttle tanker collision, which should include the following aspects:

- Resistance to drive-off
- Robustness of recovery.

These principles were outlined in Sect. 10.4.4, which referenced the most important parameters to be distance between FPSO and shuttle tanker, DP class of shuttle tanker and FPSO position and heading keeping beyond anchoring system. Chen [17] suggests a relatively detailed model which could be used directly. A detailed model however, is rarely useable in early concept phases, but should be performed during pre or detailed engineering.

The assessment presented here uses research data for the period 1996–2003, and is a relatively coarse assessment based on historical incidents, where the shuttle tanker DP class (part of resistance to drive-off) is the only parameter which is explicitly modelled. This assessment distinguishes between DP1 and DP2 class.

This is done because DP class is considered to be an important parameter for the resistance to drive-off.

Frequencies may be expressed per 'installation year' or per 'off-loading operation'. There is an extensive variation between installations from around a dozen off-loading operations per year to fields where off-loading occurs more than twice per week (i.e. between 100 and 150 times per year).

It is therefore considered that on a 'per off-loading' basis is the most appropriate, in order to reflect differences between field characteristics.

10.4.12.1   Average Collision Frequency

Based on the data reported in Sect. 10.4.11, the following average collision frequency for UK and Norwegian sectors in the period 1995–2003 is calculated:

- 1.0×10^{-3} per offshore loading (1 collision per 989 offshore loadings)

This value reflects five collisions and about 4950 offshore loading operations. Based on data reported in Sect. 10.4.11, the following collision frequency for DP1 tankers in UK and Norwegian sectors in the period 1995–2003 is calculated:

- 1.51×10^{-3} per offshore loading (1 collision per 662 offshore loadings)

This value reflects four collisions and about 2650 offshore loading operations.
Based on the data reported in Sect. 10.4.11, the following collision frequency for DP1 tankers in UK and Norwegian sectors in the period 1995–2003 is calculated:

- 4.4×10^{-4} per offshore loading (1 collision per 2300 offshore loadings)

This value reflects one collision and about 2300 offshore loading operations.

10.4.12.2   Trends in Collision Frequencies

The data basis for the assessment is relatively sparse, as can be clearly seen from Sect. 10.4.11. Thus it is questionable whether trends should be calculated or not, because they could be rather uncertain.

On the other hand, the last collision in the UK sector occurred a number of years ago (September, 1998), and there has been more than 4000 off-loading operations since then resulting in near-misses only. If accumulated values for UK sector since 1995 are computed for each year in the period, the result is as shown in Fig. 10.18. DP1 and DP2 tankers are included simultaneously.

A similar diagram may be computed for DP1 tankers, taking experience data both from UK and Norwegian sectors; see Fig. 10.19.

A further perspective may be created from integrating collisions and near-misses. If we consider DP1 tankers alone, the last collision occurred in 1998,

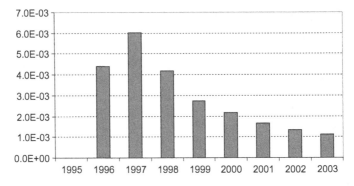

Fig. 10.18 Accumulated collision frequencies for UK sector since 1995

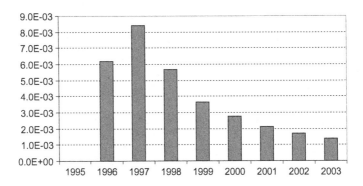

Fig. 10.19 Accumulated collision frequencies for DP1 tankers since 1995

but there have been near-misses each year since then. The accumulated frequencies for collisions and near-misses are shown in Fig. 10.20.

All of the diagrams presented suggest a falling trend. The question is which frequency should be considered representative for future operations. Consider the following predictions:

- Average collision frequency, DP1 tankers, 1995–2003: 1.51×10^{-3} per off-shore loading
- Average collision frequency, DP1 tankers, 1998–2003: 4.4×10^{-4} per offshore loading.

There has been one collision for DP2 tankers in the period 1996–2003, but two when the longer period is considered. No trend is sensible to calculate. It should further be noted that the differences are not statistically significant, due to the limited extent of data. This is clearly shown in Fig. 10.21.

It would thus appear that the frequency for DP1 tankers based on the period 1998–2003 is not representative. If a second accident occurs, that would double the

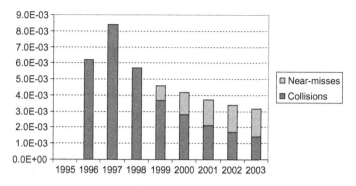

Fig. 10.20 Accumulated collision and near-miss frequencies for DP1 tankers since 1995

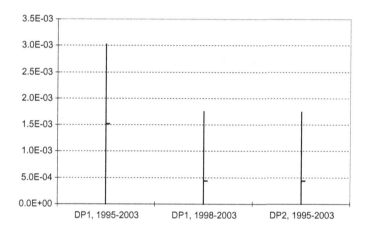

Fig. 10.21 Comparison, calculated values and prediction intervals, DP1 and DP2 tankers

frequency, if the frequency from 1 January 1998 to date was considered. The frequency for the entire period 1995–2003 is on the other hand probably too high to be representative, and a value between these two values should be used.

10.4.13   Accident Frequencies—1996–2011

Figure 10.22 presents the trend in collision frequency for NCS over the period 2000–2011, based on 10 year rolling average.

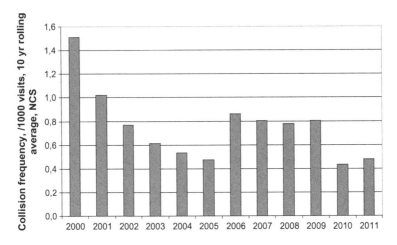

Fig. 10.22 Collision frequency as 10 year rolling average, NCS, per 1000 visits

10.4.14   Updated Frequencies

The latest update of historical shuttle tanker collision [with FPSO] risk was performed by Lundborg in his MSc thesis in 2014 [36], for the period 2004–2013. The
values from Ref. [36] are presented in Table 10.6. The values are mainly the same
as the ten year averages reported for the period 2002–2011.

10.4.15   Analysis of Timelines in Drive-Off Incidents

Dong et al. [37] has investigated timelines for nine anonymous incidents, some
resulting in impact with FPSO or loading buoy, others only resulting in drive-off
near-miss. Figure 10.23 presents one example timeline for one of the nine incidents
analysed.

Figure 10.24 presents the summary of timelines for the nine anonymous incidents, relevant for FPSOs and loading buoys. Loading buoys are included in order
to increase the data basis. It is considered that collision with loading buoys are just
as relevant as collision with FPSO for the purpose of drive-off and their timing.

Table 10.6 FPSO—Shuttle tanker collision risk, 2004–2013 [36]

Country	Drive-off probability	Recovery failure probability	Collision probability
Norway (NCS)	1.4×10^{-3}	0.33	4.6×10^{-4}
UK (UKCS)	4.4×10^{-4}	1.0	4.4×10^{-4}

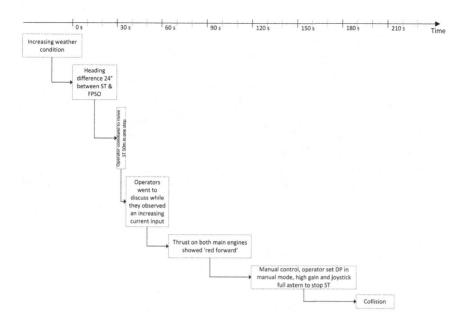

Fig. 10.23   Example timeline for anonymous collision (adapted from Ref. [37])

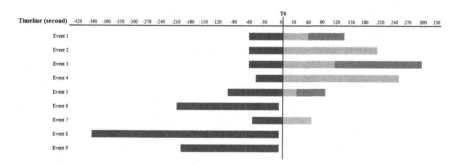

Fig. 10.24   Summary of timelines for nine anonymous collision (adapted from Ref. [37])

Figure 10.24 uses three colours according to the different phases the DP system may be in. T0 in the diagram refers to when the position (or angle) of the ST leaves the green zone. The zones are as follows:

- Green zone: Normal deviations within normal operating zone
- Yellow zone: Deviations outside green zone, within yellow warning limits
- Red zone: Deviations outside yellow zone, within red warning limits.

It should be noted that abnormal conditions (such as contaminated fuel) may exist for a long time, and that the crew may be aware of such conditions. The timeline analysis [37] considers the time after arrival on the field, and disregards whatever took place before.

It is seen that a minimum of 60 s is usually available from the start of abnormal behaviour until T0. The time is shorter (typically down to 40 s) in a few cases, and also considerably longer in other cases. A period of 60 s is really quite short in order for the DP operator to appraise the situation, decide on the best course of actions and implement these actions. Such short time (mainly dependent on the distance) is the main source of insufficient robustness (see also Fig. 10.13). A lack of robustness is most likely the main contributing factor to DP operator (DPO) failure to respond sufficiently early, as discussed in Ref. [21].

Improvement of decision-making in these circumstances is the main objective of the 'on-line risk modelling' concept see below.

10.4.16   On-Line Risk Modelling

The traditional approach to DPO notification of potential problems is focused on alarms as means for informing operators about any deviations in systems outside the operating envelope. Alarms are not proactive and often leave only a short time for operators to react properly. 'Alarm overflow' is very often a significant aspect, which makes the alarm function virtually useless. For the operator to achieve full situation awareness and the knowhow on how to act, the prediction of incidents, near-misses and accidents is necessary.

Developments in wireless and sensor technology are opening up the possibility of increased use of on-line measurements and automation for operator decision support. The amount of information a person can comprehend and utilize depends on the available time in operation and the method of data representation. DP operations are characterized by a need to rapidly understand changes in the state of complex processes from real time sensor data and video.

The development of on-line risk monitoring and decision support for safer marine operation involves risk assessment and modelling, data models and representation, sensors and communication technology, visualization, human-machine interface (HMI), and organization theory, as well as system integration. On-line decision support systems leads to improved functionality in safety critical software-based systems, for example, in terms of increased automation and autonomy in the systems, better informed operators, less manual operation and intervention, and longer response time if manual intervention becomes necessary.

A risk monitoring and decision support system needs to be separated from other systems and provide functionality to constitute an independent barrier function. More details are described by Vinnem and Utne [38], Hogenboom et al. [39] and Dong et al. [40]. On-line risk monitoring will supply the operators in marine operations with a real time risk picture and pre-warnings of possible system deviations. Such a system is illustrated in Fig. 10.25 and consists of a number of modules and layers, such as risk models, data models, human–machine interface

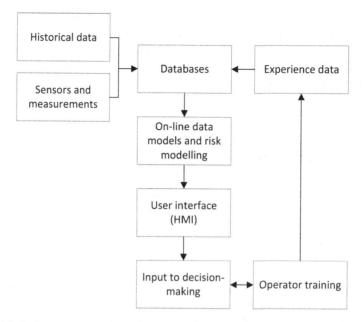

Fig. 10.25 Outline of on-line risk modelling (adapted from Ref. [38])

(HMI), as well as decision support. Input data are historical data, on-line sensor data and experience data.

The on-line risk models build upon data from different sources and the data models may range from empirical models based on historical or on-line data to physics-based models. Without on-line data, the models' effectiveness with respect to supporting operational decision-making is quickly lost. The risk models can be based on quantitative risk assessment, fault tree analysis, event tree analysis and Bayesian belief networks.

Current risk methods do not take the time aspect into account sufficiently well. Risk should be modelled as a function of time (in terms of precursor warnings of allowable windows to avoid loss of control). The timelines (see Sect. 10.4.15) suggest how frequently the updating of the risk models has to take place. The risk models set requirements for data input and information. For efficient decision support, uncertainty has to be addressed because it is essential for generating probabilistic risk information. Further, the on-line system may require improved sensors and data measurements, to provide better state representation and more advanced analysis of existing data. This may be important, for example, for giving early warning and preparing operators for rapid manual actions in those cases where the decision support system does not behave autonomously. Part of the input may also be based on simulations of possible scenarios for training of the operators.

The performance of FPSO–ST tandem off-loading systems has improved over the past 10–15 years, but the statistics show that considerable improvement is still needed. If the on-line risk monitoring and decision support system is to constitute a barrier function it has to be independent of other barriers, including the DP system. The advisory functionality should enable the operator to make better and more timely decisions, including independent automatic avoidance manoeuvring possibly as a last resort. Ongoing research focuses on how to model risk and how to give the most effective input to decision-making.

10.5   Loss of Buoyancy Due to Gas Plume

Another hazard which was considered in depth in the RABL programme, is the possible loss of buoyancy due to a gas plume in the water, typically from a subsea gas blowout. Such occurrences have been reported on a few occasions for drill ships.

Model testing was carried out with an extensive programme during the course of the RABL project [41], in which very extensive gas blowout rates were simulated. In no case could any loss of buoyancy be registered, the only effect was in fact a net upward force with the highest gas flow rate s, due to the high velocity upwards flow of water caused by the gas plume (see also Sect. 11.5).

Riser rupture just below the pontoons was also simulated, with the same result. The conclusion from the tests was therefore that loss of buoyancy due to a gas plume in the sea is not a feasible hazard for a semi-submersible installation.

Sandvik [18] refers to a survey conducted, in which 11 incidents were reported in relation to loss of buoyancy, eight incidents involving barge or ship type structures, and three with semi-submersibles. The major conclusions were:

- The sinking could not be related to plume density reduction in any of the cases where the vessel sank (about one third). Hull damage due to explosion and down-flooding of open compartments were the major factors.
- Some apparent loss of freeboard and a list or heel angle into the boil were observed in most cases, especially on ships or barges.
- Low freeboard ships or barges were most prone to sinking. They experienced large amounts of water on deck. This effect was not observed on semi-submersibles. The following phenomena are the main contributors to ship flooding:

  - Elevated water surface and froth layer in the central boil
  - Heeling moment caused by the interaction of the flow force on the hull and the anchor system with mooring lines from deck level, such as indicated in Fig. 10.26.

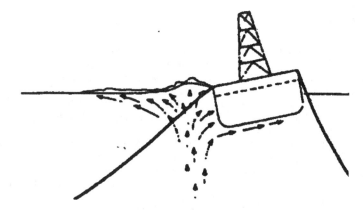

Fig. 10.26 Single hull vessels are pushed to one side, and the mooring line forces can cause heeling into the plume

10.6 Accidental Weight Condition

An accidental weight condition may in theory lead to capsize of a semi-submersible installation, for instance unsymmetrical weight conditions, or partial anchor line failure. The RABL project concluded [42] that the probability of these events was very low, and may be disregarded.

It should be noted that the failure of two anchor lines on Ocean Vanguard (see Sect. 4.28) in the Norwegian Sea in December 2004 caused a list of some 8–10°, which is a significant list, but not at all critical.

10.7 Tow-Out and Installation Risk

Risks associated with tow-out and installation of offshore structures and modules include risks during transportation and during installation. Installation risk is mainly a dropped object phenomenon (see Sect. 11.1) where an incident during installation is briefly reviewed in the introduction.

Hazards during transportation include collision, grounding, loss of buoyancy or stability; these hazards have been considered earlier in this chapter. The analysis of risks during tow-out is discussed by Trbojevic et al. [43].

One case of failure is known from the North Sea in the late 1970-ties, where during tow-out of a jacket structure, partial loss of buoyancy occurred just a few hundred meters prior to reaching the location, and the structure was lost without the possibility for recovery.

References

1. PSA (2011) Framework regulations; petroleum safety authority, Norwegian pollution control authority and the Norwegian social and health directorate; Stavanger
2. PSA (2011) The management regulations; petroleum safety authority and the Norwegian social and health directorate; Stavanger
3. Kvitrud A, Nilsen LR, Næss TI, Dalsgaard LJ, Vinnem JE (2006) In service Experiences 1996–2005 in Norway, and risk analysis of anchoring systems. In: Soares C (ed). Proceedings ESREL 2006. Balkema: London
4. Vinnem JE, Kvitrud A, Nilsen LR (2006) Stability failure for installations in the Norwegian sector (In Norwegian only). Stavanger; PSA. http://www.ptil.no/Norsk/Helse+miljo+og+sikkerhet/HMS–aktuelt/6_stabilitet_risikoanalyser.htm
5. PSA (2015). Trends in risk levels, main report 2014 (In Norwegian only), Petroleum Safety Authority; 24.4.2015
6. Vinnem JE, Haugen S (1987) Risk assessment of buoyancy loss—introduction to analytical approach, paper to City University International conference on mobile units, London 14–17 September 1987
7. Lotsberg I, Olufsen O, Solland G, Dalane JI, Haver S (2004) Risk assessment of loss of structural integrity of a floating production platform due to gross errors. Mar Struct 17:551–573
8. HSE (2005) Accident statistics for floating offshore units on the UK Continental Shelf (1980–2003). London; HMSO; 2005; RR353
9. MMS (2005) Multiple cable transit failures, safety alert No. 235, minerals management service, Washington; 2005
10. Nilsen LR (2005) Marine Systems in risk analyses for mobile units (In Norwegian only). University of Stavanger; M.Sc. thesis
11. Haugen S (2005) Risk analysis of maritime systems. Trondheim; Safetec; 2005 Sep. Report No.: ST–20649–RS–3–Rev00
12. Sklet S, Aven T, Hauge S, Vinnem JE (2005) Incorporating human and organizational factors in risk analyses for offshore installations. In: Kolowrochi K (ed) Proceedings ESREL 2005. Balkema, London
13. Vinnem JE, Seljelid J, Aven T, Sklet S (2006) Analysis of barriers in operational risk assessment—a case study. In: Soares C (ed), Proceedings ESREL 2006. Balkema, London
14. Ersdal G, Friis–Hansen P (2004) Safety barriers in offshore drill rigs derived from accident investigation. In: Proceedings of the 23rd international conference on offshore mechanics and artic engineering; Vancouver, Canada
15. PSA (2012) Trends in risk levels, main report 2011 (In Norwegian only), Petroleum Safety Authority; 24.4.2012
16. Næss TI, Nilsen LR, Kvitrud A, Vinnem JE (2005) Anchoring of installations on the Norwegian continental shelf (In Norwegian only). Stavanger; Petroleum safety authority. http://www.ptil.no/NR/rdonlyres/B186B607–98EB–4F25–8B8D–71447322B48B/10306/2005RapportForankring.pdf
17. Haver S, Vestbøstad TM, (2001) The storm outside mid Norway 10–11 November, 2001 (In Norwegian only). Stavanger; Statoil; 2001 Nov. Report No.: PTT–KU–MA–024
18. Denton N (2002) FPSO mooring system integrity study, Rev 02; Noble Denton, London; 2002 Report No.: 2002\A3792–02
19. PSA (2014) Trends in risk level in the petroleum activity, summary report—trends 2013, petroleum safety authority, 24.4.2014
20. Lloyds Register Consulting (2005) Safety of DP operations on mobile offshore drilling units on the Norwegian Continental Shelf. Lloyds Register Consulting; 2005 Dec. Kjeller, Norway; Report No.: 27.740.114/R5
21. Chen H, Moan T (2005) DP incidents on mobile offshore drilling units on the Norwegian Continental Shelf. In: Kolowrochi K (ed) Proceedings ESREL 2005. Balkema, London

22. Chen H (2006) Summary, safety of DP drilling operations. Kjeller; Lloyds register consulting; 2006 Jun. Report No.: 61.000.000
23. Chen H, Moan T, Verhoeven H (2006) Barriers to prevent loss of position for dynamically positioned mobile offshore drilling units on the Norwegian Continental Shelf. In: Soares C (ed) Proceedings ESREL 2006. Balkema, London
24. Chen H, Moan, T Verhoeven H (2006) Critical DGPS failures on dynamically positioned mobile offshore drilling unit on the Norwegian Continental Shelf. In: Soares C (ed) Proceedings ESREL 2006. Balkema, London
25. Vinnem JE (2000) Operational safety of FPSOs: initial summary report. OTO Report No.: 2000:086; HSE, 2001
26. Vinnem JE, Hauge S, Huglen Ø, Kieran O, Kirwan B, Rettedal WK et al (2000) Systematic analysis of operational safety of FPSOs reveals areas of improvement. In: SPE international conference on health, safety, and the environment in oil and gas exploration and production; 2000 June 26–28; Stavanger, Norway
27. Vinnem JE, Hokstad P, Saele H, Dammen T, Chen H et al (2002) Operational safety of FPSOs, shuttle tanker collision risk, main report. Trondheim; NTNU; 2002 Oct. Report No.: MK/R 152
28. Vinnem JE, Hokstad P, Dammen T, Saele H, Chen H, Haver S et al (2003) Operational safety analysis of FPSO—shuttle tanker collision risk reveals areas of improvement, OTC paper 15317. OTC conference 2003; Houston, USA
29. Tveit O (2003) Private communication with O Tveit, Statoil
30. Safetec Nordic AS (1996) Coast database, Rev. 1. Trondheim; Safetec
31. Wiik O (2003) Private communication with Olav Wiik, Navion
32. PSA (2011) Activity regulations; petroleum safety authority, Norwegian pollution control authority and the Norwegian social and health directorate; Stavanger
33. Oil & Gas UK (2002) FPSO tandem loading guidelines. Oil & Gas UK, London; 2002
34. Vinnem JE (2003) Operational safety of FPSOs, shuttle tanker collision risk, summary report, RR113. HMSO: London
35. Chen H (2002) Probabilistic evaluation of FPSO—tanker collision in tandem off–loading operation. Trondheim; NTNU. During thesis
36. Lundborg ME (2014) Human technical factors in FPSO-shuttle tanker interactions and their influence on the collision risk during operation in the North Sea, MSc thesis, NTNU, June 2014
37. Personal communication with PhD student Anna Dong, NTNU
38. Vinnem JE, Utne IB (2015) Risk reduction for floating offshore installations through barrier management. In: Proceedings of the ASME 2015 34th international conference on Ocean, Offshore and Arctic Engineering, OMAE2015, May 31-June 5, 2015, St. John's, Newfoundland, Canada
39. Hogenboom S, Rokseth B, Vinnem JE, Utne IB (2018) Human reliability and the impact of control function allocation in the design of dynamic positioning systems, reliability engineering and system safety. https://doi.org/10.1016/j.ress.2018.12.019 (in press)
40. Dong Y, Vinnem JE, Utne IB (2017) Improving safety of DP operations: learning from accidents and Incidents during Offshore Loading Operations. EURO journal on decision processes, vol 5
41. Sandvik P (1988) Hydrodynamic effects from subsea gas blowouts. Safetec Nordic, Trondheim; 1988 Feb. Report No.: ST–87–RR–007–02. Appendix 2 to RABL project report No 1
42. Vinnem JE (1988) Risk assessment of buoyancy loss, summary report. Safetec Nordic, Trondheim, Norway; 1988; Report No.: ST–87–RF–024–01
43. Trbojevic VM et al (1994) Methodology for the analysis of risks during the construction and installation phases of an Offshore Platform. J Loss Prev Process Ind 7(4):350–359

Chapter 11
Risk Due to Miscellaneous Hazards

11.1 Crane Accidents

Accident and incident experience may be used in order to illustrate the risk picture. If we start with dropped objects, the main characteristics in the North Sea are as follows:

- Several fatalities have been caused when the entire crane has toppled overboard, but this was before 1990.
- Equipment damage has been caused by falling load impact on the deck.
- Subsea wellheads have been damaged especially as a result of BOPs falling during exploration drilling.

Fatal crane accidents were quite frequent in the period 1975–1985, with approximately one such accident per year in the North Sea. However, only one accident has occurred since 1985. This occurred in 1988, when the crane hook caught on a vessel due to heavy swells and the crane was dragged overboard. All of the fatal accidents have occurred in the British sector of the North Sea.

One somewhat special crane accident occurred on 3 December 1998 during installation of the production deck for the compliant tower platform on the Petronius field in the US Gulf of Mexico, in some 530 m water depth, where one of the two deck modules for Texaco's Petronius project fell into the Gulf of Mexico during installation as it was being lifted into place by J. Ray McDermott's DB50 barge. The north module was hoisted by the DB50 into place on the compliant tower structure earlier the same day, and was in the process of being secured. The north module weighed 3,876 tons and contained wellbay, power and compression equipment. The south module which fell into the sea weighed 3,605 tons. The module contained the production equipment, water flood facilities and crew quarters.

© Springer-Verlag London Ltd., part of Springer Nature 2020
J.-E. Vinnem and W. Røed, *Offshore Risk Assessment Vol. 1*,
Springer Series in Reliability Engineering,
https://doi.org/10.1007/978-1-4471-7444-8_11

The south module was in the process of being lifted when it suddenly broke from its support. The module struck the transport barge as well as the DB50 before falling to the sea floor. Both barges sustained some damage.

It is claimed sometimes that less serious crane accidents are quite frequent, but it appears that systematic recording is not performed. A detailed study of causes of dropped loads has been performed for PSA, but is only available in Norwegian [1].

11.1.1 Modelling of Dropped Object Impact

The modelling of risk associated with dropped objects is often formulated as follows:

$$P_{FDI} = \sum_I N_i P_{Di} \sum_J P_{Hij} P_{Fij} \tag{11.1}$$

where

P_{FDI} probability of equipment failure due to dropped object impact
N_i number of lifts per load category, i
P_{Di} probability of load dropped from crane for load category i
P_{Hij} probability of equipment j being bit by falling load in category i, given that the load is dropped
P_{Fij} probability of failure of equipment j given impact by load in category i.

The probability of hitting equipment in particular requires further modelling depending upon the type of equipment that could be hit, the process equipment on deck, the support structure above or below water, and/or the arrangement of subsea installations.

In addition to the probabilities of hitting equipment, the energy of the impacting load and the energy transfer both have to be established as the probabilities of failure due to impact are very dependent on energy levels. Each of these aspects is discussed separately below. The physical modelling of the fall is considered first.

11.1.1.1 Crane Load Distributions

Cranes are of vital importance to the operation of an offshore platform. During the drilling period in particular the cranes are operated almost continuously. Even during normal production operations the cranes are used regularly.

The loads handled with cranes differ in weight from light loads to multiple drill collars with weight to up to about 30 tons. In addition blowout preventers (BOPs), which can weigh up 150–220 tons, are handled with the derrick drawworks. The mass of an object and its velocity determine the energy it will gain through a free fall, and thus the damage it might cause. It is also likely that the probability of crane

Table 11.1 Load distribution for different phases of production

Load categories	Load distributions (%)	
	Simultaneous drilling and production	Normal production
Heavy or multiple drill collars	22.2	0.0
Other heavy (>8 tons)	0.3	0.7
Medium heavy (2–8 tons)	27.1	33.6
Light (<2 tons)	50.5	65.7
Number of lifts/year	20,884	8,768

failures increase with increasing weight. Therefore, it is important to obtain statistics on the load distribution for crane activity. Table 11.1 presents two load distributions that are considered to be representative.

This table shows a typical number of crane operations per crane during one year for a production installation, both for simultaneous drilling and production, and for normal production operations. Other surveys have given a range from 2,700–30,500 lifts per year, depending on the number of cranes (from one to four cranes).

BOPs are mainly moved during drilling or workover, and usually not by the crane. During the drilling of one well, the BOP may be moved 1–5 times, by special lifting/transporting equipment and derrick drawworks. There are several known instances where a BOP has been dropped during such movements. This may cause damage to the BOP itself, but also cases of damage to the subsea installations are known. A fatality has also occurred in one instance when a BOP was dropped.

In addition to falling loads from a crane, there have been various cases of boom fall and crane fall in the North Sea. The first type of accident arises when the crane boom (typically 25 tons) falls from the crane, and the second when the entire crane structure (typically 60 tons) breaks loose from its base.

The main hazard associated with the fall of a crane structure or boom is that the crane driver may not be able to escape in time, and thus be dragged under the water with the crane.

11.1.2 Physical Aspects of Falling Loads

There are principally two cases that need to be considered separately in modelling the fall of a load from a crane. These are:

- Loads that are dropped onto equipment/structures on the deck or otherwise above the sea surface.
- Loads that are dropped over the sea with the possibility to hit structures in the water or on the sea bottom.

The first case has only one phase, whereas the second case has three phases, the fall through air, the impact with the sea surface, and the fall through the water. The following discussion is focused on these three phases which, implicitly address the first case as well.

11.1.2.1 Fall Through Air

Friction loss during the fall through air is negligible, due to the high specific weight of loads and thus a falling object will accelerate towards the sea surface in accordance with the force of gravity. The sideways movements will be determined by possible movements of the platform (applicable to floating units only) and the crane hook. Typically the dropped object will hit the sea at an angle within $3°$ of the angle it was positioned when on the crane hook.

11.1.2.2 Impact with Water

A falling object will hit the sea surface with the velocity v_1, and proceed through the water with the velocity v_2. These two velocities are given by the following equations:

$$v_1 = \sqrt{2gh} \tag{11.2}$$

$$v_2 = v_1 - \int_0^t \frac{P(t)}{m_{f_0}} dt \tag{11.3}$$

where

g gravity acceleration
h height from which the drop occurs
$P_{(t)}$ impact force
m_{f_0} object's mass.

The integral represents the loss of momentum during the impact with the water surface. It is shown that this integral is a function of:

- the density of the object
- the impact angle with the water surface
- the mass of the falling object
- the density of water.

After the impact the object will accelerate towards its terminal velocity, v_t, given by:

$$v_t = \sqrt{\frac{2(W-O)}{C_d \cdot A \cdot \rho}} \qquad (11.4)$$

where

W   gravity force (in air)
O   buoyancy force
ρ density of water
A   cross-section area
C_d shape coefficient of the object depending on the Reynolds number

It is also known that an object will tend to oscillate sideways during the fall through water. These oscillating movements are determined by the impact angle with the water surface and the external shape of the object. 'Barlike' objects and objects with large surface areas will oscillate more than massive and spherical objects. An oscillating object will have a lower terminal velocity than a non-oscillating object.

The path of the object through the water is also influenced by the currents that are present. After passing the sea surface, the object will move a distance s in a horizontal direction where s is given by the equation:

$$s = \int_o^t v_0 \frac{Xt}{1+Xt} dt \qquad (11.5)$$

$$X = \frac{\rho \cdot CAv_0}{2m_{f_0}} \qquad (11.6)$$

where

v_0 current velocity
C   drag coefficient

The drift caused by the currents has to be taken into consideration when calculating the most probable landing point on the seabed of a falling object.

11.1.3 Probability of Dropped Loads

The probability of dropped loads during crane operations is considered to be dependent on the characteristics of the load and environmental conditions (when floating installations are involved). It is however, unusual to have sufficient data to discriminate between these differences. Typically, only one average frequency may be estimated, for instance, an average drop frequency per lift or per crane year.

WOAD® [2], probably the most commonly used data source for incidents involving dropped loads, falling crane boom or failure of the crane base itself. It is considered that events such as the failure of the crane boom or base, are unlikely to occur without being noticed and reported. When it comes to loads falling during handling, it is quite likely that these may not be reported if there is no subsequent damage. It is therefore quite possible that the frequency of falling loads based on WOAD® reported events is an underestimate.

The typical frequency of dropped loads per crane is in the order 10^{-5}–10^{-4} loads dropped per crane per year, it could even be up to one order of magnitude higher. For critical lifting operations, particular emphasis is sometimes placed on adhering to strict procedures and this is sometimes called a 'procedure lift'. The frequency of dropped loads may under such conditions be typically 30–70% lower than the value for a 'normal' crane operation.

11.1.4 Probability of Hitting Objects

It is useful to distinguish between different types of objects that may be hit, mainly on the basis of the potential worst case consequences of such occurrences:

- Topside equipment: May cause loss of integrity of hydrocarbon containing equipment possibly causing a process fire.
- Subsea installations: May cause loss of containment of production (HC containing) equipment, possibly causing a significant spill.
- Structural components: May cause structural failure or loss of stability or buoyancy.

11.1.4.1 Dropped Loads on Topside Equipment

The probability of hitting topside equipment is usually based on geometrical considerations reflecting the areas over which the lifting is performed.

Lifting over process areas is usually not permitted by operational procedures unless special restrictions are implemented. If a load is dropped under such conditions, it may be a critical event. The probability of being hit may be expressed as follows:

$$P_{Hij} = \frac{A_{lij}}{A_{criti}} f_{crit} \tag{11.7}$$

where

A_{lij} area of equipment j over which loads in category i may occasionally be lifted

A_{criti} total area of hydrocarbon equipment over which load category i may be lifted

f_{crit} ratio of critical area to total area over which lifting is performed

11.1.4.2 Probability of Impact on Subsea Installations

The probability of hitting subsea installations is also usually based on geometrical considerations, which will then reflect the areas over which the lifting is performed.

When lifting over subsea installations lifting or lowering is frequently performed with a horizontal offset, in order to avoid damage to the subsea facilities if the load is dropped. The probability of the subsea equipment being hit may be expressed as follows:

$$P_{Hij} = \frac{A_{lij}}{A_{subsi}} f_{subs} \tag{11.8}$$

where

A_{lij} area of equipment j over which loads in category i may occasionally be lifted

A_{subsi} total area of subsea equipment over which load category i may be lifted

f_{subs} ratio of area over subsea installations to total area over which lifting is performed

11.1.5 Consequences of Impact

11.1.5.1 Consequences for Topside Equipment

The principles are the same as for subsea equipment, outlined in the following section. It should be noted, however, that the probability of loss of containment and subsequent fire is often considered in a simplified manner.

11.1.5.2 Consequences for Subsea Equipment

The most important installations which may be subjected to falling objects, are underwater production systems (UPS) and pipelines.

Underwater production systems are mechanical equipment units, consisting of pipework, valves and controls, mounted on a frame or in an enclosure on the seabed. Their purpose is to connect the wells to the pipelines or risers which convey the well fluids to the process module. Typical UPS-modules are Xmas trees, made to control and shut down the wellstream, and control modules. UPS-modules normally have masses up to 30 tons. A Xmas tree has a height of about 4 m, and covers a horizontal area of about 8 m². Pipelines on the seabed for oil and gas, have diameters up to 40' inner diameter.

The actuators on the Xmas trees are among the most vulnerable subsea components, and these are considered as an illustration of damage to subsea equipment.

The actuators are intended to operate the valves on the X-mas tree. A typical actuator consists of a stem which keeps the valve open. Hydraulic pressure is used to keep the valve open against a spring. Thus, if problems lead to a loss in the hydraulic pressure, the actuator will operate the valve, which will isolate the well. The actuators have a length of about 1 m, and they are often mounted in relatively unprotected positions on the Xmas tree.

The consequences of an impact are dependent on how a falling load actually hits subsea equipment i.e., the velocity of the falling load, where the subsea equipment is hit, the impact angle, the impact time, and the contact area. For a specific load, these values are difficult to estimate, and concentrates on the amount of energy which is transferred between the objects, and the deflection this energy causes on the equipment. For an actuator the deflection is:

$$y = P\frac{l^3}{3}E \cdot I_0 \tag{11.9}$$

$$P = 2m_{fo}\frac{v}{t_d} \tag{11.10}$$

where

l	length
E	elastic tension module
I_0	moment of inertia
m_{fo}	mass of falling object
v	velocity of impacting object (before contact)
t_d	duration of energy transfer during impact.

Some calculations have been made for ideal situations. These indicate that a falling load with a mass of 2 tonnes could easily damage an actuator, and for heavier loads a blowout would be a most probable consequence. The same loads applied to a pipeline may cause damage and leakages.

11.1.5.3   Consequences to Structural Components

Loads due to falling objects and equipment are a result of the impact energy, direction and geometry of the contact area. Hence, it is natural to distinguish between loads due to long cylindrical objects (pipes) and loads from bulky objects because they have different drop rate, trajectory/velocity in water, and effect on the structure. The impact loads on the following structural components are of interest:

- topsides
- module support beams
- supporting structure
- buoyancy compartments.

The elements supporting structure and the buoyancy compartments will often need to be subdivided further, due to strength variations, different hit probability, etc. In principle, the probability of an impact by a falling load should be based on:

- Frequency of lift operations.
- Frequency of dropped loads as a function of lifting procedures, precautions, etc.
- Conditional probability of drop location and height.
- Conditional probability of a particular dropped object hitting a particular structural component, given the drop location and height. For underwater parts of the structure due account needs to be taken of the behaviour of the load in water (which depends upon the object's angle with the water surface when hitting the water, its shape etc.).
- Conditional probability of impact geometry (e.g. the angle between the axis of a pipe and the impact surface), given a dropped object and a hit.
- Conditional probability of velocity, with a given dropped object and a hit.

Studies are often based on several simplifications, due to lack of data.

11.1.6 Impact Energy Distributions

The energy distributions may be calculated based on geometrical distributions, frequencies, and probabilities of failure and hit. Three examples are presented below, for columns of a floating production vessel, module supports beams, and topsides modules. Table 11.2 applies to impact on columns and the top of the column, while Table 11.3 applies to impacts on cantilevered structures and exposed beams.

RNNP has collected an extensive amount of dropped object incidents on production installations.

Figure 11.1 presents the conditional exceedance probability distribution for impact energy for the dropped objects. The curve applies to all types of dropped objects, not only from cranes.

Table 11.2 Cumulative hit frequencies for cylindrical objects and columns

Object and target	Energy class, kJ					
	0	100	200	400	800	1600
Cylind. objects on columns	$2.0 \cong 10^{-3}$	$5.8 \cong 10^{-4}$	$2.6 \cong 10^{-4}$	$8.5 \cong 10^{-6}$	$6.8 \cong 10^{-6}$	$3.1 \cong 10^{-6}$
0 to −2 m	$1.1 \cong 10^{-3}$	$3.2 \cong 10^{-4}$	$6.0 \cong 10^{-5}$	$3.8 \cong 10^{-6}$	$2.0 \cong 10^{-6}$	$7.7 \cong 10^{-7}$
−2 to −52 m	$4.8 \cong 10^{-4}$	$1.4 \cong 10^{-4}$	$2.6 \cong 10^{-5}$	$1.7 \cong 10^{-6}$	$8.9 \cong 10^{-7}$	$3.4 \cong 10^{-7}$
−52 to −64 m						
Container on top of column	$8.9 \cong 10^{-4}$	$8.9 \cong 10^{-4}$	$6.9 \cong 10^{-4}$	$6.9 \cong 10^{-4}$	$2.8 \cong 10^{-7}$	$1.6 \cong 10^{-7}$
Sum columns	$4.5 \cong 10^{-3}$	$1.9 \cong 10^{-3}$	$1.3 \cong 10^{-3}$	$7.0 \cong 10^{-4}$	$1.0 \cong 10^{-5}$	$4.4 \cong 10^{-6}$

Table 11.3 Cumulative hit frequencies and energies for dropped objects on module support beam

Target	Energy class, kJ				
	0	200	400	800	1600
Cantilever	2.5×10^{-4}	2.5×10^{-4}	2.5×10^{-4}	3.0×10^{-6}	2.9×10^{-6}
Module beam	1.0×10^{-4}	1.0×10^{-4}	6.4×10^{-5}	6.4×10^{-5}	1.9×10^{-6}

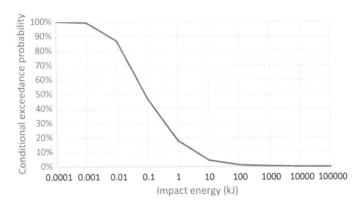

Fig. 11.1 Impact energy distribution for dropped objects on production installations, 2013–2017, NCS

11.1.6.1   Impact Energy Exceedance Curves

Figure 11.2 shows an example of an impact distribution of loads dropped from a crane. The annual impact frequencies shown relate to hits on a subsea installation, as a function of the impact energy.

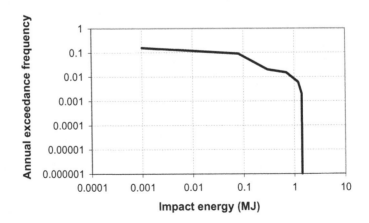

Fig. 11.2 Impact distribution for dropped objects from crane

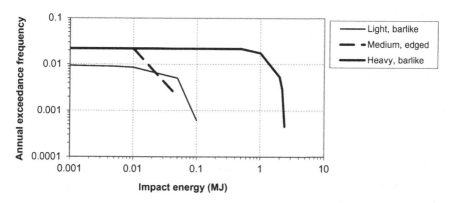

Fig. 11.3 Impact distributions for different objects

For design purposes, this would not be sufficient information, as the type of object, its velocity etc, would also be needed.

The next diagram, Fig. 11.3, presents separate impact energy distributions for three different objects, two 'barlike' objects (light and heavy) and medium 'boxlike' (edged) objects. The distributions correspond to the maximum energies shown in Table 11.4 below.

11.2   Accidents During Tow

Accidents during tow are particularly relevant for jack-up platforms. An overview of accidents to jack-up platforms for the period 1980–1993 revealed that 11 accidents out of a total of 69 accidents were due to towing problems [2].

Serious accidents during towing of jack-up platforms have also occurred in the North Sea, in August 1990. The West Gamma jack-up was being towed between two locations in the North Sea, when it capsized during severe weather conditions. No fatalities occurred.

11.3   Man-Overboard Accidents

Man-overboard accidents may be considered a subcategory of occupational accidents, but are an important subcategory, particularly because a person falling overboard is in need of assistance from emergency response services, by means of a Fast Rescue Craft (FRC), or Man-Overboard (MOB) boat. Lately, Daughter Crafts are also being used, i.e. a large MOB boat, with a small steering cabin that also serves as a shelter for crew and survivors, and dual propulsion systems.

Table 11.4 Energies of falling objects at seabed level

Load type	Weigth, tons	Speed, m/s	Hit energy, kJ
Light, barlike	1	10.3	53
Light, edged	1	3.1	4.6
Medium, barlike	5	9.0	200
Medium, edged	5	4.0	41
Heavy, barlike	19	12.8	1540
Heavy, edged	19	5.1	250

The emergency resources may be installed on the installation itself, and/or on the standby vessel. There is a requirement in Norwegian legislation for production installations that two independent systems be installed. A daughter craft will satisfy this requirement due to its dual propulsion systems.

Installations in North-west European waters have a large freeboard, due to high wave heights in extreme conditions. This implies that a person falling over board may fall up to 30–40 m before hitting the sea. Persons falling over board may be injured due to hitting structural elements in the fall, or when hitting the water. They may also suffer from hypothermia if left in the water for a long period without protective clothing, and may also drown. It is therefore important to rescue such persons within a short period.

The North Sea practice is that a person should be rescued out of the water within 8 min from the alarm is sounded. This implies that MOB boat crews must be available rapidly when performing activities that may lead to man-overboard accidents, such as when erecting scaffolding over the side of the installation.

11.3.1 Frequency of MOB Accidents

The Risk Level project [3] is the source of overview of occurrences of MOB accidents in the Norwegian sector. In the UK, HSE has published overview of accidents and incidents on production and mobile installations. Since recent UK data have not been available, only data until 2003 are included for UK.

Figures 11.4 and 11.5 presents the available statistics for the Norwegian and UK sectors. It should be noted that the majority of the MOB incidents in Norwegian waters have occurred from attendant vessels. These are not covered in the UK statistics. The frequencies are therefore not directly comparable.

There has been one fatality due to MOB incidents in Norwegian waters in the period 1990–2017. This occurred at Maersk Interceptor in 2017 where a person fell to the sea as a consequence of a dropped load when working over sea. If we restrict consideration of the UK sector to 1990–2005, there have been two fatalities, in 1990 and 1996, the former from a mobile installation, the latter from a production installation.

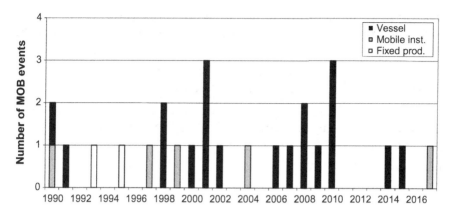

Fig. 11.4 Occurrence of MOB accidents in Norwegian sector, 1990–2017

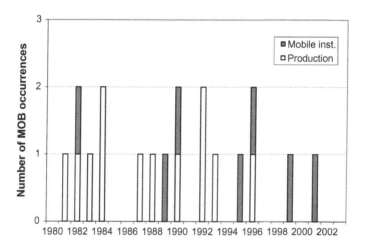

Fig. 11.5 Occurrence of MOB accidents in the UK sector, 1980–2003

In order to compare the two sectors, we restrict the consideration to production and mobile installations in the period 1990–2003:

- UK: 10 incidents
- Norway: 5 incidents.

It should, on the other hand, be observed that more incidents have occurred on attendant vessels in Norwegian waters, compared to production and mobile installations. When it is considered that there are more UK installations than Norwegian, and that the average annual number of manhours in the UK industry is about 50% higher than in Norway, the ratio between the number of cases may not be very different, when normalised according to the manhours. If we calculate the incident frequency, the values are:

- UK: 1.5 incidents per 100 million manhours
- Norway: 1.3 incidents per 100 million manhours.

In the UK sector we may calculate a FAR value, based on the two fatalities that have occurred. The value is based on the period 1990–2003, as an average for production and mobile installations:

- UK: FAR = 0.30.

11.3.2   Scenarios Involving MOB Accidents

The Norwegian data is, as noted above, the most detailed, which gives the best opportunity to consider scenarios where man-overboard incidents have occurred. Figure 11.6 shows an overview of the scenarios where such incidents have occurred in the Norwegian sector, whereas Fig. 11.7 shows the same for the UK sector.

It is shown that work over [open] sea is completely dominating (six of nine) for the man-overboard occurrences in the UK sector, whereas the opposite is the case for the Norwegian sector (two of 23). If vessels are excluded, the Norwegian ratio becomes two of six occurrences, which is still considerably less than in the UK.

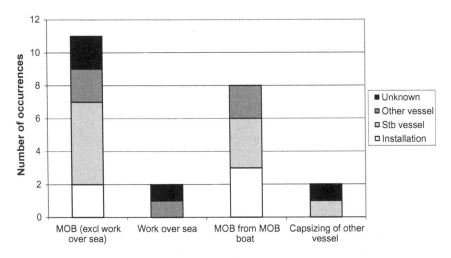

Fig. 11.6 Scenarios for MOB accidents in Norwegian sector, 1990–2011

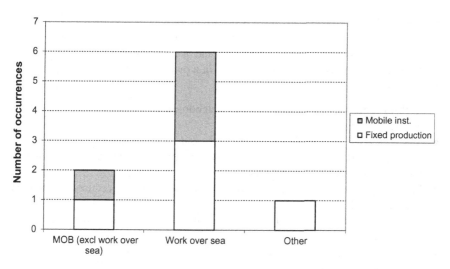

Fig. 11.7 Scenarios for MOB accidents in UK sector, 1990–2003

11.4 Structural Failure

Structural failure has been one of the difficult aspects when it comes to risk quantification. It has generally been omitted in QRA studies, but this is unfortunate with respect to presenting a complete risk picture. It should therefore be the aim to include risk due to structural failure in the risk results. It must at the same time be acknowledged that risk analysis is probably not the applicable source in order to identify risk reduction measures and their effects. With respect to the main sources of probability of structural failure it is practical to group the sources as follows:

- Probability of failure of the structure resulting from statistical variations in loads and structural loadbearing capacities
- Probability of failure due to accidents
- Probability of failure due to a gross error during design, fabrication, installation and operation of the structure.

The first element is identical with the usual scope of structural reliability studies. Risk in this respect should be controlled by appropriate design standards with specified load and resistance coefficients. A low probability of failure (i.e. $<10^{-4}$ per year) is aimed for when design standards are developed. Thus, this probability of failure usually is small compared with the other risks for structural failure.

Probability of failure due to accidents reflects systems failures, such as ballast system failure, anchor system failure, collision impact, falling objects, etc. These mechanisms are addressed in other sections of this book.

The probability of failure due to a gross error is the most difficult to handle, partly because it is outside the scope of structural reliability studies, and partly because such gross errors are impossible to analyse with a normal risk analysis approach. Normally, a detailed plan for verification and quality assurance of important items in the design, fabrication and installation process is required in order to keep the probability of gross errors in a project at a low level. Gross errors are understood to be [4]:

- Lack of human understanding of the methodology used for design,
- Negligence of information,
- Mistakes such as calculation errors (this can be input errors to the analysis programs used and also errors in computer software that are used for design),
- Lack of self-check and verification,
- Lack of follow-up of material data testing, welding procedures, inspection during fabrication, etc.,
- Mistakes resulting from lack of communication or misunderstanding in communication,
- Lack of training of personnel onboard the installation that may lead to maloperation of ballasting systems,
- Errors in systems used for operation of the installation.

Thus, gross errors are understood to be human errors. The nature of the failures as listed above is such that all these scenarios should be possible to detect and rectify in time. Gross errors have been a significant contributor to the failure of structures, and a focus on these issues is considered to be important in order to ensure project success. Two examples of gross errors are the sinking of the Sleipner GBS structure during construction [5], and the overpressure of the cargo system on an FPSO [6].

The approach used by Lotsberg et al. [3], as outlined in Fig. 10.1 may be useful in order to indicate the risk contribution from gross errors.

11.5   Subsea Gas Release

The possible sources of subsea gas leaks are subsea gas wells, as well as subsea leaks from risers and pipelines. Subsea oil leaks are not considered in this context. The special aspect associated with a gas plume in the water is that a flammable gas cloud may be formed above the sea surface. The possible consequences for buoyancy of floating objects are discussed in Sect. 10.5.

A subsea gas leak may be observed on the surface, if the leak rate is significant, such as in the Snorre Alpha subsea gas blowout (see Sect. 4.9). Other leaks may be difficult to observe, and thus hard to detect, except with ROV. Visually the following parameters may be observed when relevant:

- The diameter of the plume on the surface
- The swell of the water within the plume
- The horizontal water speed.

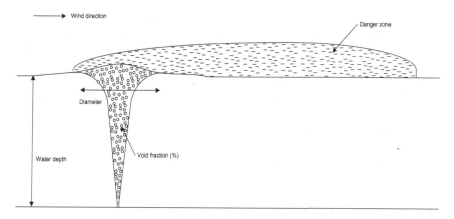

Fig. 11.8 Subsea gas leak which may be observed on the surface

There may also be a gas cloud formed above the gas plume in the sea. This may have concentrations above LEL, depending on the size of the gas leak and other parameters. An idealised representation of the gas plume and associated gas cloud is shown in Fig. 11.8.

For one 1000 kg/s leak at a depth of 100 m, the void fraction may be around 40%, the water rise may be up to 4 m, and the diameter may be around 250 m. The following factors will, in general, affect the behaviour and shape of the plume:

- Release rate (kg/s)
- Gas density (kg/Sm^3)
- Depth of release
- Diameter of release opening
- Release direction
- Currents
- Vertical sea temperature and salt variation.

A gas cloud from a subsea gas leak may cause ignition, if an ignition source is present within the zone with concentration above LEL. The most obvious source of ignition could be a vessel entering the cloud. A passing vessel could enter the zone without knowledge of its presence, or a vessel engaged in emergency response actions if proper safety zones have not been established. See also Sect. 4.9.1, pipeline rupture at the Jotun field, in which case emergency actions were delayed due to the need to establish the extent of the danger zone. In this event, a plume with approximate diameter 100 m was observed by the standby vessel. The water depth at the location of the leak was 126 m. The gas flowrate was calculated to be initially 25–30 kg/s, later dropping to 3–5 kg/s.

In the case of the Jotun leak, the incident was detected due to the effect on the export line pressure on the Jotun FPSO, and the leak location was detected visually by the standby vessel, when following the line in order to search for leaks.

An important source for modelling of the gas plume in the water is Fanneløp [7]. When the behaviour of the gas plume is established, CFD simulation may be used in order to model the behaviour of the gas cloud above the sea surface, and thus the dimensions of the danger zone.

The extension of the danger zone is strongly dependent on the wind speed. For a 1000 kg/s gas leak, the danger zone has a typical downwind extension of around 200–300 m at sea level, with a low wind speed. With moderate wind speed (8–10 m/s) the extension will exceed 1000 m. The zone will reach 30–40 m above sea level.

PSA organised a joint modelling effort in the period 2006–2008, from which two reports have been published [8, 9].

References

1. Drangeid SO, Grande BV, Skriver J (2005) Analysis of causal relationships for unwanted occurrences with offshore cranes (In Norwegian only). Stavanger, Scandpower, 2005 Jun. report no.: 33.790.007
2. DNV (1998) WOAD. Worldwide Offshore Accident Database, DNV, Høvik
3. PSA (2018). Trends is risk levels, main report 2017, (In Norwegian only), Petroleum Safety Authority, 26 Apr 2018
4. Lotsberg I, Olufsen O, Solland G, Dalane JI, Haver S (2004) Risk assessment of loss of structural integrity of a floating production platform due to gross errors. Mar Struct 17:551–573
5. Trbojevic VM et al (1994) Methodology for the analysis of risks during the construction and installation phases of an offshore platform. J Loss Prev Process Ind 7(4):350–359
6. Vinnem JE (2000) Operational safety of FPSOs: initial summary report. OTO report no.: 2000:086; HSE, 2001
7. Fanneløp TK (1994) Fluid mechanics for industrial safety and environmental protection. Elsevier, London
8. Fanneløp TK, Bettilini M (2007) Very large deep-set bubble plumes from broken gas pipelines, Petroleum Safety Authority, report no.: 6201, 18 Nov 2007
9. Tveit OJ, Huser A (2008) Risk associated with subsea gas release (in Norwegian only), Petroleum Safety Authority, 20 May 2008

Chapter 12
Fatality Risk Assessment

12.1 Overview of Approaches

Analysis of fatality risk is one of the crucial elements of a full Quantified Risk Analysis. There is insufficient amount of data available for important parts of this analysis, and there is extensive uncertainty about some of the results. It is therefore an important element requiring thorough discussion, and consideration of the various options available to assess the risk of fatalities.

12.1.1 Why Fatality Risk?

The presentation of risk to personnel is virtually without exception focused on fatality risk, in spite of the fact that the accident statistics from PSA shows that the ratio of fatalities to injuries for exploration and production during the last 10 years is about 1:1400. It might be argued that the primary focus should be on injuries, these being in much higher number.

But, as this chapter will demonstrate, modelling fatalities is very complex and is as such a sufficiently large challenge. Including the assessment of injuries in the QRA would be extremely difficult and would probably be counterproductive.

There is also some justification for omitting injuries. QRA is primarily focused on major hazard s, and in these cases there is probably a relatively fixed average (at least for each accident type) relationship between the number of injuries and fatalities, although this ratio is unknown. If we can assess the fatality risk and focus on risk reducing measures, then implicitly we will probably also reduce the risk of having injuries although perhaps not to the same extent.

When we address the fatality risk, particularly in the context of major hazard, it is impossible to distinguish rigidly between fatalities and severe injuries, and it is probably no point in doing so either. Sometimes it appears that the term 'casualties'

© Springer-Verlag London Ltd., part of Springer Nature 2020
J.-E. Vinnem and W. Røed, *Offshore Risk Assessment Vol. 1*,
Springer Series in Reliability Engineering,
https://doi.org/10.1007/978-1-4471-7444-8_12

is used, in order to cover both fatalities and injuries. Fatalities calculated in a fatality risk assessment might be more correctly considered as 'fatalities or serious injuries'. This may not be very crucial for the risk assessment but may have some implications for emergency planning.

Throughout this chapter the term 'fatalities' is used, recognising that the distinction between fatalities and injuries sometimes is not very clear.

A final point is related to the use of fatality risk results. Sometimes it is claimed that fatality risk results are very unsuitable for analysis of effects of risk reducing measures. This may to some extent be correct but is often a reflection of how fatality risk is used. The effect of risk reducing measures is too often only expressed in terms of the reduction of overall average FAR value for the entire installation and its crew. Then it is no surprise that few proposals have large effect. Effect of proposals should be quantified for more limited aspects of fatality risks in order to see significant effects.

12.1.2 Statistical Analysis

Statistical analysis of fatality risk may be used when the statistical database is sufficiently extensive. The extent of the database has to be seen in relation to the objectives of the analysis. What may be sufficient in one analysis may be insufficient in other contexts. A statistical analysis is insufficient if the detailed causes of accidents are of interest.

As an illustration, fatality statistics relating to Norwegian production installations are adequate to allow calculation of the average fatality risk to employees, arising from occupational hazards. The results were presented in Sect. 2.2. If differences between different trades on the installation are required, then the fatality statistics is insufficient in order to allow the calculation of occupational risk for particular groups of workers (trades).

Uncertainties may be less extensive in a statistical analysis, if past operations are directly relevant to the planned operations i.e., no new hazards are introduced. Regrettably, this is not always the case. Calculation of fatality risk based upon statistical analysis is often used for occupational hazards, diving hazards and for helicopter transportation hazards.

12.1.3 Phenomena Based Analysis

Analysis of accident phenomena implies that the mechanisms of accident causation are modelled. It could also be called analysis based on physical modelling. This usually involves a stepwise analysis, typically involving a sequence of events such as causes of fire, fire loads, responses, and effects on personnel from fire loads.

With this modelling approach, the behaviour of a person during a major accident may be described as a process of several steps in series, as indicated in the block diagram below.

This figure shows the various steps a person has to go through in order to preserve his or her life in the event of a major accident. The relevant parts of the analysis where possible fatalities will be assessed are indicated below the blocks of the diagram.

Uncertainties are usually associable with each step of the analysis. This may lead to a higher level of uncertainty with this type of analysis as compared to statistical analysis. In fact, uncertainty could in principle be used in order to determine the level of detail in the analysis. The uncertainty due to simplification may be extensive if it is only split into a handful of steps in this event, but adequate data may be available for the analysis. In the opposite case, if the task is broken into perhaps two dozen steps or more, one may end up in a situation where there is no data for most of the steps, thus again increasing the uncertainty. Some level of modelling between these two extremes may produce an analysis with the least uncertainty.

Major hazards are usually analysed by means of an approach reflecting the accident phenomena, such as fire, explosion, collisions and other structural or external impacts, as well as marine accidents related to capsize or sinking. Fatality Risk Assessment for major hazard s has the following main elements:

- Assessment of the fatalities in the immediate vicinity of an accident. This assessment is usually closely coupled with the event trees and/or the terminal events in these trees.
- Assessment of fatalities during escape, mustering, evacuation and pick-up. This assessment is usually done by a separate analysis whereby different categories of accidental events are considered.

A phenomena based approach usually has to be adopted for fatality risk assessment for major hazard s, because the statistical base is far too small. The second aspect is that an average fatality rate would be completely unrepresentative for a specific installation, as the fatality risk levels associated with major hazards are strongly dependent on the probabilities of major accidents, and these are certainly installation specific values.

Analysis of fatalities in the immediate vicinity of an accident and fatalities during escape, mustering, evacuation and pick-up require the development of dedicated models. Such models are introduced briefly below.

12.1.3.1 Analysis of Immediate Fatalities

The term 'immediate fatalities' usually has a double meaning, in the sense that 'immediate' may be interpreted as:

- Immediate in **space** i.e., fatalities caused at the location of the initiation of the accident, for instance caused in the module by the initial jet fire. These fatalities will usually **also** be immediate in time.
- Immediate in **time** i.e., fatalities caused at the time of the initiation of the accident, for instance caused in the Shelter Area by an initial explosion (if very strong). These fatalities may **not** be immediate in space.

Usually both types of immediate fatalities are included in a fatality analysis. There are at least two different approaches to assess of immediate fatalities. These are:

- Analysis coupled with the nodes in event trees
- Separate scenario modelling based on terminal events.

The number of fatalities in an accident is dependent on the manning level in the individual modules and areas at the time of accident occurrence. The analysis should therefore distinguish between day-shift and night-shift and should take account of the manning level in the different areas.

12.1.3.2 Analysis of Escape, Evacuation and Rescue

The entire process of emergency abandonment of an installation is often called EER; Escape, Evacuation and Rescue. A platform emergency abandonment process consists of three major phases (see also Fig. 12.1):

- Mustering phase i.e., the escape to TR or lifeboat stations [and temporary sheltering].
- Evacuation phase, involving primary, secondary and in extreme cases tertiary evacuation means.
- Rescue phase i.e., the rescue of survivors from the evacuation to a place of safety.

Sometimes an additional phase precedes emergency abandonment, namely the precautionary evacuation of all non-essential personnel, upon detection of a hazardous situation. The entire process of emergency abandonment should be evaluated, from the time the alarm sounds until all personnel are safely brought to shore

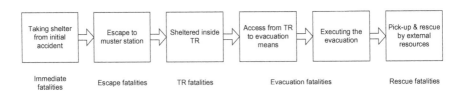

Fig. 12.1 Block diagram of the total process involved in phenomena based modelling of fatalities

or taken onboard a vessel, rescued from the sea by helicopters or have been accounted for as fatalities.

A comprehensive model to cover all these phases will be quite complex, mainly because there is such a large number of possible outcomes, depending on platform conditions as well as external and environmental conditions. A statistical simulation method is often adopted in EER analysis, based upon ranges of environmental conditions and failure conditions, which are used to generate particular scenarios.

The starting point of the EER modelling is to determine under what conditions a decision to evacuate would be expected, in order to use this as a basis for the simulation. Such information could be available from the emergency preparedness manuals, but experience shows that this is seldom the case. Typical scenarios that usually will require evacuation are the following:

- All blowout scenarios (irrespective of ignition).
- Process fires with escalation to other modules.
- All large and medium sized riser leaks (irrespective of ignition).
- Ignited riser leaks, irrespective of leak size.
- Majority of pipeline leaks (irrespective of ignition), in the vicinity of the platform.
- Major structural failure as well as serious critical deformation of critical members.

Helicopters are often the preferred means of evacuation, and will be used as long as there is sufficient time and other conditions allow the helicopter to be used. Precautionary evacuations are frequently undertaken by helicopter, because the time pressure is much less severe in these cases.

Helicopters will be used in a majority of the evacuation situations, for unignited events. Bad weather and/or risk of a gas cloud near the helideck are the main factors preventing the use of helicopters for evacuation in such situations.

For ignited events the fraction of helicopter evacuation s is significantly lower. Shorter available time and the severity of the scenario are the main reasons for this. For structural events evacuation will normally be by means of lifeboats, due to short available time in these events.

Large variations are often observed in the outcome (i.e. success or failure) of lifeboat evacuation. This is due to variation in the reliability and robustness of evacuation means, the weather conditions, the available time and the accident scenarios.

12.1.4 Averaging of FAR Values

Fatality risk values are often averaged over separate groups of personnel or over the entire crew of the installation. Another typical averaging is over all personnel who

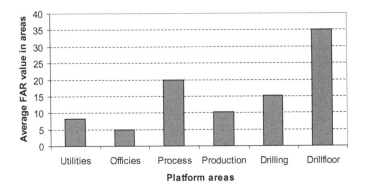

Fig. 12.2 FAR values in different platform areas

are working in the so-called hazardous areas. Such averaging will inevitably mask quite extensive variations between individuals and smaller groups.

It is also possible to calculate average fatality risk levels in different areas on the installation as shown in Fig. 12.2.

FAR levels may vary considerably from one platform area to another. Consequently personnel working in different areas will be subjected to different FAR levels.

12.1.5 Variations Between Installations

Fatality risk level s have over the years been analysed for most of the installations in the North Sea. These levels show considerable variations, due to differences such as the date when the installation was designed, the extent of process and well systems, the availability of a separate accommodation platform is available or not. FAR values calculated for different installations are presented in Fig. 12.3.

The FAR values presented in this diagram, are annual average values for the total platform crew without the contribution from transport of personnel to and from the platform. The following contributions are included:

- Production, well and process activities, including EER effects
- Accidents during off-duty time in the accommodation area
- Occupational accidents.

It should be pointed out that the values in the diagram are strictly speaking only illustrative in the sense that some of the platforms were analysed more than 25 years ago and have had considerable modifications since that time. These modifications are not reflected in Fig. 12.3. Nevertheless, some of the variations

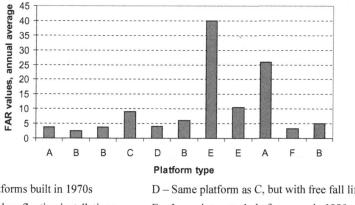

A – Platforms built in 1970s   D – Same platform as C, but with free fall lifeboats

B – Modern floating installations   E – Large integrated platforms, early 1980s

C – Early 1990s steel jacket platform, F – Integrated GBS platform, early 1990s
    conventional lifeboats

Fig. 12.3 Comparison of FAR values for some North Sea platforms

that exist between risk level s of different platforms are indicated. It should also be pointed out that the choice of platforms shown was completely random.

Some of the differences presented in Fig. 12.3 may be due to differences between the organisations performing the QRA studies. There is some experience that demonstrates that two organisations performing studies of the same installation may differ in results up to half an order of magnitude (a factor about 3). These differences are due to modelling differences, differences in assumptions and analytical premises, simplifications and possibly data. The largest differences in Fig. 12.3 are more extensive than such variability between organisations.

It is shown that the availability of free fall lifeboats has a considerable effect on the risk level as shown by the difference between platforms (C) and (D). It may further be noted that the oldest platforms (Type A), had at the time of analysis conventional lifeboats. For the two cases illustrated here, both actually have a limited need for boat evacuation due to being bridge connected installations. The platforms of Types (E) and (F) have free fall lifeboats.

It may be observed that even platforms designed roughly at the same time can be quite different from a risk point of view. The question may therefore be raised as to what is socially acceptable. The only known reference to such a value, is the value implied by UK Health and Safety Executive, in a number of publications which indicate that 10^{-3} per year is the upper tolerability limit for average individual fatality risk. When this is transposed to a FAR value (see Eq. 2.6 in Chap. 2), only one of the values in Fig. 12.3 exceeds this limit. Social tolerability is outside the scope of the present discussion.

12.2   Occupational Fatality Risk

Occupational fatality risk is usually calculated based on statistical data. There is, in fact, virtually no alternative, because there are no available models which may be used in order to model explicitly the occupational accident phenomena.

Section 2.2 has detailed the statistical data for occupational accident s, which are available for production installations, mobile drilling units and offshore related vessels. The available data is, however, most suitable for calculation of average values for the entire industry, due to the limited data set. Thus, distinctions between different installations and different trades may not usually be possible purely based on statistics.

One way which may be used to establish such distinctions, is to assume that there is a relationship between injuries and fatal accidents, and then use the accident statistics for a particular platform or group of works as the basis for calculating fatalities. The FAR value from occupational accident s may then be derived as follows:

$$FAR^i_{occ} = R_{av} \cdot H^i_{occ} \qquad (12.1)$$

where

FAR^i_{occ} FAR value for installation i
R_{av} ratio between FAR value and injury as average for the industry
H^i_{occ} injury rate for platform i.

This equation is based upon the assumption that there is a constant relationship between injuries and fatal accidents on every installation. When considering injuries, care must be taken in order to have the same extent of injuries covered both in the average and in the platform specific injury rate. One solution is to use details of injuries that need to be reported to the authorities as the basis for a particular platform or a particular operator, and then to use the authority average statistics for the R_{av}. The reporting threshold for injuries in the Norwegian offshore sector is consistent throughout this industry. This is somewhat unique; other Norwegian industries do not have consistent reporting thresholds between the different companies.

An example of this is presented in Table 12.1 below. This may be used in lieu of a comparison of fatality statistics. If no occupational fatalities have occurred over the years at the field, no direct comparison of fatality rates is possible. One way would then be to use the triangle of fatalities, serious, minor accidents. This is not very accurate but may give some indication of the occupational FAR if all else fails. The use of a triangle would be parallel with, for instance, Heinrich's triangle [1].

Table 12.1 Example, operator specific FAR value

	Industry average	Operator average
Injury rate (per 1 million working hours)	27.2	31.4
FAR value (Average industry/Adjusted operator fatality rate)	2.8	3.2

The comparison shows that the operator's statistics were 15% above industry average in the ten year period considered. The same adjustment is therefore made in the calculation of fatality risk, which also is increased by 15%. As noted above, the condition which needs to be satisfied in order for this to be valid is that the ratio between injuries and fatalities is the same for the operator in question and the industry average.

An alternative to this approach is to replace all the injuries that are reportable to the authorities with the so called 'serious injuries', the advantage is that the definition of serious injuries is very distinct, and is actually stated in the regulations. Due to this it is unlikely that there shall be any significant under-reporting. The disadvantage is that there are fewer such injuries each year. During a five year period for all production installations in Norway, 2006–2010, there were the following number of fatalities and injuries:

- Fatalities: 1
- Serious injuries: 117
- All reportable injuries: 1405.

An example may illustrate the use of such data. An installation in the Norwegian sector had a fatal occupational accident about 20 years ago, has not had fatal accidents since, and had seven serious occupational injuries in the period 2006–2010, with an exposure of 4.96 million manhours. The total for the Norwegian sector is 117 serious injuries, with 143.6 million manhours. This implies that the frequency of serious injuries is 1.41 per million manhours for the actual installation, in contrast to 0.81 per million manhours as an average for the Norwegian sector. If the average FAR value for the Norwegian sector in this period is 0.70, the adjusted FAR value for the installation is 1.21 fatalities per 100 million manhours, according to Eq. 12.1.

It has also been suggested that the approach indicated by Eq. 7.2 may be used to determine trade specific fatality rates on a specific installation, as follows:

$$FAR_{occ}^{i,j} = FAR_{occ}^{i} \frac{H_{occ}^{i,j}}{H_{occ}^{i}} \qquad (12.2)$$

where

$FAR_{occ}^{i,j}$ FAR value for installation i and trade j

$H_{occ}^{i,j}$ injury rate for platform i and trade j.

This equation assumes that the ratio of the fatality rate for a specific trade in relation to the average fatality rate, is the same as the ratio between the injury rate for that occupation in relation to the average injury rate on the installation. It is doubtful whether this is a justifiable assumption. Consider the difference between drilling personnel and administration personnel. It is unreasonable that the ratio between injuries and fatal accidents will be the same for these two groups.

If such ratios were available from a larger statistical base, then the approach could be used in an adjusted form. However such ratios are not available for all personnel groups, even if all activities on the Norwegian Continental Shelf for the last 20 years are considered. The last option is to use expert judgement to create these ratios.

12.3 Immediate Fatality Risk

12.3.1 Overview

It was noted in Sect. 12.1 that the term 'immediate' relates to fatalities in both time and space. The modelling of immediate fatality risk is mainly associated with accidents that are initiated in the topsides areas, most typically fires and explosions. When external impacts on the support structure are involved, e.g. collision, the immediate fatalities and the EER fatalities become inseparable, to the extent that it is not fruitful to distinguish between them.

When we focus on the topside events, it becomes both meaningful and necessary to distinguish between immediate fatalities, and the EER fatalities that often occur much later. Consider for instance the Piper Alpha disaster where probably only two or three fatalities resulted immediately from the initial blast, whereas most of the fatalities occurred more than an hour later, when those who had escaped to the muster area (living quarter) were either asphyxiated or died in their attempts to get away from the platform.

There are two somewhat different approaches to the modelling of immediate fatalities:

- Subjective modelling
- Modelling based on physical effects.

The subjective modelling is a coarse approach which is quick to carry out, builds on assumed values by the analyst, and is not particularly traceable. The modelling based on physical effects has much better traceability, and is more dedicated in the sense that effects of risk reducing measures may be quantified explicitly. Both approaches are discussed in the following subsections.

12.3.2 Subjective Modelling

Subjective fatality modelling is directly based on the event trees, with fatalities considered for relevant nodes in the event tree, for instance where explosions, fires or escalations occur. Figure 12.4 is an example of this approach, where the event tree for the medium gas leak shown in Chap. 6, is used as the illustration.

Usually there is no separate EER analysis carried out when this approach is used, as all fatalities during immediate and later stages are included in the subjective assessment. This is usually seen in the event tree in the sense that escalations imply several fatalities. This approach is shown in Fig. 12.4.

The subjective assessment may, however, be limited to immediate fatalities alone, and in this case a separate EER analysis will be needed based upon an alternative approach. The high number of fatalities arising from escalation as

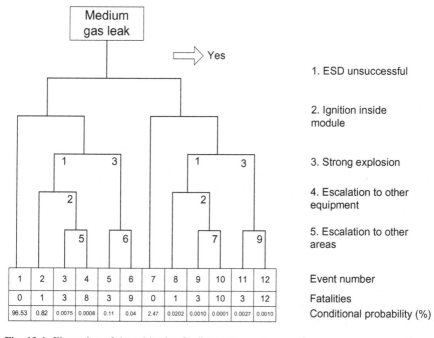

Fig. 12.4 Illustration of the subjective fatality assessment approach

detailed in Fig. 12.4 would therefore be eliminated, because the escalation occurs after some time, when personnel have had sufficient time to escape from the hazardous areas. The calculation of PLL from the simplified approach is quite straight forward, with Eq. 12.2 as a starting point:

$$pll_j = \lambda_i \cdot \prod_K p_k \cdot \sum_K fat_k \tag{12.3}$$

where

pll_j contribution to total PLL from end event j
λ_i frequency of initiating event in the tree
p_k conditional probability of branch k
fat_k fatalities implied by node k
K set of branches that defines the path from initiating event to end event j.

The values for each terminal event in Fig. 12.4 are a summation of the following:

- $\sum_K fat_k$
- $\prod_K p_k$.

The total contribution to PLL from the event tree is then calculated according to the following equation:

$$PLL^i = \sum_K \lambda_j \cdot pll_j \tag{12.4}$$

where

PLL^i PLL contribution from event tree i
J total number of terminal events for event tree i.

Subjective fatality modelling is quick and easy to carry out, and can also easily be implemented in a spreadsheet if the risk calculations are done in this format. The approach is therefore suitable to obtain a rough calculation of the fatality risk.

The main disadvantage with the approach is that it is entirely subjective, and is usually undocumented in the sense that the factors used to calculate fatalities for each node are not documented. It is therefore not traceable, and its repeatability is a problem, because even the same analyst will probably not arrive at identical results if performing the assessment more than once. Due to this lack of documentation and repeatability, the approach is not suitable for sensitivity studies, when assessing the effect of risk reducing measures.

There are a number of disadvantages associated with this approach, but it should be realised that the approach is still quite popular, due to the speed with which it may be completed.

12.3.3 Modelling Based on Physical Effects

12.3.3.1 General Modelling Aspects

The shortcomings of the subjective approach may be overcome by the use of modelling based on the physical effects arising from accident scenarios i.e., the geometrical considerations, heat loads, explosion loads and impacts (either by fragments, equipment or structural elements). The objective of setting up a model based on these physical effects is to:

- Reflect the physical processes and their effects on personnel as realistically as possible.
- Achieve a model formulation which is traceable and repeatable.
- Enable model verification in relation to experimental data and/or statistical data.
- Achieve a model formulation which is effective in assessing the possible effects of risk reducing measures.

Some of the factors that need to be addressed in relation to physical modelling of fatalities are related to the general shortcomings in event tree modelling. Some of the factors that are proposed below for inclusion in the fatality model may already be included in the accident scenario modelling, if a very detailed event tree or more advanced techniques are used. Fatality modelling should consider factors such as:

- Heat loads on personnel, considering the probabilities that flames are directed towards personnel in the area.
- Shielding against the effect of an accident arising from equipment, structures, etc.
- Probability that escape from the initiating area is completed prior to ignition.
- Occurrence of secondary explosions, BLEVEs.
- Immediate effects in neighbouring areas.
- Probability of immediate effects in accommodation spaces and other safe areas.
- Personnel distributions for each area, considering variations according to time of day, operational phase, etc.

The personnel distribution may be a particularly difficult aspect to model realistically. The distribution of primary interest is that when the accident occurs, not the average distribution during the day or shift. It is often claimed that accidents are strongly correlated with human activity. For instance is it often postulated that 80–90% of all accidents are caused by personnel, although 'causation' has to be taken in a wide sense in order to defend such ratios.

Using the average distribution of personnel in an area is likely to result in an under-representation of the risk to personnel, if personnel are involved in causing accidents. A higher concentration of personnel close to the origin of the initial accident may be considered, but there are aspects against such an assumption, in relation to a gas leak.

- Only when ignition occurs within a few seconds of the leak starting, would personnel have been prevented from escaping once the leak is detected.
- Incidents have shown that some people may move towards the source of the leak, in order to attempt to control the leak.
- People may sometimes be ordered to verify that an alarm is genuine, not a false alarm.

There is also another aspect which should be reflected in the event tree, if the over-representation of personnel is considered valid. If people are present, then the likelihood of manual detection of the leak may be set to 1.0.

It may be amazing, but repeated accounts of accidents refer to people being injured because they are running towards the source of the leak (either without reflection or in an attempt to control the fire) in order to try to limit the leak or stop it. Probably the majority of personnel immediately injured or killed have suffered just because of this. (The author can vividly remember how a platform manager once described how he had entered a room to isolate a leak in an instrument. The gas concentration was so high that it was virtually impossible to breathe!) Such tendencies strongly support the assumption that there will be a higher than average concentration of people in the immediate vicinity of the accident.

On balance it is probably valid to assume that there is a higher concentration of people in the vicinity of an accident, compared to the average personnel distribution during a shift. The difficult aspect is to determine what the likely increase would be. Sometimes a simplistic assumption is used in which it is assumed that all personnel in the area at the time of ignition are killed. It is, however, contended that such an assumption represents an overly conservative assessment.

12.3.3.2 Modelling Principles

It is probably only the limitations of imagination which limit how advanced a model of fatality causation due to fire and explosion effects may be. At the same time the model needs to be realistic and have a reasonable computing time, if it has to be repeated thousands of times. The type of modelling often included in a QRA may be illustrated by some examples.

Consider a small gas leak of about 1 kg/s, which is ignited at a certain point in the module. The resulting fire is for simplicity assumed to be a diffuse gas fire. We assume that an event tree has been generated. This example will consider a specific branch k and that radiation levels have been calculated in relation to this end event and a known or assumed origin (leak source).

The fatalities in branch k may now be expressed as follows:

$$fat_k = \sum_P \sum_{X_m} \sum_{Y_m} p(x,y) fat_{heat}(x,y) A^*(t) \tag{12.5}$$

where

fat_k	number of fatalities for branch k
$p(x, y)$	position of persons
$fat_{heat}(x, y)$	probability of person in position (x, y) dying from heat loads
$A^*(t)$	adjustment factor, time dependent
P	number of persons in module
X_m, Y_m	coordinates of module

The value of the probability could for instance be determined as follows:

$$fat_{heat}(x,y) = \begin{cases} 1, & if\ (x,y) \in A^j_{R1} \\ linear\ increase\ 0-1, & if\ (x,y) \in A^j_{R2} \\ 0, & if\ (x,y)\ outside\ A^j_{R1} \end{cases} \qquad (12.6)$$

where

A^j_{R1} area in inner circle around the origin with $R1$ (highest) heat load
A^j_{R2} area in doughnut outside inner circle with $R2$ (lower) heat load.

The heat loads $R1$ and $R2$ are determined on the basis of human tolerability limits (see Sect. 15.10.5). Typical heat loads for $R1$ and $R2$ could be 50 and 20 kW/m^2.

The time dependent adjustment factor $A^*(t)$ may tie in with a time dependent ignition probability, in order to reflect both that persons in the area may have had the time to escape, and others may have arrived to try to isolate the leak.

The modelling of fatalities due to explosion is often done in the same manner, using Eqs. 12.5 and 12.6, just replacing 'heat' with 'blast'. The limits $R1$ and $R2$ are now to be understood as blast loads. These should in principle be determined on the basis of human tolerability limits. This is not as straight forward as for heat. The blast loads required to kill people are quite high, typically 3–4 bar overpressure. But fragments and structural failure s may affect people indirectly, such that they are crushed, or not able to escape from a subsequent fire. These loads are therefore often based on subjective evaluations. Typical values may be:

- $R1 = 0.5$ bar
- $R2 = 1.0$ bar (see Ref. [2]).

12.3.3.3   Suitability of Physical Modelling

It is usually not very easy for different analysts (or even the same analyst at a later time) to recreate a subjective analysis associated with nodes in the event trees. The scenario based analysis should certainly be easier to recreate, because most, if not all, factors are assessed based upon explicit formulations and physical consequence aspects and loads. However, a slight disadvantage is that it may be more difficult for a reviewer to trace it in all details, as it is more complicated. This should not be seen as a real disadvantage of the physical modelling, if the analysis is fully documented.

12.3.4 Is There a Need for Benchmarking?

There has been some improvement of individual models used in QRA over the last ten to fifteen years. This has included fire and explosion load modelling in particular, but also in the modelling of leak frequencies, ignition and so forth. Despite this improvement, the modelling of fatalities appears to be a subject which has attracted no interest. One gets the impression that most QRA professionals see fatality modelling as a 'numbers game' which need not be taken very seriously, because the uncertainty is high in any case.

It is probably true that most experts view fatality risk modelling as the most uncertain element in a QRA study. Most offshore operators, however, have their risk tolerance criteria connected to fatality risk values, expressed either in terms of FAR values, AIR values or an f–N distribution. It may not often be recognised by QRA consultants that considerable costs in risk reduction measures may result due to the over-representing of fatalities implied by the coarseness of the fatality models. A brief and limited comparison exercise is summarised below, in order to appraise the extent of this problem.

12.3.4.1 Summary of Immediate Fatalities in Major Accidents—Explosions

A brief statistical analysis of immediate fatalities arising from fires and explosions on production installations in the North Sea has been undertaken [3] for the period 1973–1997. This has been based upon an analysis of explosions, where the following number of accidents was identified:

- Explosions with overpressure < 0.2 bar: 23 cases
- Explosions with overpressure 0.2–1 bar: 7 cases
- Explosions with overpressure 1–2 bar: 2 cases.

Figures 12.5 and 12.6 are based upon an analysis of these events.

First of all the number of cases should be considered. 23 events were identified with overpressures below 0.2 bar, seven in the 0.2–1 bar category. In both these categories, there are sufficient cases to be statistically significant (on the border line in the second category). In the highest overpressure category there are only two events, one occurred during normal operation (no fatalities, four injuries), and the other case occurred during a shut-down state under very special circumstances (no fatalities or injuries). There is very limited statistical significance in the values from the highest category. This category is therefore virtually eliminated from the following discussion. A summary is presented in Fig. 12.7.

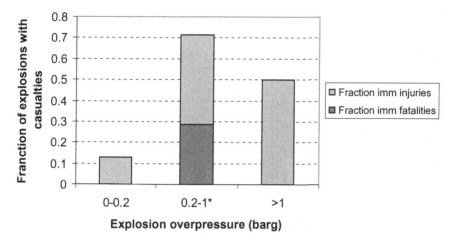

Fig. 12.5 Fractions of explosions with immediate fatalities and injuries

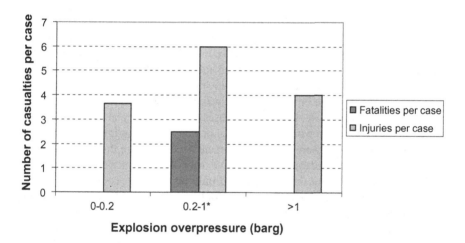

Fig. 12.6 Number of immediate fatalities and injuries per case with casualties

It may be somewhat surprising to find that fatalities are only found in the category 0.2–1 bar overpressure. One of the two cases with overpressure about 1 bar was rather special in several respects. The ignition of the gas cloud occurred several hours after the leak occurred, and just by sheer luck nobody was hurt in the blast.

One of the cases in the 0.2–1 bar category is the Piper Alpha disaster. It is assumed, based on the indications from various sources that probably two people perished in the initial blast (or at least were so badly injured that they could not escape from the subsequent fire). These two fatalities are included in the analysis. No information is available about injuries, and thus no injuries from the Piper Alpha disaster have been included in the analysis.

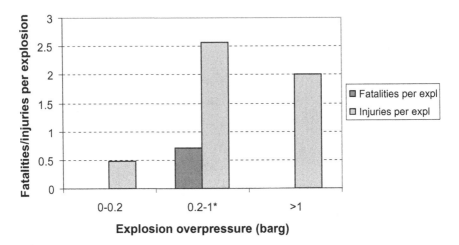

Fig. 12.7 Average number of fatalities and injuries per explosion

It is noteworthy that in only 29% of the cases with explosion overpressure in the range 0.2–1 bar did fatalities occur. If both fatalities and injuries are counted together, it can be seen that casualties occurred in 71% of explosions. When fatalities occurred there were on average 2.5 fatalities and when injuries occurred there were on average six injuries. For the lowest category, no fatalities occurred, the fraction of cases where injuries occurred was 13% (total of three cases), with on average 3.7 injuries per case. One of these cases was rather special in the sense that the gas migrated back to a workshop (unclassified area) and was ignited by an open flame. All seven persons were injured. The event occurred more than 20 years ago, and should not be considered representative of current standards. The average number of injuries without this event would only be two per case with injuries.

The analysis referenced here [3] is quite old, but it has not been found worthwhile to repeat it. Some comments relating the Macondo accident may nevertheless be appropriate. The investigation reports do not appear to focus on the explosion and fire and how fatalities and injuries have occurred. This is obvious to some extent, because the installation sank and is not recovered. The testimonies by survivors on the other hand focused quite some attention on these issues [4]. It is in any case obvious that the blast overpressure has been over 1 barg. 11 people have died, but it is unknown how they died. Similarly, there is no documentation of how people were injured (17 people with serious injuries). It is in any case obvious that the data from the Macondo accident do not fall within the experience data summarized in Figs. 12.5, 12.6 and 12.7. This tends to underline that the Macondo accident was quite extreme.

Table 12.2 Summary of personnel injuries and fatalities in fire events, 1973–1997

Damage category		Fires	Fatalities	Injuries	Immediate fatalities	Immediate injuries
Local	Events	27	0	3	0	3
	Fatalities		0			
	Injuries			5		
One module or more	Events	7	2	1	2	1
	Fatalities		9		9	
	Injuries			7		7

12.3.4.2 Summary of Immediate Fatalities in Major Accidents—Fires

The analysis conducted for fire events is somewhat more simplified, but the conclusions are easier to draw. It can be observed first of all that fires are separated into two categories:

- Local i.e., limited to the equipment where it started
- Fires involving escalation to one module or more.

The identified number of fires in these two categories was reviewed in order to eliminate those events that had started with explosion. It was found that 13% of the local fires resulted from an initial explosion, whereas 53% of the escalated fires started with explosions. The cases shown below are those that did not start with explosion. (Table 12.2)

In the case of local fires, three of 27 events caused 'immediate' injuries, no fatalities were caused, either as 'immediate' or during evacuation. Two of these fires are not necessarily representative of topside hydrocarbon fires, since one of the cases was an electrical fire in 1995, and another was the fire in the galley in 1988. The only relevant topside hydrocarbon fire with casualties is the crude oil export pump on a Norwegian platform in 1989, where injuries were actually quite minor.

For fires that escalated to the entire module (or further), two of the seven cases caused immediate fatalities. Both of these were special however, in the sense that the fatalities were caused inside a concrete shaft. The cases are the fire on Statfjord A in 1978 with five fatalities (during start-up preparations), and a similar fire on Brent B in 1984 with four fatalities and seven persons injured. None of these casualties are really representative of topside fire events. It could therefore be argued that even for escalated fires, no events have occurred causing casualties due to topside hydrocarbon fires. For calibration purposes, all fires will have to be considered as one category.

12.3.4.3 General Observations

The following overall observations may be made, relating to fire incidents:

- Surprisingly few casualties have resulted from ignited hydrocarbons leaks on production installations in the North Sea.
- It would have been interesting to extend the analyses to other areas and include the most recent accidents, but this would not be possible generally, due to lack of information
- There are indications that the average fraction of cases where casualties occur is considerably higher in the US GoM than in the North Sea, the difference probably being statistically significant.
- For risk analysis modelling, both the probability of fatalities and injuries should probably be considered as more or less equal. It will probably be somewhat random whether an injury is so severe that a person will perish or not.
- For minor explosions, it appears that about 15% of the cases result in casualties. The average number of casualties per event is 3.7, but this is strongly influenced by the type of area/room where ignition occurs. In these situations it may be argued that injuries are more likely than fatalities.
- For stronger explosions, it could be argued that casualties have occurred in about 33% of the cases, although the data basis is quite limited. The average number of casualties cannot be calculated, but would probably be in the order of 10 (or more!), and would be strongly influenced by Piper Alpha. Fatalities appear to be more probable than injuries, but the fatalities in the majority of the cases are probably mainly caused by fires following the explosions.
- For fires that are not preceded by an explosion, casualties have occurred in about 10% of the cases that are relevant for topside hydrocarbon areas. The average number of casualties is 1.7 per fire. Such events are most likely to result only in injuries although there is theoretically (not practically, because none have occurred) a small probability of fatalities.

12.3.4.4 *Comparison with QRA* Results

Table 12.3 presents a comparison of average number of fatalities per case of explosion from some randomly selected QRA studies, as well as the average value found from accident statistics in the North Sea in the period 1973–1997.

Table 12.3 Number of fatalities per explosion event from some QRA studies, compared with accident statistics

QRA study	Average fatalities per explosion incident
Study i	1.38
Study ii	2.19
Study iii	1.95
Study iv	0.057
Average from North Sea, 1973–1997	**0.15**

The comparison is limited to immediate fatalities from explosions, but it is probably difficult to distinguish between fatalities from the initial blast, and those that may be injured by the initial blast to the extent that they cannot escape the effects of the subsequent fires. There are in fact indications that all actual fatalities in explosions in the North Sea have been caused by the subsequent fires.

It is clearly demonstrated that these QRA studies are quite conservative in the calculation of fatalities for explosion scenarios; when compared with actual event statistics.

12.4   Analysis of Escape Risk

12.4.1   Overview

The term 'escape' sometimes has different interpretations. The interpretation chosen here is in line with that adopted by the NORSOK guidelines for Risk and Emergency Preparedness analysis. This states that the escape phase is the withdrawal of all personnel from their working place when the accident or incident occurs, to their designated muster station (see also Fig. 15.5). The muster station is usually either a lifeboat station or a place inside the Temporary Refuge/Shelter Area (according to British or Norwegian nomenclature).

Success in escape to the muster area depends on a number of factors, which include:

• location of the accident	• wind direction
• size of the accident	• heat and smoke protection
• heat load and smoke	• capacity of escape ways and stairs
• duration	• alternative routes on the particular deck as well as to decks above or
• escalation	below

This evaluation of escape is mainly focused on fire events, including explosions. In the following it is always assumed that there is a fire between some of the personnel groups and the muster area. If this is not the case, the escape analysis becomes trivial.

When an explosion occurs, the time of ignition determines the conditions pertaining during the escape phase. Sometimes escape can take place prior to ignition, and in this event escape analysis is also trivial. If this is not the case, then an explosion may be followed by fire or not. If there is no fire then escape ways may still be blocked by the effect of the explosion. When there is no fire, there is no urgency to escape, and it should be possible to escape successfully even though there may be some blockage of the escape ways.

If the explosion is followed by fire, then probably the fire is the main challenge for escape, but blockage by structural effects may make the escape phase even more difficult.

Other accidental events such as collisions and excessive environmental loads, are of such a nature that either precautionary escape and evacuation can take place, or the effect of the accident is over when evacuation is performed. Following a collision, there will be a possibility of collapse of the platform. It may then be decided to evacuate the platform and it is of vital importance that usable escape routes still exist, particularly for transportation of injured personnel.

Structural events are somewhat special in the sense that the topside is either left intact or severely impaired by overall structural collapse. Escape is therefore not normally prevented by damage to topside equipment, but rather as a consequence of severe structural impact to the substructure. These phenomena are such that either all or none of the escape ways are impaired simultaneously. Consequently, if the structural damage is less than critical, escape to muster areas should be possible. Any injuries or fatalities will thus depend on how emergency evacuation is carried out. Thus the following discussion is limited to escape following fire and explosion events. The escape analysis is usually performed in the following manner:

- Qualitative engineering evaluation of escape possibilities from each area on each deck, following different accidental scenarios.
- Provision of a summary table to show where escape problems exist.
- Formulation of quantitative input to a fatality risk study, based on the qualitative analysis.

Another approach uses a quantitative escape time study, whereby times are simulated for the movement of personnel from the different areas to the TR. This does not give an indication of where personnel may be exposed to unacceptable conditions but it does give an identification of possible bottlenecks.

It is recommended that the sequence involving a qualitative evaluation followed by a quantitative study should be followed at least when engineering new installations. All the following subjects are therefore discussed separately below:

- Analysis of escape times
- Impairment analysis
- Escape fatality analysis.

12.4.2 Escape Time Analysis

The 'escape time' is the entire period from when the alarm is sounded until all personnel have arrived at the mustering stations (see also 'muster time' below). There is software available [5] which is able to simulate the movement of personnel in an escape situation on a rather detailed level. This may be useful for installations or vessels with very high manning level, but may not be particularly useful for an installation with less than 50 persons, which is often the case on modern installations. For coarse calculations, the following may be used:

- Walking speed in corridors is often considered to be in the order of 1 m/s.
- For stairs with normal elevation the walking speed is calculated to be 0.7 m/s along the stairs.

It may be necessary to distinguish between 'escape time' and 'mustering time'. The mustering time includes the escape time as well as the time needed to identify any missing personnel (often called 'confirmed POB'). Such a time may typically be in the order of 15 min, but even for drills, more time may occasionally be required to investigate why some persons are missing. In the following, an average mustering time of 20 min is used.

For a large integrated installation, it is often claimed that search for and rescue of injured persons, may take up to 30 min to complete. It should be realised that such a time can never be confirmed and calculated precisely and will always remain an assumed value. This is another reason why a very sophisticated and 'exact' calculation of the time it takes for people to move back to the Shelter Area/Temporary Refuge is of limited value.

Some platforms have installed automatic movement registration, especially where there may by movement of personnel between bridge connected platforms. These systems should be capable of providing information about where a person was last registered in case of being missed from a 'POB count' at the muster station. Some credit in terms of reduced search time, may be taken when such systems are in place as the search area is better defined.

The minimum periods before search and rescue may be completed will be about 50 min if no automatic movement registration is applied, and 35–40 min when such facilities are available. These values establish the minimum periods of intactness required for the escape ways. These are inputs to the impairment analysis described below.

12.4.3 Impairment Analysis

The impairment probability for escape ways takes into account those circumstances where heat, explosion load, or poor visibility due to smoke prevent personnel away from the original accident, reaching the Shelter Area. If the wind is blowing from the process/wellhead areas towards the muster area, smoke impairment of the escape ways may be a problem. This is particularly serious if oil is ignited on the sea surface. Then large amounts of smoke may surround the platform, causing problems in using external escape ways as well as hindering the evacuation itself.

The impairment probability will take the wind direction into account in relation to the calculated heat loads. Impairment of the shelter area occurs when personnel cannot stay inside the shelter area for a period long enough to perform a safe and complete evacuation.

It is common practice that each main area shall have two independent escape routes, such that a blockage of one route is not critical. This must be taken into consideration when the impairment probabilities are calculated.

The approach to modelling of impairment can follow the same principles as the quantification of fatality risk during escape (see Sect. 12.4.4 below), apart from replacing the probability of fatalities by impairment probabilities.

12.4.3.1 Qualitative Scenario Analysis

The qualitative analysis considers the possible escape routes from any area to the muster area(s). An evaluation approach based on tables is suggested, where three table formats are proposed; for movement deck by deck, a summary table, and a table with the input to the quantitative analysis. The objectives of the evaluation are to:

- Ascertain that there is at least one safe escape route from any area of the platform.
- Ascertain that there is a usable alternative to this route.
- Ascertain that the escape to a safe area is as short as possible.
- Point out details of the present arrangement which could be improved to meet the goals above.

A safe escape route implies that the route will be passable for a sufficiently long period after the event to allow escape of all personnel to the muster area. A usable alternative (secondary) route requires that the route at most is partly exposed, although complete exposure may be tolerable if the route is very short.

In the qualitative analysis, one table describes the possible escape routes from **one deck** to the muster area(s). A sample table is shown in Table 12.4.

The use of some type of form is considered to be essential in the documentation of an analysis of escape routes and fatalities. This avoids ending up with an analysis which is extensively based on subjective and non-traceable evaluations made by the analyst.

A number of abbreviations and code words are used in the tables. These are as follows:

Primary Route	The first choice for a person being in the particular area described in the table. The primary route should preferably be one which is not exposed to fire loads even when the running distance to the muster area is longer than the shortest way (see below).
Secondary Route	The best alternative to the Primary Route when this is unusable. The Secondary Route should also offer a relatively high success probability.

Table 12.4 Example, qualitative evaluation of escape ways

| Escape from module | Primary route | | | Secondary route | | | Alternative mustering | |
	Identification	Distance	Protection	Identification	Distance	Protection	Identification	Protection
PXX.Y	Walkway E or W side of WD	Short	Exposed	Stairs E and W down to MD. Walkways E & W side of MD	Long	Partly exposed	Not relevant	–
PZZ.W	Walkway E or W side of WD	Short	Exposed	Stairs E and W down to MD. Walkways E and W side of MD	Long	Partly exposed	Not relevant	–

Abbreviations WD—Weather Deck, UPD—Upper Production Deck, MD—Mezzanine Deck Level, PD—Production Deck, LPD—Lower Production Deck

Alternative Mustering   Mustering for evacuation in areas other than the mustering
                         station.

A muster station is defined as a place at which personnel may gather in a
relatively safe environment prior to evacuation or abandonment of the installation.
If there are parallel escape ways on both sides of the platform these should be
counted as **two alternative primary** escape ways (not as primary and secondary),
due to the fact that escape on one side may easily be prevented by heat loads,
depending on the fire origin and the wind direction.

The distance to move is a function of the platform size and design. The following abbreviations are used to characterise the distance to escape:

Short   If the deck has direct access to the muster area, the shortest way is the
          walkway on the sides of the deck being considered.
Medium   When it is necessary to descend or ascend one deck level to avoid the
          fireFire loads, the distance to run is denoted medium.
Long   When it is necessary to descend or ascend two or more deck levels to
          avoid the fireFire, the distance is denoted long.

A medium or even a long route may sometimes be preferred (Primary Route) to
avoid the consequences of fire. The protection against fire loads during escape is
therefore important. The following classifications are used to classify the escape
ways protection:

Protection

Exposed   Lack of shielding by deck or bulkhead for the entire length of the
                 escape way.
Partly exposed   Lack of protection on part of the length of the escape way.
Good   Escape way shielded for the entire length.

The individual scenario tables for each module and each hazard category are
summarised into an overall table. A sample table is shown in Table 12.5. This table
uses the same abbreviations as shown for the individual modules.

12.4.3.2   Input to Quantitative Analysis

The qualitative evaluation provides the input to the quantitative escape study. This
input consists of the following distributions:

- Number of personnel in muster area (input to Evacuation study).
- Number of personnel at secondary evacuation station (input to Evacuation
  study).
- Number of personnel trapped in other areas (input to Fatality risk study).

An example format is presented in Table 12.6.

Table 12.5 Summary table for qualitative assessment of escape routes

ESCAPE FROM	Escape distance P = Primary S = Secondary	Accidental event in							
		P1 Explosion/ Fire	P2 Explosion/ Fire	P3 Explosion/ Fire	P4 Explosion/ Fire	P5 Explosion/ Fire	P6 Fire	Explosion & Fire	P7 Explosion/ Fire
P7 West	P: Medi./long	PE	NE	NE	NE	NE	NE/PE	NE/PE	–
	S: Long	NE	NE	NE	NE	E	PE/NE	PE/NE	–
East	P: Medi./long	NE	NE	NE	NE	E	PE/NE	PE/NE	–
	S: Long	PE	NE	NE	NE	PE	NE/PE	NE/PE	–

Abbreviations NE = Not exposed, PE = Partly exposed, E = Exposed, PD = Partly damaged, D = Damaged

Table 12.6 Summary table for input to evacuation analysis

Accidental scenario (main categories)	Number of personnel in muster area	Number of personnel at Secondary evacuation station
Blowout, wellhead area	80	15
Blowout, drill floor	90	5

12.4.4 Escape Fatality Analysis

Escape fatality analysis is usually performed for the terminal events in the event trees. The basis of the proposed modelling is the same as in the physical modelling of immediate fatalities (see Sect. 12.1.3.1).

This outline considers only one escape route on one deck, $ER(x,y)$, without vertical movements, and one terminal event in the event tree, j. The fatalities in branch k may now be expressed as follows:

$$fat_{ER,j} = P_{ER} fracfat_{ER,j} A^+(t) \tag{12.7}$$

where

$fat_{ER,j}$	number of fatalities during escape for end event j
P_{ER}	Number of persons needing to use escape way ER
$fracfat_{ER,j}$	fatality fraction for ER for end event j due to heat loads
$A^+(t)$	adjustment factor, time dependent, to cover amongst other factors, the possibility that ignition occurs after escape has been completed.

The value of the probability could for instance be determined as follows:

$$fat_{ER,j} = \begin{cases} 1, & if\ \dot{q}(x,y) \geq 50\,kW/m^2 \\ linear\ increase\ 0-1, & for\ area\ with\ \dot{q}(x,y) \geq 20\,kW/m^2 \\ 0, & if\ \dot{q}(x,y) \leq 20\,kW/m^2 \end{cases} \tag{12.8}$$

where

$\dot{q}(x,y)$	heat load on $ER(x,y)$
$R1$	lowest heat load limit for no survival
$R2$	highest heat load limit for complete survival
X, Y	coordinates of escape route.

The heat loads $R1$ and $R2$ are determined on the basis of human tolerability limits, similar to those mentioned in Sect. 6.9.5. Typical heat loads for $R1$ and $R2$ could be the same as for immediate fatalities, 50 and 20 kW/m^2.

12.5 Analysis of Evacuation Risk

12.5.1 Overview of Evacuation Means

A realistic evaluation of the evacuation options is required prior to an evacuation risk analysis. In the beginning of this chapter the use of helicopters for emergency and precautionary evacuation was briefly introduced. A helicopter will normally be preferred for evacuation when time and conditions allow. The following discussion illustrates the considerations and evaluations that are performed for an actual installation. There are two options for helicopter assistance in an emergency situation:

- Helicopters located on the field
- Helicopters from an onshore base.

There are also two alternative sources of onshore based helicopters:

- SAR (Search and Rescue) helicopters
- Normal helicopters for personnel transport.

Before year 2000 there were few fields in the North Sea which had a helicopter stationed offshore permanently. None of them were full SAR helicopters, but some had partial SAR facilities. Whether SAR capabilities are essential or not depends entirely on the accidental scenario.

Conditions are somewhat changed in the Norwegian sector. Today there are several full SAR helicopters stationed offshore. This has resulted from much more extensive cooperation between fields and operators relating to emergency preparedness. The overall picture is that standby vessels have been reduced in number, and SAR helicopters have been increased. Typically, five to ten installations within an area are covered by one SAR helicopter permanently stationed on one of the installations. The same installations may also share one or two standby vessels.

An analysis of risk level on the Norwegian Continental Shelf [6] considered two assumed field locations, and the possible helicopter response times for these two fields. Table 12.7 is extracted from that work showing the distances and minimum response times for SAR related services.

The two cases considered, are an FPSO installation (50 persons crew) at the following locations:

- Norwegian North Sea Block 30/8, 15 km South West of Oseberg Complex.
- Norwegian Sea Block 6507/6, 30 km North East of Heidrun, 35 km South West of the Norne field.

The onshore based SAR services have Sea King helicopters which can take 18 passengers, and have a cruising speed of about 190 km/h. If transporting stretchers, it is possible to load seven stretchers with injured personnel. The Sea King is also fully equipped for search and rescue operations during times of darkness and low visibility.

Table 12.7 SAR services and response distances and times for two field locations

'SAR' resource	Block 30/8	Block 6507/6
Distance and response times for onshore based SAR services		
• Sola	185 km, 70 min	570 km, 185 min
• Ørland	400 km, 135 min	180 km, 70 min
• Bodø	735 km, 175 min	265 km, 95 min
Distance and response times for offshore based SAR helicopters		
• Statfjord	85 km, 40 min	435 km, 145 min
• Ekofisk	330 km, 115 min	Not applicable

The offshore helicopters are actually more modern and have to some extent better equipment, and may have a higher cruising speed. They are with one exception so-called 'all weather SAR' helicopters. Other helicopters may be in the area when the accident occurs, but it is obviously not possible to depend on this.

The desired means of evacuation is normally by helicopter, if conditions allow. However, helicopters will not arrive until some time after the accident, and the evacuation means that are immediately available will be those installed on the installation. The use of helicopters or lifeboats is, however, not only a question of availability, but will be dependent upon:

• Whether there are fires on the installation
• Weather conditions
• Exposure of helideck
• Capacity
• Response time.

A helicopter should be useable in riser fires which do not expose the helideck (if gas or smoke is blown away from it) as well as structural impacts and failures. It is assumed that lifeboats will be the safest evacuation means in the following scenarios:

• process fires
• blowouts not burning on the sea surface
• blowouts before ignition.

The muster area should be protected against heat loads from sea fires and be located at a reasonable distance from mustering stations for both lifeboat and helicopter evacuation. Any muster areas at the opposite end of the platform from the main muster and evacuation facilities should only be used for those who are unable to reach the main muster area.

12.5.1.1 Experience from Performed Evacuations

Table 12.8 gives details of previous emergency evacuations in the Norwegian sector, excluding those cases where evacuation is just across a bridge to a separate accommodation platform or flotel. Precautionary evacuation is also excluded.

It can be observed that the only two fixed platform emergency evacuations that have taken place using capsules or lifeboats are from the late 1970s. Since then three platform evacuations from mobile drilling units have been performed. In the case of the capsize of Alexander L. Kielland and West Gamma some people had to be rescued from the sea, as both accidents occurred rapidly in severe weather conditions. Evacuation with conventional lifeboats was successful in the three cases which occurred in good weather conditions.

It is also noteworthy that although many installations have had free fall lifeboats installed for over 30 years, no actual lifeboat evacuation has taken place involving free fall lifeboats. There are also free fall lifeboats installed on ships, but there is no known case of deployment on ships either.

Table 12.8 Summary of platform evacuations by lifeboat in Norwegian North Sea

Installation	Year	Accident type	Total fatalities	Injury associated with evacuation means	Persons in sea
Ekofisk A	1975	Ignited riser leak	3 fatalities (during evacuation)	3 injured inside damaged capsule	0 persons escaped to sea
Ekofisk B	1977	Unignited blowout	0 fatalities	0 injured persons in capsule	0 persons escaped to sea
A. Kielland	1980	Capsize	Unknown number of fatalities on platform and in sea (123 in total)	> 20 persons in damaged lifeboats	≈15 (?) escaped to sea
West Vanguard	1985	Ignited shallow gas blowout	1 fatality on drill floor	0 injured persons in 2 lifeboats	0 persons escaped to sea
West Gamma	1991	Capsize during tow	0 fatalities	No lifeboats usable	≈35 persons escaped to sea

12.5.1.2 Lifeboat Evacuation

Some events will prevent some personnel groups from reaching the main shelter area and these persons might be evacuated by a lifeboat installed at a secondary lifeboat station. It is assumed that the escape and launching time for this group will be less or the same as for persons at the main shelter area.

Another group of personnel may reach the main shelter area, but may not be able to evacuate using the lifeboats here because of a failure in the lifeboat launching. There is usually at least one extra lifeboat at the main lifeboat station to allow for this eventuality.

It is recognised that launching of vertical free fall lifeboats may be carried out rather quickly, say in the order of 5 min. It is assumed that it will take approximately 10 min to embark and launch another lifeboat in case of release failure. Launching of conventional lifeboats is considered to take about twice the time for free fall lifeboats.

Based on the above factors the total duration of escape and evacuation for individual groups will be from 25 to 60 min, without including the time needed to assist injured persons or to try to combat the accident. It is considered relevant to allow some time for these activities. If an additional 30 min is applied for these actualities then the shelter area, the evacuation system, the control room, and the platform main structure must remain intact for a period of approximately 90 min. Most of the personnel will, however, be evacuated long before that time.

The probability for release failure of more than one free fall lifeboat, which would result in insufficient lifeboats being available, is negligible. The probability of launch and/or release failure of conventional lifeboats is significant, and strongly dependent on the weather conditions and sea state.

12.5.1.3 Helicopter Evacuation

The calculated time to evacuate a total platform complement of 115 persons by means of helicopters is shown in Table 12.9. The case is constructed for a specific location and with specific onshore SAR helicopter stations in mind. The actual

Table 12.9 Helicopter evacuation time

Activity	Duration (min)		
	Airport 1	Airport 2	Neighbour platform
	Sea King		Bell 412
Time from request to the helicopter crew have mobilised	60	60	5
Time required to evacuate all personnel (assuming one helicopter, and shuttling to neighbour platform)	(7 flights) 195	(7 flights) 195	(9 flights) 255
Total time required for evacuation	315	310	275

locations are not relevant for the example, and the required periods for evacuation of the total complement are illustrative.

There may be more than one helicopter available to perform the evacuation. Two or more helicopters will obviously shorten the time required to perform such an evacuation. However, it is probably not possible to operate safely more than three helicopters simultaneously, due to air space considerations.

Some persons may not reach the main muster area, and cannot be evacuated by helicopter, but should be able to use the lifeboat at the secondary shelter area.

Another group of persons may be those who reach the main muster area, but because of the event development, or some other reason, are unable to be evacuated by helicopter. Most likely they will have the possibility to use the main lifeboats. Such an evacuation scenario is however, not considered in this discussion.

There is also a possibility that some people are not able to reach any of the defined muster areas. This might happen if the escape ways are impaired by heat, smoke or obstructions. These personnel will have to use other means of evacuation if possible. It should be emphasised that this problem will exist also when considering lifeboat evacuation.

12.5.1.4 Summary of Required Intact Times

The times that the Main Safety Functions need to remain intact are summarised in Table 12.10, based on the conservative values in the preceding sections.

The times in this table are the periods which the personnel must be able to remain safely on board for completion of a safe evacuation. These periods equal the required periods of intactness for the Shelter Area, Control Room and Main/hull Structure safety functions.

Lifeboats will still be the preferred evacuation means when time is crucial, often even if a helicopter is located in the field. The minimum periods are therefore established by the lifeboat option. Consequently the following are the minimum periods of intactness to be used in safety evaluation:

Table 12.10 Summary of required evacuation periods

Evacuation means	Required time to evacuate (min)				
	Escape ways	Main shelter area	Evacuation system	Control room	Main/hull structure
Helicopter (Airport 1)	50	315	315	315	315
Helicopter (Airport 2)	50	310	310	310	310
Helicopter (Neighbour Field)	50	275	275	275	275
Lifeboats	50	90	90	90	90

Escape Ways: 50 min
Shelter Area: 1 h 30 min
Evacuation system: 1 h 30 min
Control room: 1 h 30 min
Platform Main Structure: 1 h 30 min.

The requirements for escape ways, shelter area, evacuation system and main structure are obvious. The control room has to be operating long enough to provide personnel with enough information to allow safe evacuation.

It has been observed in several accidents that the evacuation time has been longer than the time observed during drills. Probably, the main reason for this is delay in the decision-making process. Even when circumstances have been such that evacuation has been urgent, considerable time has passed until evacuation has been initiated. The reason is probably a tendency to devote too much time to try to combat the accident.

Therefore, decision-making appears to be an important factor in addition to the escape way layout, with respect to mustering and evacuation time.

12.5.2 Impairment Analysis

The required periods of intactness in relation to shelter area, evacuation system and platform main structure are governed by the possibilities for safe evacuation, as shown above. In most accidental events, some kind of pre-warning is received, such that all non-essential personnel may already have been evacuated prior to critical conditions. A blowout will almost always give pre-warnings, which may initiate precautionary evacuation. This will probably reduce the risk related to a burning blowout.

Impairment of the main structure resulting from heat loads on the structure also cause impairment of the shelter area if global collapse occurs. The impairment of safety functions is therefore a function of the period of intactness required for safe escape and evacuation of the platform.

The impairment analysis becomes to a large extent an exercise relating to probabilistic survival times of:

- Shelter Area (Temporary Refuge)
- Command/control centre
- Main structure.

The dimensioning loads for the Shelter Area (disregarding structural failure, which is treated directly) are often related to smoke ingress into the muster area. Studies of smoke ingress therefore have to be done in a probabilistic manner, in order to give the required probabilities. The failure of the command and control centre due to smoke ingress is often handled in the same manner.

The probability of failure of the main structure may be assessed by means of structural analysis tools, relating to the accident loading in question.

12.5.3 Evacuation Fatality Analysis

There are several models available for evacuation analysis. The model outlined here is based on an event tree approach. The following tasks would be included in an evacuation analysis:

1. Assessment of failure probabilities for optional evacuation means under different environmental conditions and scenarios by fault tree analysis.
2. Evaluation of each evacuation means concept.
3. Overall analysis of evacuation efficiency and success by event tree analysis.
4. Formulation of input to analysis of rescue fatalities.

Another option for evacuation fatality analysis is the use of a statistical simulation technique, such as Monte Carlo analysis. With this approach all factors are described by means of statistical distributions and the probability of different outcomes determined based upon consideration of many randomly selected conditions. The simulation will replace only Step 3 in the list above, whereas the other steps would be as described. The evacuation simulation will usually be integrated into an overall simulation of the rescue phase.

12.5.3.1 Failure Probabilities for Evacuation Means

The failure probabilities for the optional evacuation means under different environmental conditions and accident scenarios are usually assessed by means of Fault Tree Analysis. This assessment will be dependent on the available type of evacuation systems.

Results from this step may be presented for all evacuation means, with separation of scenarios as shown in Table 12.11 below. The evaluation of individual

Table 12.11 Summary of failure probabilities for evacuation means

Failure probabilities							
Evacuation means	Evacuation in good environmental conditions						Evacuation in severe weather condition
	Blowout burning on platform	Blowout burning on sea level	Riser fire	Process/ utility fire	Collision	Severe structural damage	
Primary	0.17	0.29	0.33	0.08	0.05	0.04	0.43
Secondary	0.29	0.37	0.30	0.21	0.09	0.07	0.53

lifeboat concepts will be based on the results of the Fault Tree Analyses, and consider the following aspects in particular:

- Risk of set back (when a free fall lifeboat hits the wave with an undesirable angle, such that it is thrown back rather than dive through the waves)
- Sea state and weather operational limits
- Reliance upon external vessels or systems
- Risk of unintended release or operation
- Total evacuation time for platform complement.

12.5.3.2 Results from Evacuation Study

The availability of evacuation means may be measured by means of fatality fractions. This is usually found to be the best way to express availability of evacuation means. The availability may be quantified by the following formula:

$$A_{e,i} = 1 - \frac{N_{evacfail,i}}{N_{escape,i}} \tag{12.9}$$

where

$A_{e,I}$ availability for alternative 'i'

$N_{evac\ fail,\ i}$ number of personnel not evacuated by all available means, alternative 'i'

$N_{escape,\ i}$ number of personnel who succeeded in escaping to TR or secondary evacuation station, alternative 'i' (maximum is total POB).

Distributions of outcomes under different accident and environmental conditions should also be presented in addition to the overall availability as outlined above. Such information will be valuable for the emergency planning.

12.5.3.3 Input to Rescue Study

If a simulation is used to determine the results from different evacuation systems being launched, the model will give quantitative input to the rescue study as specified in Table 12.12.

12.6 Analysis of Risk Associated with Rescue Operations

The rescue analysis is the last step of a complete EER (Escape, Evacuation and Rescue) analysis. The rescue analysis is not only dependent on the installation's own resources, in fact most of the rescue resources will be external. The standby

Table 12.12 Summary of evacuation fatality study

Accidental scenario	Expected numbers of personnel in different consequence groups						
	Evacuated successfully	Evacuated craft damaged	Launching failure	Remaining on the platform	Survivors seaborne	Not evacuated from platform	Casualties, seaborne
Blowout	120	45	0	5	5	0	2
Burning blowout	90	48	0	8	12	8	11
Riser leak	60	78	0	4	15	12	8

vessel may play an important role in the rescue/pick-up of personnel, but the track record of the standby vessel in this role is not very impressive. In the following accidents the installation's standby vessel was unable to rescue a single person:

- Alexander L. Kielland capsize (Norwegian North Sea, 1980)
- Ocean Ranger capsize (Canadian North Atlantic, 1984)
- West Gamma capsize/sinking (North Sea, 1990, attended by Norwegian standby/tug vessel).

It should however, be added that the weather conditions were quite severe in all these cases. People were rescued by other vessels in the first case mentioned, and in the last case all people were saved by another vessel. Only in the Ocean Ranger case were none saved because the standby vessel failed to rescue the persons from the lifeboats. There are, however, several cases where the standby vessel has been in a position to rescue all persons from the lifeboats.

Tables 12.13 and 12.14 present experience data from rescue operations in the North Sea, from platform accidents and from helicopter accidents near the installations [7]. Three cases of rescue operations from helicopter ditching occurred in 2012, but the details are not known at the end of 2012.

Table 12.13 Overview of rescue experience in the North Sea from helicopter accidents

Date	Accident type	Installation	Position	No of pers involved	Rescue by	Rescue time (mins)	Weather cond.
15.07.88	Helicopter controlled ditch	Norwegian North Sea	70 miles from coast	18 pers	SAR helicopter	approx 60 min from ditch	3–4 m waves
25.07.90	Helicopter crash on deck and sea	Brent Spar		7 survivors	MOB	Short time	Calm
14.03.92	Helicopter crash into sea	Shuttle flight Corm A to flotel		12 survivors, 5 died inside wreck, 6 died in water due to injuries	MOB, SAR helicopter	20 min response time, additional 70 min to rescue 6 survivors	27 knots, 15 m waves
19.01.95	Controlled ditch	Brae	120 nm from Aberd	18 pers	FRC from standby vessel	approx 60 min from ditch	30 knots, 7 m waves
18.01.96	Helicopter controlled ditch	Flight to Ula/Gyda	41 nm from Sola	18 pers	SAR helicopter	60 min from ditch	Calm
18.02.09	Controlled ditch	Flight to ETAP field	300 m from install.	18 pers	SAR helicopter & Daughter Craft	approx 120 min from ditch	Calm

Table 12.14 Overview of rescue experience from the North Sea, excluding helicopter accidents

Date	Accident type	Installation	Position	No of pers involved	Rescue by	Rescue time (mins)	Weather cond.
01.11.75	Riser rupture, explosion & fire	Ekofisk A			66 persons transferred by work boat		5 m/s, 0.5 m w
27.03.80	Capsize	Alexander L. Kielland		89 survivors	59 in 2 LB, 47 in 1 LB by SAR hel	Few hours	Gale force
					12 in 2 LB by supply v	Few hours	
					16 in lifer, 9 by SAR hel, 7 by supply v	Unknown	
					7 from sea by supply v	< 30 min	
					7 by Edda basket	Unknown	
15.02.82	Capsize	Ocean Ranger	Off New Foundland	84 pers	31 in 1 LB crushed against SBV, all drowned, remaining not sighted		
06.10.86	Burning shallow gas blowout	West Vanguard	Haltenbanken	79 survivors, 77 in 2 LB rescued by SBV, 2 from sea by MOB, 1 fatal from explosion	MOB, SBV	< 30 min for 2 survivors, 1.5–2 h for 77 in LBs	Low wind, 2–3 m waves
06.07.88	Gas explosion & escalating fire	Piper Alpha	North Sea	62 survivors	MOB, SBV	22 in 20 min, 39 in 50 min, 63 in 120 min	
22.09.88	Gas explosion & fire from blowout	Ocean Odyssey	Fulmar area	66 survivors, 8 from sea, 58 from 2 LBs	MOB, SBV	Unknown	8 m/s wind

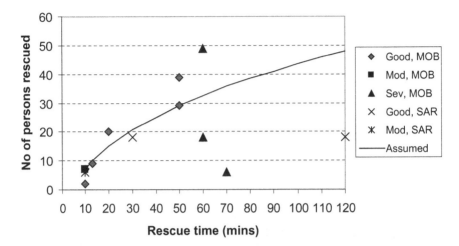

Fig. 12.8 Relationship between number of persons to be rescued and required rescue time, for MOB boats and SAR helicopters

The diagram in Fig. 12.8 presents a summary of experience from actual accidents as summarised above [7] with updates), where both the number of persons to be rescued and the time required to rescue them are presented. Three states of environmental conditions are considered, 'good', 'moderate' and 'severe' conditions. An assumed average distribution for good weather conditions is also presented.

12.6.1   Rescue Time Analysis

An analysis of risk levels on the Norwegian Continental Shelf [6] has considered the possible helicopter response times for two assumed field locations. Figure 12.9 is extracted from that study and shows the time dependent pick-up success probabilities for platforms located in the Northern North Sea and the Norwegian Sea ('Haltenbanken'). The rescue response times are determined for the following conditions and capacities, as detailed below:

- One person in sea, Blocks 30/8 and 6507/6
- Five persons in sea, Blocks 30/8 and 6507/6
- 15 persons in sea, Blocks 30/8 and 6507/6
- Two persons on platform, Blocks 30/8 and 6507/6.

Figure 12.9 presents these categories in a way which allows comparison between the two locations. There is a clear difference between the conditions of Blocks 30/8 (50 min) and 6507/6 (80 min) with respect to the maximum time required to complete pick-up in all scenarios. The differences may be explained as follows:

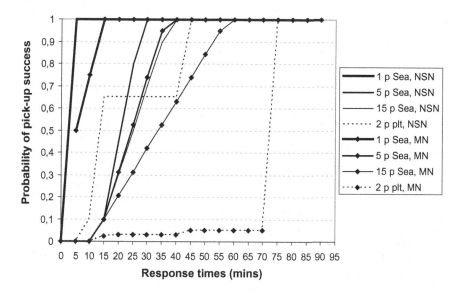

Fig. 12.9 Comparison of response times and probabilities for case studies in Northern North Sea (NSN) and Norwegian Sea (MN)

- For scenarios involving persons in sea, lifeboats or rafts, the FRC from the installation or standby vessel provide the quickest response, unless weather conditions prevent this.
- When the sea state exceeds $H_s = 6$ m, the arrival of the first SAR helicopter will be the deciding factor with respect to time required to complete rescue operations.
- For scenarios requiring personnel on the platform to be ferried to onshore medical care as rapidly as possible, the presence of a helicopter on the installation, or on a neighbour installations, will give the most rapid response.
- The arrival of the first onshore SAR helicopter will determine the maximum response time.

12.6.2   Rescue Capacity

Rescue capacity is dependent on both time and environmental conditions. Wind, waves, visibility and daylight conditions will all be crucial. There is little data available for such capacities, and the examples below should be regarded as an indication of what should be established, rather than what is established in terms of data.

Table 12.15 presents some considerations of important factors relating to the pick-up capacity of different rescue means. This table does not however, consider

Table 12.15 Assumed rescue capacities for some rescue means

Pick-up mode	Pick-up from lifeboats	Pick-up from life rafts	Pick-up from sea
Directly onto standby vessel Capacity,	Usually not possible	Can be done in good weather, and bad, if special equipment	Requires special equipment (net or similar)
Good weather		5–6 life rafts per hour	Up to 50 per hour
Bad weather		3–4 life rafts per hour	Up to 20 per hour
Directly onto other vessels	May be possible under ideal conditions	May be possible under ideal conditions	Not possible, unless it has a FRC
Standby's FRC, 1 boat, 1 crew (including transfer to standby vessel) Capacity,	Feasible up to $H_S = 6$ m Will be demanding on crew	As for lifeboats	Feasible up to $H_S = 6$ m, but dependent on locating survivors in sea. Will be very demanding on crew
Good weather	30 persons per hour	30 persons per hour	15 persons per hour
Bad weather	15 persons per hour	15 persons per hour	<10 persons per hour
Standby's FRC, if 2 crews (including transfer to standby vessel) Capacity,	As above, but crew changes may tire them less	As for lifeboats	As above, but crew changes may tire them less
Good weather	50 persons per hour	50 persons per hour	25 persons per hour
Bad weather	25 persons per hour	25 persons per hour	15 persons per hour
Standby's FRC, 2 vessels and 3 crews (including transfer to standby vessel) Capacity,	As above, but crew changes may tire them less	As for lifeboats	As for lifeboats
Good weather	70 persons per hour	70 persons per hour	35 persons per hour
Bad weather	35 persons per hour	35 persons per hour	25 persons per hour
SAR helicopter (assuming 20 min delay when emptying helicopter) Capacity,	Feasible up to $H_S = 8$ m Limiting factor will be time required to empty helicopter when full	As for lifeboats	
Good weather	17 persons per hour (helicopter capacity)	15 persons per hour	10 persons per hour
Bad weather	10 persons per hour	8 persons per hour	5–6 persons per hour

Note These are assumed values, based on subjective evaluations

the pick-up of one or two persons from the sea, due to having fallen from the platform during work over the platform's side.

The effect of environment conditions on rescue capabilities are dealt with in this table, which is mainly focused on the conditions during daylight and good visibility conditions.

There will nevertheless be quite considerable limitations on the rescue capacities, if the survivors have evacuated to the sea. Another factor which may limit the effective capacity even further, is if the survivors do not manage to stick together (standard practice is to bond together by rope or similar), and thus time must be spent locating the survivors first. Helicopter passengers in the UK and Norwegian sectors today routinely wear transponders, in order to eliminate delays for locating survivors. It may be argued that personnel in an undamaged lifeboat will not need to be rescued immediately in good weather conditions or at least, there will be no urgency about it. These conditions are, however, the least demanding, and even if the standby vessel has one FRC and only one crew, the rescue capacity is high.

It should be noted that the only helicopter considered to have a potential pick-up function is the fully equipped SAR helicopter, either from shore based or offshore locations, which however, will have a significant mobilisation time.

Other helicopters are not considered to have a real rescue capability in these circumstances, even if they have a personnel winch installed. Only if they have a dedicated and specially trained rescue person, who can assist the survivors in the boat, raft or the sea, would they be considered capable of rescue operations.

It is relatively rare that the total rescue system is exercised (and even rarer that the system gains practice from accidents), it is therefore quite interesting to note the experience from a full scale exercise in the North Sea in 1998:

- Premises:

  - 55 'persons' (i.e. dolls) in survival suits to be picked up within 120 min.
  - Primary standby vessel with nine-men crew, one FRC and one 'Sea Lift' (equipment to assist in lifting personnel out of the sea).
  - Secondary standby vessel with nine-men crew and one FRC.
  - 10 knot wind and minor swell.

-
  Resulting capacity:

  - All 55 'persons' rescued within 65 min.
  - Pick-up capacity per MOB boat was roughly 2 min per 'persons' (including the transfer to standby vessel).
  - Pick-up capacity by SAR helicopter was roughly 3 min per 'persons' (excluding transfer to vessel or installation).

- Important observations:

  - A minimum of two persons required on the bridge of standby vessel when simultaneous operation of FRC and Sea Lift.

- Increasing wind speed (15–20 knots) and sea state (2–4 m) would reduce the rescue capacity, most significantly for FRC.
- A minimum of two persons required on the bridge of standby vessel when the master is acting as 'on-scene commander' and communicating with the onshore emergency management team.

Additional data regarding helicopter pick up capacities have become available as a result of the cooperation between operators with respect to emergency preparedness for several installations in an area.

Experience data has been collected from rescue of personnel at sea by onshore based SAR helicopters [8]. This shows that rescue of personnel from life raft is quickest, whereas it is more time consuming to rescue persons from a lifeboat or ditched helicopter, due to the structures which may snag the recue wire during the operations. It has been demonstrated that the effect of wind is minimal, but significant movements of lifeboat or helicopter will complicate the rescue operation. The average time from the experience data is 1–2 min per person in life rafts.

Further illustration is provided by data from two exercises that have been conducted, using dummies ('dolls') to be picked up from the sea by platform based SAR helicopter.

In one exercise, the rescue of seven persons in severe weather conditions was tested in 45–55 m/s wind. Figure 12.10 presents the comparison of what was calculated and what the exercise demonstrated, again using dummies to be picked up from the sea. In the calculation, it was estimated that the 7 persons could be transferred to the nearest installation in 50 min. In the exercise, the operation was performed in two stages, whereby five persons were shuttled to the installation first, followed by two persons in the second stage. The second stage was completed in 53 min.

Figure 12.11 shows a comparison of actual performance in the exercise with what had been calculated as expected performance, based on experience data. The

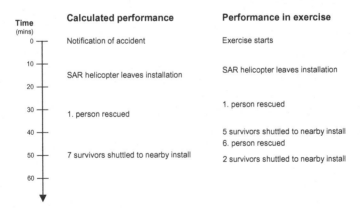

Fig. 12.10 Comparison of calculated performance by SAR helicopter with data from exercise in severe weather conditions

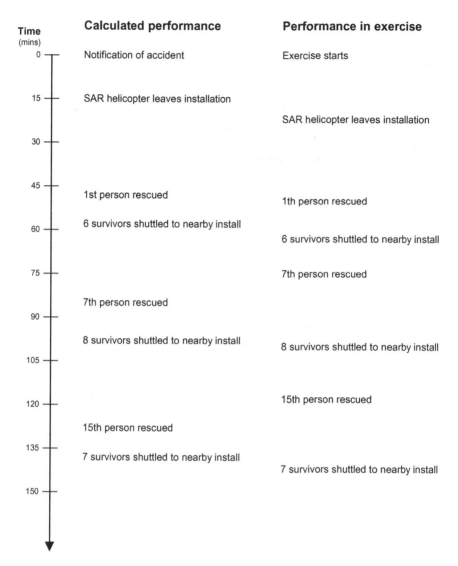

Fig. 12.11 Comparison of calculated performance by SAR helicopter with data from exercise in good weather conditions

exercise took place in good weather conditions. In the exercise, it took 143 min to rescue 21 persons, compared to 137 min in the calculation.

12.6.3 Rescue Fatality Analysis

12.6.3.1 Analytical Basis

The success probability of a rescue pick-up operation is dependent on a number of conditions and circumstances which may give very different results. This has been demonstrated by accidents in the past where the pick-up success probability has varied from 0 (Ocean Ranger) to 1.0. A success probability equal to 1.0 represents all survivors being rescued in time to survival. The following aspects should be incorporated in the rescue fatality assessments:

- time for vessel or helicopter to arrive at the scene of the accident
- availability of FRCs for pick-up
- capacity of rescue means as indicated above
- day or night
- visibility conditions
- weather conditions, sea state
- sea temperature
- use of survival suits and life vests
- size of area where survivors may be picked up
- total number of persons to be rescued and their distribution.

A Monte Carlo simulation approach is sometimes used for the rescue analysis. This uses a method based on random statistical simulations, reflecting assumed probability distributions to consider possible combinations of factors, conditions and circumstances. Timing of sequences as well as success/failure probabilities may also be simulated, based on the applicable conditions.

12.6.3.2 Results from Rescue Study

The availability of the rescue operation will be measured by means of fatality fractions. The availability will be quantified by the following formula:

$$A_{r,i} = 1 - \frac{N_{rescue\,fail}}{N_{evac}} \qquad (12.10)$$

where

$A_{r,i}$ rescue availability for alternative i
$N_{rescue\,fail}$ number of personnel not rescued by all available resources
N_{evac} number of personnel who succeeded in being evacuated.

Distributions of outcomes under different accident and environmental conditions should also be presented in addition to the overall availability as outlined above. Such information will be valuable for emergency response planning.

12.6.3.3   Synthesis

The final synthesis consists of tying the different steps of the analysis together, in order to produce the final results i.e., overall availabilities. The results may be calculated according to the formula:

$$A_{tot} = 1 - \frac{N_{escapefail} + N_{evacfail} + N_{rescuefail}}{N_{total}} \qquad (12.11)$$

where

A_{tot} total availability for alternative i
$N_{escape\ fail}$ number of personnel not being able to escape
$N_{rescue\ fail}$ number of personnel not rescued by all available resources
N_{tot} total number of personnel on the platform.

In addition to the overall availability and partial distributions presented for each of the phases, distributions of the overall values for different accident and environmental conditions should also be presented. Such information will be valuable for the emergency response planning.

12.7   Diving Fatality Risk

Diving fatality risk is usually calculated on the basis of a statistical analysis. There were several fatalities among divers in the 1970s and early 1980s. No further fatalities have occurred since then in the UK and Norwegian sectors. The use of divers has been reduced in the last 25 years, as the use of Remote Operated Vehicles (ROVs) has expanded. The values presented in Sect. 2.2 for air diving and saturation diving are nevertheless used for risk assessments purposes.

12.8   Fatality Risk During Cessation Work

Decommissioning of offshore installations involves risks that are not often present during the operations phase. Also some risks that are associated with drilling and production operations may become more critical during dismantling of the installations, e.g. the possibility of failure of the platform structure.

Risk management during decommissioning activities will require a combined qualitative and quantitative approach. It is considered that the qualitative techniques of preliminary hazard analysis, HAZID and HAZOP analysis are satisfactory to identify hazards and qualitatively assess risk. In addition, risk-reducing measures, barriers, may be identified and monitored.

Table 12.16 Summary of FAR values applicable to decommissioning [9]

No	Work task	FAR value
1	Rope access	10.3
2	Lifting operations—platform cranes	26.8
3	Lifting operations—external cranes	$1.1 \cdot 10^{-5}$ [a]
4	Lifting onshore	26.8
5	Scaffolding	5.5
6	Equipment decommissioning operations—offshore	1.9
7	Deconstruction operations—offshore	4.1
8	Prefabrication and construction—onshore	10.4
9	Demolition—onshore	12.3
10	Marine operations—Supply	18.1
11	Marine operations—Standby	3.3
12	Marine operations—Anchor handling	37.4
13	Marine operations—Tugs	13.2
14	Marine operations—Crane barges/vessels	5.5
15	Marine operations—Diving Support	7.5
16	Diving—Saturation	97
17	Diving—Air	685
18	Helicopter	32/97[b]
19	Management and administrative activities	0.4
20	Off-duty time	0.2

[a]Fatal accident rate per lift [b]Values for take-off/landing and cruise respectively

The quantification of risks for personnel engaged in decommissioning operations is more difficult due to the absence of suitable historical accident data for appropriate activities. In addition, occupational accident s are likely to have greater significance in the decommissioning phase than during production operations.

A Joint Industry Project (JIP) was carried out, aiming to provide a better basis for quantification of risk to personnel during decommissioning and removal operations of offshore installations [9]. Focus has been mainly on occupational accidents, on the basis that the lack of available data has been a bigger problem for analysis of this type of accident than for major hazards. Table 12.16 presents a summary of the FAR values associated with various activities recommended by the study.

A simple verification exercise was performed in the study, where coarse values for the total manhours consumption in all UK and Norwegian decommissioning projects were calculated and an average FAR value was calculated on basis of the existing accident records.

The number of fatalities which has been considered to be relevant to include for the period 1994–2003 is three. The total number of manhours was difficult to calculate, but a value of 11.4 million manhours was calculated, with an 80% confidence interval from 5 to 15 million manhours. On this basis, the average experienced FAR value in decommissioning projects in the North Sea in the period 1994–2003 was 26, with a variation range from 20 to 60 taking into account the uncertainty in the manhour volumes.

The historical FAR value indicates that the FAR values which have been calculated, and which generally are based on average industry statistics, may underpredict the risk associated with decommissioning projects. There may be a number of reasons for this, but in particular the uncertainty regarding the condition of structures, equipment etc, during decommissioning may introduce additional risk. Some of the fatal accidents which have occurred may be considered to be due to such uncertainty.

References

1. Kjellén U (2000) Prevention of accidents through experience feedback. Taylor and Francis, London
2. IOGP (1996) Quantitative risk assessment, datasheet directory. IOGP, London. Report No.: 11.8/250
3. Vinnem JE (1998) Blast load frequency distribution, assessment of historical frequencies in the North Sea. Preventor, Bryne, Norway. Report No.: 19816–04
4. Skogdalen JE, Khorsandi JD, Vinnem JE (2012) Escape, evacuation, and rescue experiences from offshore accidents including the deepwater horizon. Loss Prev Process Ind 25(1):148–158
5. Soma H (1995) Computer simulation for optimisation of offshore platform evacuation. In: Proceedings of the 14th international conference on offshore mechanics and arctic engineering. ASME Press, New York, June 18–22 1995
6. Vinnem JE, Vinnem JE (1998) Risk levels on the norwegian continental shelf. Preventor, Norway. Report No.: 19708–03
7. Vinnem JE (1999) Requirements to standby vessels. phase 1—Survey of current practice (In Norwegian only). Norwegian oil and gas, stavanger, Norway
8. Norwegian Oil and Gas Association (2000) Guidelines for area based emergency preparedness (In Norwegian only). Stavanger
9. Haugen S, Myrheim H, Stemland E (2004) Quantitative risk analysis of decommissioning activities. Trondheim, Safetec, Report No.: ST–20447–RA–1–Rev02

Chapter 13
Helicopter Transportation Fatality Risk Assessment

13.1 Overview

When offshore operations started in the North Sea, there were three severe accidents within a few years (1973, 1977 and 1978) in the Norwegian sector, with 34 fatalities. This created a high awareness level in the Norwegian offshore industry, as well as in unions and among employees. The UK sector had a series of fatal accidents in the early 1980s, which culminated with an accident with 45 fatalities when a Chinook crashed just before landing in Sumburgh on Shetland in November 1986. This accident caused the complete abandonment of the Chinook helicopter in offshore operations in UK the and Norway.

The high attention in Norway on the risk during helicopter transportation to offshore fields has led to several initiatives over time. The first initiative was the initiation of a series of Helicopter Safety Studies (HSSs), of which the latest is HSS–3 [1]. These studies are conducted once every 10 years. A new report was published in 2017, in order to reflect the fatal accident outside Bergen in 2016, this has been "HSS-3b" [2].

Helicopter transportation safety in offshore flying was also the topic of an official Norwegian White Paper in 2002 [3], which proposed ambitions and actions for a significant reduction in risk levels. A safety advisory group for helicopter safety on the NCS has also been formed [4] with representatives from supervisory and air traffic control authorities, helicopter operators, oil companies and unions for offshore employees.

Helicopter operators in Norway have been quick to replace old helicopters with new models as soon as they have become available. This is believed to be to some extent because of the high attention on these issues, but also the fact that some of the largest oil companies have requested modern helicopters in bidding for their transportation contracts. Norwegian companies have also employed more sophisticated preventive maintenance schemes. This is believed to have had a positive effect on accident statistics, with the exception of Airbus H225LP (previously

© Springer-Verlag London Ltd., part of Springer Nature 2020
J.-E. Vinnem and W. Røed, *Offshore Risk Assessment Vol. 1*,
Springer Series in Reliability Engineering,
https://doi.org/10.1007/978-1-4471-7444-8_13

known as Eurocopter EC225, in use from 2008), where a new helicopter model has had two full-crew fatal accidents. This helicopter model has not been in use in Norwegian and UK sectors, in spite of having been allowed by European Aviation Safety Authority (EASA).

There have been two helicopter accidents in Norway associated with the offshore transportation of personnel since 1978, when a Super Puma crashed into the Norwegian Sea in September 1997 as well as the Airbus H225LP which lost its rotor and crashed fatally on a small island just few minutes before scheduled landing at Bergen airport in April 2016. It should be noted that the accident in August 1991 when a helicopter crashed into the sea in the Ekofisk field is not counted as a transportation accident, as the helicopter was used in the maintenance of a flare tip on one of the Ekofisk installations, with fatal outcome for the three persons on board.

The accident in 1997 involved people being shuttled daily between accommodation onshore and the FPSO installation during the commissioning phase, because of insufficient accommodation capacity in this phase. This accident put considerable focus on the significant risk increase employees were being exposed to, if shuttled between offshore installations or between an offshore installation and onshore on a daily basis. The volume of shuttling has been reduced significantly since then.

The accident in 2016 occurred with a helicopter in normal transit from the Gullfaks B platform to Bergen airport, with 11 passengers and 2 pilots, operated by CHC Helikopter Service AS. The helicopter had just descended from 3,000 ft and had established stable cruising at 140 knots at 2,000 ft for about two minutes when the main rotor detached, without any prewarning, and with a completely normal flight up until that point. The helicopter crashed on a small island east of Turøy. All 13 persons on board perished. The investigation [5] has shown that the accident was a result of a fatigue fracture in one of the eight second stage planet gears in the epicyclic module of the main rotor gearbox. The fatigue had its origin in the upper outer race of the bearing (inside of the gear), propagating towards the gear teeth.

The risk levels in helicopter transportation of personnel to offshore installations has improved considerably since 1970s. This can still be maintained in spite of the tragic accident in 2016. Safety standards have not improved correspondingly in UK offshore helicopter operations, and there are significant statistical differences between the UK and Norway when it comes to FAR for passengers.

Fatal accidents have continued to occur outside Norway during the past 15 years, and there have been recent fatal accidents during the transportation of personnel offshore in both the UK and Canada. We therefore propose different fatality rates for the UK and Norway in this chapter.

However, it should be emphasised that helicopter transportation is not risk free in the Norwegian sector either. There have been several near-misses the past 15–20 years, for instance when one of the main blades was almost 75% fractured from a foreign object in 2002 and the helicopter was lucky enough to find a nearby tanker onto which it could make emergency landing. The values in Chap. 17 also demonstrate that helicopter transportation risk is still the highest contributor to offshore employees' risk levels in Norway.

This chapter builds on previous work, such as Refs. [6, 7], in addition to the following studies: HSE [8], SINTEF [1, 2] and Heide [9].

13.2 Accidents and Incidents—Offshore Northwest Europe

This section provides a brief overview of the helicopter accidents and incidents in the UK and Norwegian sectors since 1990.

Table 13.1 presents the accidents and incidents that have occurred, mainly based on Refs. [1, 2]. There is one difference with respect to the HSS-3 study, namely precursor events; i.e. when flights could return to land or the installation, which are not included. Experience from the Risk Level project [10] has demonstrated that the list of precursor events included in HSS-3 is not complete. The two occurrences in Holland were found on company webpages.

An accident in offshore Newfoundland, Canada in 2009 can be added to the incidents and accidents in Table 13.1. A warning light for main gearbox lubrication failure in the S92 helicopter came on 13 min after levelling off at a cruising altitude of 9,000 feet. The crew declared an emergency, started to return and descend to 800 feet, believing they had 30 min of emergency lubrication available. Ten minutes later they crashed with high force in the sea. The investigation of the 2016 accident in Norway [5] showed that if the causes of the accident in 2009 had been investigated more thoroughly leading to preventative measures, the accident in 2016 might have been avoided. Two pilots and 15 passengers drowned, and one passenger survived 80 min in the sea before being rescued with severe injuries. None of the emergency locator beacons in the helicopter, on the rafts and personal locator beacons worn by crew and passengers had activated in this accident. There was also a problem with the personal locator beacons in the 2009 controlled ditching in the UK sector, but this was a different problem, involving interference. The personal locator beacons were therefore withdrawn for some months, while these problems were solved.

An important observation from Table 13.1 is that all the accidents during cruising seem to have technical failures as their main causes. During take-off, landing and approach there are six occurrences, of which five are associated HOFs and one is technical.

All the accidents and incidents during take-off, landing and approach in Table 13.1 have occurred in the UK sector. In fact, even if the helicopter accidents in the 1970 and 1980s are included, no accident has occurred during take-off, landing and approach in the Norwegian sector. This difference is not statistically significant, because of the low number of accidents and incidents, but it is still noteworthy.

Table 13.1 Overview of helicopter accidents and incidents, UK, Holland and Norway, 1990–2012

Year	Country	Helicopter type	Fatalities/ injuries/ survivors	Location	Short description	Primary cause	Flight phase
1990	UK	S-61 N	6/0/7	Brent Spar	While manoeuvring to land on the helideck, the tail rotor struck a crane. Aircraft descended onto the helideck and fell into the sea where it sank rapidly	HOF	Landing
1992	UK	AS332 L1	11/1/5	Near Cormorant A	Aircraft taking pax from platform to flotel 200 m away. Access Bridge had been lifted because of adverse weather. Aircraft departed and then turned downwind with insufficient airspeed and descended rapidly into the sea and sank	HOF	Take-off
1995	UK	AS332/L2	0/0/18	Cruising to Brae A	Controlled emergency landing caused by lightening damaged the tail rotor	Environmental	Cruising
1996	Norway	S-61 N	0/0/16	Crusing to Ula/ Gyda	Controlled emergency landing caused by strong vibrations when main rotor blade failed	Fatigue cracks	Cruising
1997	Norway		12/0/0	Cruising to Norne	Failure of shaft to MGB caused helicopter to disintegrate	Technical	Cruising
1997	Holland	AS332/L2	0/0/8	Cruising from L7 to Den Helder	Controlled emergency landing (20.12.97)	Unknown	Cruising
2002	UK	S-76	11/0/0	Leman field	Rotor blade failure during approach to platform. Aiicraft went out of control and crashed into the sea	Main Rotor Blade Failure	Approach
2002	Norway	AS332/L2	0/0/16	Crusing from Sleipner to Sola	Strong vibrations because of severe damage to main rotor blade, caused emergency landing on nearby tanker	Main Rotor Blade damage	Cruising
2006	Holland	AS332/L2	0/0/17	SAR flight from Den Helder Apt	Controlled emergency landing because of abnormal engine indications	Technical	Cruising

(continued)

Table 13.1 (continued)

Year	Country	Helicopter type	Fatalities/ injuries/ survivors	Location	Short description	Primary cause	Flight phase
2006	UK	SA365 N	7/0/0	Close to North Morecambe platform	When preparing to land on the North Morecambe platform in the dark, the helicopter flew past the platform and struck the surface of the sea. The fuselage disintegrated on impact and the majority of the structure sank	HOF	Landing
2008	UK	SA365 N	0/0/7	During landing on helideck	Aircraft tail hits crane during landing on helideck	HOF	Landing
2009	UK	H225LP	0/0/18	ETAP field	Controlled ditching close to installation. All on-board escaped safely, rescued by a SAR helicopter and a 'Daughter craft'	HOF	Approach
2009	UK	AS332L2	16/0/0	Cruising from Miller to Aberdeen	Catastrophic failure of the planet gear in MGB because of fatigue	MGB technical failure	Cruising
2012	Norway	H225LP	0/0/21	Cruising from Halten bank to Kristiansund	Controlled emergency landing when a warning light for low hydraulic pressure came on	Unknown technical	Cruising
2012	UK	H225LP	0/0/14	30 miles east of Aberdeen	Controlled emergency landing when a warning light for low hydraulic pressure came on	Unknown technical	Cruising
2012	UK	H225LP	0/0/19	32 nm south of Shetland	Controlled emergency landing when several alarms started to indicate low hydraulic pressure	MGB technical failure	Cruising
2013	UK	AS332L2	4/0/14	Just outside Sumburgh airport, Shetland	Fog and low visibility. Autopilot Altitude Acquire modus was not used during descent, which it should have been according to procedure	Procedure not followed	Approach
2016	Norway	H225LP	13/0/0	Small island outside Bergen	Catastrophic failure of planet gear in the MGB because of fatigue	MGB technical failure	Cruising

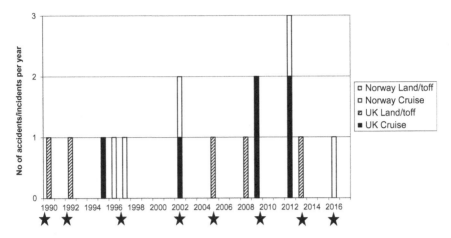

Fig. 13.1 Fatal (marked with star) and non-fatal helicopter accidents in the UK and Norwegian sectors, 1990–2012

Accidents that occur on the helideck while the helicopter is parked are not included in Table 13.1 or in the discussion in this chapter.

HOFs dominate for accidents and incidents during take-off, landing and approach. This implies that accidents and incidents because of HOFs have a major contribution to the occurrences in the UK sector, but not in the Norwegian sector. This is somewhat surprising, as the qualification and training requirements are based on international standards, and thus should be the same. However, the differences are not statistically different, as noted above. Actually, if the accidents in the Norwegian sector in the 1970s were included, this would have changed somewhat, as HOFs played strong roles in at least two of these accidents.

Figure 13.1 presents an overview of non-fatal as well as fatal accidents. Fatal accidents are marked with a star below the year in question. Five such fatal accidents are in the UK sector and only one in the Norwegian sector.

It is noteworthy that only four out of the 14 accidents and incidents shown in Fig. 13.1 are from the Norwegian sector. The number of person flight hours was in 2011 41% higher in the UK compared with in Norway (see also Sect. 13.5). The difference is even more significant since 2000, but this is not statistically significant because of the low number of events.

The number of accidents and incidents associated with MGB has been increasing; in fact, all the accidents in 2012 and one in 2009 as well as a fatal accident in Canada in 2009 and the accident in 2016 were all caused by MGB problems.

13.3 Risk Modelling

13.3.1 Assumptions and Premises

When the risk of the helicopter transportation of personnel was initially assessed back in 1970s, it was assumed that the main factors would be the same as in fixed wing flying, namely that non-technical causes would contribute 70–80% and that the majority of accidents would be associated with take-off, approach and landing. However, the majority of accidents initially occurred during transit (cruising), and therefore these two key assumptions had to be reconsidered.

Another aspect is also different. If engine or gear box failure occurs during cruising altitude, the helicopter is supposed to be able to make a controlled emergency landing, because of main rotor autorotation. The crew and passengers should be unhurt in such circumstances, especially as they are provided with survival suits, personal emergency beacon, and liferafts in the helicopter. Experience has shown that the helicopter sometimes disintegrates in the air, thus making controlled emergency landing impossible. In other cases, like in the Turøy accident in 2016, the rotor and/r gear box failure mode prevents autorotation. There have in fact also been three cases of controlled emergency landing without fatalities in 2012.

A further aspect to consider is shuttling between two installations, which often has a short duration, and thus an entirely different relationship between take-off, approach and landing, compared with the cruising phase.

The UK and Norwegian sectors have traditionally been considered together, without any difference, when calculating accident and incident statistics. The risk levels presented in this chapter are predicted separately for the UK and Norwegian sectors, as the experience during the past 20 years has been quite different.

Finally, with respect to helicopter operations in the Norwegian sector, most helicopters in operation in the beginning of 2013 were new models, mainly Sikorsky S92 and Airbus H225LP (formerly Eurocopter Super Puma EC225). These new models had at that time been involved in only one fatal accident (offshore Newfoundland, Canada, 2009), whereas all other fatal accidents are with older models. The Super Puma AS332L2 by contrast, had been involved in several incidents, as shown in Table 13.1. The improvement considered in 2013 to be implied by these new models would be taken into account when calculating accident frequencies. Since then, the Airbus H225LP has been subject to the 2016 accident which was basically a repeat of the 2009 accident in Scotland. This was an AS332L2 helicopter with the same main gearbox as the H225LP helicopter, and all confidence with the Airbus H225LP has been lost in UK and Norway.

All these aspects need to be considered when developing the risk model. Many risk models in the literature fail to address some of these aspects.

13.3.2 Risk Model

The FAR values for personnel on an installation are usually expressed as the number of fatalities per 10^8 exposure hours (see Sect. 2.1.4). It is customary to express the FAR values for helicopter transport as the number of fatalities per 10^8 person flight hours.

The modelling is based on the same principles as those adopted by Heide [9], namely that risk during helicopter transportation is function of the flight time during the cruise and approach phases as well as a function of the number of landings/take-off during landing and take-off. This can be expressed as follows:

$$FAR_{Hel} = FAR_{Hel}^{cruise} + FAR_{Hel}^{landing} \tag{13.1}$$

where:

FAR_{Hel} FAR value (per 10^8 person flight hour) for flying from onshore airport to the offshore helideck or back

FAR_{Hel}^{cruise} FAR contribution from the cruising phase

$FAR_{Hel}^{landing}$ FAR contribution from the take-off and landing phases.

The FAR for the helicopter transportation of personnel can be expressed as follows:

$$FAR_{Hel}^{cruise} = \frac{Fat_{Hel}^{cruise} \cdot 10^8}{Person\,flight\,hours} = \frac{Fat_{Hel}^{cruise} \cdot 10^8}{Pass_{av} \cdot Fl.hrs} \tag{13.2}$$

where:

Fat_{Hel}^{cruise} number of fatalities in helicopter accidents during cruising in the applicable period

$Pass_{av}$ average number of passengers

$Fl.hrs$ total number of flight hours in the applicable period.

It is further usual to include pilots in the calculation of fatalities, although separate FAR values can be expressed for them.

The accident rate for helicopters, AR_{Hel}^{cruise}, can be expressed as follows:

$$AR_{Hel}^{cruise} = \frac{N_{acc,\,Hel}^{cruise}}{Flight\,hours} \tag{13.3}$$

where:

$N_{acc,\,Hel}^{cruise}$ number of accidents during cruising the in applicable period.

The FAR for helicopters, $FLAR_{Hel}^{cruise}$, can be expressed as follows:

$$FLAR_{Hel}^{cruise} = \frac{N_{acc,Hel}}{Flight\ hours} \cdot \frac{N_{F,acc,Hel}}{N_{acc,Hel}} \qquad (13.4)$$

where:

$N_{F,acc,Hel}$ number of fatal accidents during cruising in the applicable period.

Equations 13.2, 13.3 and 13.4 should also be repeated for the take-off and landing phases:

$$FAR_{Hel}^{landing} = \frac{Fat_{Hel}^{landing} \cdot 10^8}{Number\ of\ landings} \qquad (13.5)$$

where:

$Fat_{Hel}^{landing}$ number of fatalities in helicopter accidents during take-off or landing in the applicable period.

$$AR_{Hel}^{landing} = \frac{N_{acc,Hel}^{landing}}{Number\ of\ landings} \qquad (13.6)$$

where:

$N_{acc,Hel}^{landing}$ number of accidents during take-off or landing in applicable period.

The FAR for helicopters, $FLAR_{Hel}^{landing}$, can be expressed as follows:

$$FLAR_{Hel}^{landing} = \frac{N_{acc,Hel}^{landing}}{Number\ of\ landings} \cdot \frac{N_{F,acc,Hel}^{landing}}{N_{acc,Hel}^{landing}} \qquad (13.7)$$

where:

$N_{F,acc,Hel}^{landing}$ number of fatal accidents during take-off or landing in applicable period.

13.4   Previous Predictions

HSS-1 [11] was carried out immediately after the period with many fatalities in UK operations, and this study calculated a high fatality rate of:

$$3.8 \times 10^{-6} \text{ per person flight hours}$$

The study conducted in 1998 [6] divided accident frequency into separate values for cruising and landing/take-off. A comparable value would, however, be:

$$1.6 \times 10^{-6} \text{ per person flight hours}$$

The HSS was updated in 1999 [12], and the statistics from SINTEF were compiled in a white paper [3] on helicopter safety in 2002. This study documented the following value:

$$1.4 \times 10^{-6} \text{ per person flight hours}$$

The white paper also proposed the objective of reducing the risk level by 50% over a 10-year period compared with the average for the period 1990–2000.

The HSS was updated in 2010 [1]. This study documented the following value for the period 1999–2009:

$$2.4 \times 10^{-6} \text{ per person flight hours (average for North Sea)}$$
$$5.6 \times 10^{-6} \text{ per person flight hours (UK sector)}$$

The HSS was updated in 2017 [2]. This study documented the following value for the period 1999–2015:

$$1.0 \times 10^{-6} \text{ per person flight hours (Norwegian sector)}$$
$$4.0 \times 10^{-6} \text{ per person flight hours (UK sector)}$$

The frequency for the Norwegian sector has artificially assumed that the Turøy accident in 2016 to be included in the period 1999–2015, for the purpose of this calculation.

The exposure hours in the UK sector in the period 1999–2009 seems to be too low value in HSS-3 (see also Sect. 13.5), and therefore the FAR value is too high. Past reductions in FAR and future objectives may seem to be very significant reductions in the fatality rate, but several factors need to be considered:

- The original SINTEF study (HSS-1) covered the period 1969–89 and during the period 1975–1986 there were more than 125 fatalities in helicopter accidents in the North Sea. Since 1986, only three fatal accidents with 39 fatalities have occurred.
- The period 1975–1986 was considered in the HSS, but the study did not attempt to consider if any trends could be identified or whether there was any basis for making distinctions between Norwegian and UK operations.
- It is an established fact that improvements were introduced in helicopter operations in the 1980s because of the high number of accidents and thus a reduction in the frequency of accidents would be expected.

Table 13.2 Helicopter statistics for UK and Norway offshore operations, 1996–2005

Area	Person flight hours (million hours)	Sources
Norway	7.090	NOU2002:17 and RNNS report 2005
UK	10.320	NOU2002:17 extended to 2005
Total	17.410	Corresponding number of flight hours: 1.348 million hours

Table 13.3 Helicopter accident statistics for UK and Norway offshore operations, 1996–2005

Aspect	Number of persons	Sources
Accidents, cruise	3	):
Accidents, take-off/landing	0	):
Fatal accidents	2	): NOU2002:17 and HSE 2004
Fatalities	23	):
Survivors	0	):

One of the deficiencies of the SINTEF studies is the lack of distinction between fatality risk during cruise and landing/take-off. This is an important distinction especially when shuttling is considered.

It might be argued that taking only the 10 year period following a period with high fatalities gives rise to an over-optimistic prediction. However, it would be impossible to define how much of the earlier period would need to be included to avoid such optimism.

Risk parameters were in the second edition of this book (2007) presented for the period 1996–2005. The data sources are presented in Tables 13.2 and 13.3.

The most up to date FAR value for helicopter transport in the North Sea was in 2007:

$$1.32 \times 10^{-6} \text{ per person flight hours}$$

This implies a similar value to that stated in Ref. [3], which was 1.4 per million person flight hours. If the corresponding value is calculated for the period 1987–2005, this becomes 1.35 per million person flight hours. In the Norwegian white paper, five year rolling averages were shown; these were naturally varying. Table 13.4 separately presents the derivation of FAR for cruising and landing on installations.

Table 13.5 separately presents the derivation of FAR for cruising and landing on installations that were included in the 3rd Edition of this book.

New predictions are made separately for the UK and Norwegian sectors in Sects. 13.6 and 13.7, because of the differences discussed above. These predictions are made for a 20-year period, owing to data limitations. When combined predictions are made (Sect. 13.5), they are limited to 10 years.

Table 13.4 Helicopter risk parameters for the cruising and landing phases

Factor	Cruising	Landing on platform	Comments
Basis in period	1996–2005	1987–2005	
Accident rate	2.22×10^{-6}	2.0×10^{-7}	
Fraction of fatalities to total number of persons exposed	1.0	0.46	
Fatal accident rate	1.48×10^{-6}	2.01×10^{-7}	
Average number of fatalities	11.3	6.0	Both values based on period 1987–2005

Table 13.5 Helicopter risk parameters for the cruising and landing phases, 2002–2011

Factor	Cruising	Landing on platform	Comments
Accident rate	1.46×10^{-6}	3.8×10^{-7}	
FAR	7.3×10^{-7}	1.89×10^{-7}	
Fraction of fatalities to total number of persons exposed	0.60	0.50	
Average number of fatalities (in fatal accidents)	13.0	9.0	Values based on period 1992–2011

13.5   Combined Prediction of Risk Levels—UK and Norwegian Sectors

Risk parameters are presented for the 20 year period 1998–2017. The data sources are presented in Tables 13.6 and 13.7.

CAA statistics provides the number of passengers and air traffic movements from relevant airports. For 2000 and 2001, these values correlated with flight hours and person flight hours, and average conversion factors could thus be established.

Table 13.6 Helicopter statistics for UK and Norway offshore operations, 1998–2017 and 2008–2017 (in parenthesis)

Area	Person flight hours (million hours)	Sources
Norway	15.362 (8.077)	NOU2002:17 and RNNP (PSA, 2012)
UK	17.663 (8.632)	NOU2002:17, CAA statistics, HSS-3b
Total	32.980 (16.709)	Corresponding number of flight hours: 2.603 million hours

Table 13.7 Helicopter accident statistics for UK and Norway offshore operations, 1998–2017 and 2008–2017 (in parenthesis)

Aspect	Number of events/persons	Sources
Accidents, cruise	7 (5)	):
Accidents, take-off/landing	3 (2)	):
Fatal accidents, cruise	3 (2)	):
Fatal accidents, take-off/landing	2 (1)	): NOU2002:17 and HSS-3
Fatalities, cruise	40 (29)	):
Fatalities, take-off/landing	11 (4)	):
Survivors, cruise	18 (18)	):
Survivors, take-off/landing	21 (21)	):

The values were also checked against the number of offshore employees in the period, and reasonable consistency was established. HSS-3b person flight hours appear to be too low values, as for HSS-3.

The FAR value for helicopter transport in the North Sea can be calculated as an average for the two periods:

$$1998-2017 : 1.55 \times 10^{-6} \text{ per person flight hours}$$
$$2008-2017 : 1.97 \times 10^{-6} \text{ per person flight hours}$$

This implies values that are somewhat above that stated in Ref. [3], which was 1.4 per million person flight hours. It also implies that these values are slightly above the values for the periods 1992–2011 and 2002–2011.

Table 13.8 separately presents the derivation of FAR for cruising and landing on installations.

The trends are shown in Fig. 13.2 for the North Sea in total, as well as for the UK and Norwegian sectors separately. The values for the North Sea are rolling 10 year average values, whereas those for the sectors are rolling 20 year average values (except in the period 1996–2007, when they build up from 10-year average to 20-year average values).

Table 13.8 Helicopter risk parameters for the cruising and landing phases, 2008–2017

Factor	Cruising	Landing on platform	Comments
Accident rate	2.16×10^{-6}	3.3×10^{-7}	
FAR	6.7×10^{-7}	7.3×10^{-8}	
Fraction of fatalities to total number of persons exposed	0.62	0.44	
Average number of fatalities (in fatal accidents)	13.3	5.5	Values based on period 1998–2017

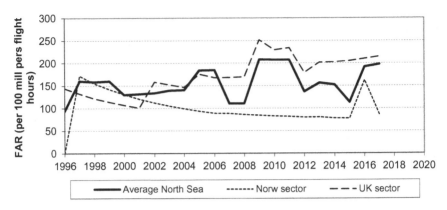

Fig. 13.2 Trends in average FAR values, North Sea and UK and Norwegian sectors, 1998–2017

It can be seen that the trend for the Norwegian sector is falling until 2015, whereas the trend for the UK sector is increasing during the period. This is the reason why the average for the North Sea is slightly increasing as an overall trend in the period. The Turøy accident in 2016 has significant influence on the values for the Norwegian sector, but obviously far less effect on the combined values.

13.6   Prediction of Risk Levels—UK Sector

There have been relatively frequent helicopter accidents in the UK sector for almost 30 years, and so there is a good statistical basis for predictions. It is nevertheless considered to be most appropriate to use an average over 20 years when considering one sector only. The FAR value for helicopter transport in the UK sector is:

$$1998-2017 : 1.53 \times 10^{-6} \text{ per person flight hours}$$

This value is somewhat above the average value for the North Sea (i.e. UK and Norwegian sectors, see Sect. 13.5). Table 13.9 separately presents the derivation of FAR for cruising and landing on installations.

The replacement of older helicopter models with new models (S92 and H225LP) is much slower in the UK compared with in Norway. It is therefore considered to be relevant to predict fatality rates for the future in the UK sector as experienced in the recent years. It has also been documented that several accidents have been caused by HOFs, which would also suggest that significant change is unlikely in the future. The last five years have been without fatal accidents, which may be an indication that the risk level has been reduced, however this is not statistically significant at the present time.

Table 13.9 Helicopter risk parameters for the cruising and landing phases, UK, 1998–2017

Factor	Cruising	Landing on platform	Comments
Accident rate	2.7×10^{-6}	4.1×10^{-6}	
FAR	8.1×10^{-7}	9.3×10^{-7}	
Fraction of fatalities to total number of persons exposed	0.60	0.34	
Average number of fatalities (in fatal accidents)	13.5	5.5	Values based on period 1998–2017

13.7 Prediction of Risk Levels—Norwegian Sector

Until the accident in 2016 there had only been one fatal helicopter accident in the Norwegian sector during a period of 30 years. The Turøy accident suggests typically in the order of 1 fatal accident per 20 years with the present volume of flight hours. It is thus considered to be most appropriate to use an average over 20 years when considering one sector only. The FAR value for helicopter transport in the Norwegian sector is:

$$1998-2017 : 0.85 \times 10^{-6} \text{ per person flight hours}$$

This value is considerably lower than that for the North Sea as well as the UK sector value (see Sects. 13.5 and 13.6), but it is slightly up from the value for the 1992–2011 period (0.82 per million person flight hours). Table 13.10 separately presents the derivation of FAR for cruising and landing on installations.

The replacement of older helicopter models is almost complete in the Norwegian sector, because oil companies have required the use of newer models when new contracts have been signed. It is claimed by experts that the S92 helicopter shows a

Table 13.10 Helicopter risk parameters for the cruising and landing phases, Norway, 1998–2017

Factor	Cruising	Landing on platform	Comments
Accident rate	3.2×10^{-6}	0	
FAR	1.08×10^{-6}	0	
Fraction of fatalities to total number of persons exposed	1.0	0	
Average number of fatalities (in fatal accidents)	13	0	

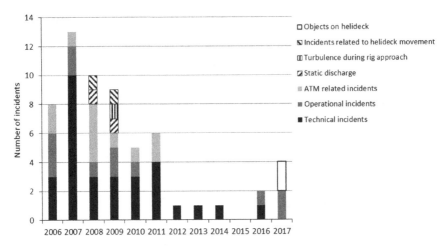

Fig. 13.3 F major hazard indicator 1 from the PSA risk level project, reflecting the major hazard potential of precursor events, Norwegian sector, 2006–2017

lower incidence rate of major failure precursors. It is therefore considered to be relevant to predict lower fatality rates for the future, compared with what has been experienced in the past. The main volume of flight hours is at present time carried out by the S92 helicopter, as explained in Sect. 13.1. This is a somewhat vulnerable situation. If severe problems were to be associated with S92 sometime in the future, there could be no helicopter which was trusted.

The Risk Level project suggests trends in potential major accidents with helicopters, based on precursor events with no (apart from 'luck') or only one remaining barrier. Figure 13.3 presents the trend for the past twelve years, suggesting a downward trend. However, the diagram also indicates that other causes may limit the reduction that is achievable, as operational (including pilot) errors and ATM errors are significant contributors, whereas helideck movement and turbulence have less importance.

It is noteworthy that the number of incidents with no or only one remaining barrier is increasing in 2016 and 2017. This is a statistically insignificant increase, but it nevertheless raises the question whether cost reductions in the last years have contributed to this increase. When the causes are considered, all of the incidents in 2017 are associated with human and organizational factors, which gives an additional indication that cost focus may be involved. See also Sect. 13.10.2 which discusses aspects related to implementation of new EU regulations, and the likely effects on safety. A study by Safetec indicated that helicopter safety already is under pressure.

The helicopter safety white paper [3] suggested a long list of improvements that together were considered to imply a reduction by 50% of the fatality frequency, according to the goal. Some of the main actions were [13]:

- Flight data monitoring
- New technology
- TCAS 1 collision avoidance system
- EGPWS, Enhanced Ground Proximity Warning System
- De icing (rotor)
- Survivability in Sea state 6.

A safety advisory group was also established [4]. Contact with two of the leading members [13, 14] of the group revealed that the majority of the actions have already been implemented for the majority of the helicopters in the Norwegian fleet [4]. There are two main features of improvement being sought through these actions:

- Reduction in the frequency of technical and operational faults that may lead to fatal accidents.
- Reduction in the consequences of such faults, i.e. reduce (or eliminate) the number of fatalities resulting from such faults.

This implies that an event that in the past could have led to serious consequences may in the future have less severe consequences. In theory this should be reflected in the criteria used to classify incidents, but it will probably take some time before a revision is made.

When all these factors are taken into account, they attempted to consider what these qualitative factors may result in with respect to the prediction of fatalities in helicopter transportation in the future. The percentage completion of recommendations was 67% at the end of 2007 and in 2012 it was considered to be 100%. It should, however, be noted that what effect these actions will have on the future incident rate is based on subjective evaluations made by a large group representing different organisations and interests.

The final aspect to consider is that helicopter accidents are so rare that some margin must be allowed for what could be called 'unexperienced events' (or unknown threats), namely mechanisms that are unknown until they occur for the first time. This may, for instance, be related to the volume of traffic in 'near arctic' conditions in the Barents Sea. Allowance has been made for such occurrences in the future. On this basis, we have subjectively considered that a representative average value for the future may be the following:

- Helicopter transport: 70 fatalities per 100 million person flight hours

This corresponds to full compliance with the 50% reduction target of the official white paper [3]. This value was considered in 2012 and is still considered to be valid.

Table 13.11 presents the predicted FAR for cruising and landing on installations separately for the Norwegian sector. For the landing and take-off values, 50% of the UK values has been applied.

Table 13.11 Helicopter risk parameters for the cruising and landing phases, Norway, future predictions

Factor	Cruising	Landing on platform	Comments
Accident rate	3.1×10^{-6}	3.3×10^{-7}	
FAR	1.34×10^{-6}	2.2×10^{-7}	
Fraction of fatalities to total number of persons exposed	0.68	0.58	
Average number of fatalities (in fatal accidents)	12	4.5	

13.8   Other Risk Parameters

13.8.1   Fatality Distribution

A distribution of fatalities occurring in helicopter accidents may be required in cases where an f–N distribution is used to express risk to personnel. This may be generated from accident statistics. Figure 13.4 presents the distribution of fatalities per fatal accident.

Most helicopters in use in the North Sea typically have 14–18 seats and therefore it is usually not necessary to distinguish between different helicopter types. The Chinook helicopter has 45 seats, but this helicopter has been out of use for North Sea activities following the accident in 1986.

Figure 13.4 shows an overview of the number of fatalities in helicopter accidents during cruising and landing. Accidents that occurred on the helideck are omitted from the presentation.

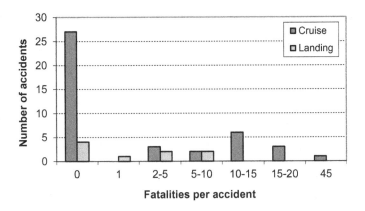

Fig. 13.4 Fatalities in North Sea offshore helicopter accidents

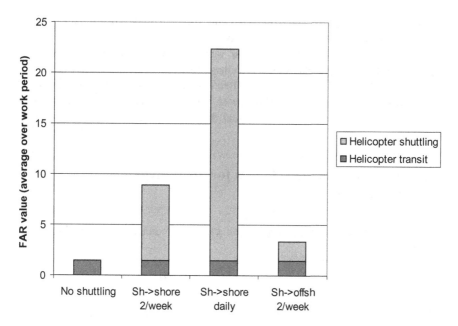

Fig. 13.5 Comparison of average FAR values for an offshore worker according to the extent of shuttling performed

13.8.2   Comparison of Risk Associated with Shuttling

The critical effect of extensive shuttling between the shore and offshore facilities on the risk levels for the persons involved was demonstrated in Ref. [6]. This is sometimes undertaken during offshore installation and/or the commissioning phase of new facilities. Figure 13.5 shows the average annual FAR value for an offshore employee, according to the extent of offshore shuttling that the person is exposed to.

The following shuttling situations are shown (abbreviations used in the diagram are also included):

1. No shuttling ('No shuttle').
2. Shuttling to shore twice per week ('Sh– > shore 2/week', 60 min each way).
3. Shuttling to shore daily ('Sh– > shore daily').
4. Shuttling to a nearby installation offshore twice per week ('Sh– > offsh 2/week', 15 min per one way trip).

All helicopter operations are included, while transport between the installation and shore at the outset and finish of a full working period (usually 2 weeks), as well as any shuttling during that working period are also included. It may thus be observed from the levels demonstrated here, that helicopter associated risk is important for the overall risk level imposed on offshore employees.

The diagram shows the considerable increase in risk to an employee who is shuttled either to shore or to another installation regularly during the offshore work period. Even for shuttling twice per week, the increase is significant, and the total risk experienced by offshore workers is doubled if being shuttled twice per week from shore. If shuttling is daily, total risk increases by a factor of almost five.

It should be noted that the total risk values presented here include transportation from shore to the installation, which is often excluded when concept or operational alternatives are compared. The influence of shuttling would obviously have been even more extensive if the initial and final flights (to/from shore) were ignored.

13.9 Prediction of Risk Levels for an Individual Installation

Risk levels for an individual installation can be predicted using the Eq. (13.7) presented by Heide [9]:

$$\theta = (T + T_1\phi) \cdot \alpha_1 \cdot \alpha_2 \cdot \alpha_3 + (1 + \phi \cdot N_1) \cdot \beta_1 \cdot \beta_2 \cdot \beta_3 \qquad (13.8)$$

where:

θ Proportion of flights where an average passenger perishes,
T Flight time directly between the heliport and offshore helipad,
T_1 Extra flight time for flights that have an intermediate landing,
φ Proportion of flights that have an intermediate landing,
α_1 Proportion of accidents per time unit for the accident rate that is dependent on flight time,
α_2 Proportion of fatal accidents per accident for time dependent accidents,
α_3 Proportion of passenger fatalities per fatal accident for time dependent accidents,
N_1 Number of intermediate landings,
β_1 Proportion of accidents per flight for the accident rate that is dependent on the number of flights,
β_2 Proportion of fatal accidents per accident for flight-dependent accidents, and
β_3 Proportion of passenger fatalities per fatal accident for flight-dependent accidents.

13.10 Other Aspects Related to Helicopter Safety

13.10.1 Can Helicopter Transportation Risk Be Reduced Further?

Personnel risk during helicopter transportation to offshore installations is still a significant contributor to the risk level for personnel working offshore, as is discussed in Chap. 17. It would therefore be essential for offshore safety to reduce further the risk levels during helicopter transportation. Even though considerable efforts have been made in Norway during many years, the Turøy accident in 2016 demonstrated that there is still considerable improvement. But we may need to pose the question is significant improvement achievable? This was the question posed in a press article just a short while after the 2016 accident [15].

As for modern fixed-wing aircraft, a large helicopter may land safely if only one of two engines is running, and they can also land on auto-rotation if both engines have failed, as long as the rotor's function is intact. But the trend in the last 15–20 years has been that the majority of the fatal accidents are caused by gearbox or rotor failure, thus preventing auto-rotation. Based on the data in Table 13.1, three out of five fatal accidents in the UK and Norwegian sectors since year 2000 have been caused by gearbox or rotor failure, to the extent that emergency landing was not possible. The rotor and main gearbox are critical components, they can-not have redundancy, and the survival of the helicopter is dependent on the integrity of these components. Modern helicopters have an emergency function to allow flying for 30 min after loss of all lubrication inside the main gearbox. This is supposed to be an emergency measure to allow safe landing. But in the case of the 2009 accident in Canada, the pilots got a warning from the main gearbox, but did not perform an immediate emergency landing, believing they still had 30 min of safe flying left (the recent requirement). Had they landed immediately on the sea level, most of the personnel would probably have survived.

The rotor and main gearbox are mechanical components with ultra-high loads and stress. Although considerable efforts are made to prevent catastrophic failures during construction as well as maintenance, it will never be possible to fully exclude failures. A new failure mechanism may be experienced any day, in the worst case leading to catastrophic failure of, for instance, the main gearbox.

The only way to completely eliminate risk from helicopter transportation is to avoid such transportation. If transportation by vessels were to replace the use of helicopters, that would eliminate such accidents. When manning from offshore is moved to shore due to digitalization or the use of normally unmanned installations (NUI), this is another way to avoid (or at least significantly reduce) helicopter transportation for the installations in question. This argument is actually used by the industry to assert that such solutions imply reduced risk levels, but when we tried to launch a study of boat transportation after the press article in 2016, the industry was not interested.

The use of vessels for personnel transportation is due to be introduced in the near future in any case, when the Oseberg H platform is started up as a NUI wellhead platform, to be visited a few times per year by personnel from a W2W-vessel. Such installations are built without helideck and vessel transportation is the only option.

But perhaps vessel transportation of personnel could be used in a few other circumstances also for installations with helideck, at least as a supplement, where logistically it makes sense and could possibly save time. This was what we wanted to study after the press article as indicated above.

13.10.2   EU Regulations

EASA is preparing new common regulations for helicopter traffic, the Common European System for Offshore Helicopter Operations (HOFO), as an amendment to EU regulation 965/2012 which creates considerable uncertainty. The implications of this regulatory framework for Norway, if implemented, are unclear, according to HSS-3b [2]:

- Threats to safety identified in HSS-3 are becoming more relevant by the introduction of HOFO. A new regulation that may decrease or all together remove the possibilities for special Norwegian regulations for offshore helicopter operations, will be a setback for the current Norwegian safety efforts.
- Norwegian oil and gas guideline No. 066 (NOROG 066) reflects the development and practical safety efforts established on the NCS over several decades. This standard is viewed by many as world leading, and the main Norwegian stakeholders consider the guideline to be an important document that needs to be preserved and further developed.
- Overall, the NOROG 066 guidelines represent a higher safety standard than the HOFO regulation. The guidelines are used voluntarily in the current contracts between the oil companies and the helicopter operators, and this may also be the case for future contracts under HOFO. However, there is a concern that NOROG 066 may become diluted and loose its position over time because of economic pressure in the industry and divergent priorities by new (or existing) actors under HOFO. Safetec Nordic was commissioned by the Norwegian Ministry of Transport and Communications to carry out a study of the consequences of potential changes to Norwegian regulations for helicopter flying [16] in 2016.
- New regulations were issued in 2018 [17] integrating the HOFO and NOROG 066, implying that the crucial requirements of NOROG 066 are embedded in the Norwegian regulations. There is also an opportunity for the helicopter operators to apply for dispensation from the technical requirements in HOFO.

13.10.3   Land Based Helicopters

The discussion in this section is limited to offshore helicopter transportation of personnel, where there is an industry guideline (NOROG 066, see Sect. 13.10.2) which is considerably stricter than the current regulations.

Safetec Nordic carried out a safety study of land-based helicopter operations in 2011–2012 [18]. The study demonstrates that the land-based helicopter transportation in Norway is significantly different from the offshore helicopter transportation. Statistically, the frequency of fatal accidents was 12 times higher for land-based helicopters.

Three out of four pilots admitted that they had recently performed flights in spite of realising that it was unsafe to fly because of their fatigue. During interviews, several pilots admitted that their knowledge about the company's weak economic position and their own job insecurity gave them incentives to push limits further in order to complete missions. This may be a good illustration of the effect of cost pressure, which may also be experienced by offshore helicopter pilots in the future.

References

1. Sintef (2010) Helicopter safety study 3 (in Norwegian only). SINTEF; 2010 Nov. Trondheim, Norway; Report No.: SINTEF A14973
2. SINTEF (2017) Helicopter safety study 3b (in Norwegian only). SINTEF; 2017 Feb. Trondheim, Norway; Report No.: SINTEF A28021
3. Norwegian official report (2002) Helicopter safety and the Norwegian continental shelf (in Norwegian only), NOU2002:17, Oslo, Norway
4. CAA (2007) Status report from cooperation forum for helicopter safety (in Norwegian only), www.luftfartstilsynet.no
5. AIBN (2018) Report on the air accident near Turøy, Øygarden Municipality, Hordaland county, Norway 29 April 2016 with Airbus Helicopters EC 225LP, LN-OJF, operated by CHC Helikopter Service AS, report SL 2018/04, July 2018
6. Vinnem JE, Vinnem JE (1998) Risk levels on the norwegian continental shelf. preventor, Bryne, Norway; 1998 Aug. Report No.: 19708–03
7. Vinnem JE (2008) Challenges in analysing offshore hazards of an operational nature, presented at PSAM9, Hong Kong, 18–23 May 2008
8. HSE (2004) UK offshore public transport helicopter safety record (1976–2002), www.hse.uk. gov
9. Heide B (2012) On helicopter transport risk for offshore petroleum personnel, presented at PSAM/11ESREL2012. Hels, Finl 24–28(6):2012
10. PSA (2012) Trends in risk level on the Norwegian continental shelf, main report, (in Norwegian only, English summery report). Pet Saf Auth Stavanger 25(4):2012
11. SINTEF (1990) Helicopter safety study. SINTEF; 1990 Nov. Trondheim, Norway; Report No.: STF75 A90008
12. SINTEF (1999) Helicopter safety study 2 (in Norwegian only). SINTEF; 1999 Nov. Trondheim, Norway; Report No.: STF38 A99423
13. Hamremoen E (2007) Personal communication with Mr Hamremoen in Statoil (now Equinor)
14. Karlsen K (2007) Personal communication with Mr Karlsen, Industry Energy
15. Vinnem JE (2016) Helicopter safety can probably not be better, time to consider boat transport (press article in Norwegian only). Stavanger Aftenblad 19(05):2016
16. Safetec Nordic (2016) Consequence study of possible regulation changes for offshore helicopter operations (in Norwegian only), Report ST-11926-2, 16.12.2016
17. Lovdata (2018) Regulations for offshore helicopter operations (in Norwegian only), FOR-2018-06-20-923, 21.6.2018, Ministry of Transportation
18. Safetec Nordic (2012) Safety study of land-based helicopters (in Norwegian only), Report ST-04215-2, 12.02.2012

Glossary

The following definitions are coordinated with NORSOK Z-013 (which reflects ISO terminology (ISO/IEC Guide 73:2002 and ISO31000:2018) where relevant, except that 'risk tolerance criteria' replaces 'risk acceptance criteria', in accordance with what is used internationally.

Accidental Event (AE) Event or a chain of events that may cause loss of life or damage to health, assets or the environment

Accidental effect The result of an accidental event, expressed as heat flux, impact force or energy, acceleration, etc. which is the basis for the safety evaluation

Acute release The abrupt or sudden release in the form of a discharge, emission or exposure, usually due to incidents or accidents

Area Exposed by the Accidental Event (AEAE) Area(s) on the facility (or its surroundings) exposed by the accidental event

Area risk Risk personnel located in an area is exposed to during a defined period of time

As Low as Reasonably Practicable (ALARP) ALARP expresses that the risk shall be reduced to a level that is as low as reasonably practicable

Average Individual Risk (AIR) Risk an average individual is exposed to during a defined period of time

Barrier element Technical, operational and organisational measures or solutions involved in the realisation of a barrier function

Barrier function The task or role of a barrier

© Springer-Verlag London Ltd., part of Springer Nature 2020
J.-E. Vinnem and W. Røed, *Offshore Risk Assessment Vol. 1*,
Springer Series in Reliability Engineering,
https://doi.org/10.1007/978-1-4471-7444-8

Barrier performance (or risk) influencing factor Factors identified as having significance for barrier functions and the ability of barrier elements to function as intended

Barrier system System designed and implemented to perform one or more barrier function

BLEVE Boiling Liquid Expanding Vapour Explosion, is defined as rupture of a hydrocarbon containing vessel due to being heated by fire loads

Causal analysis The process of determining potential combinations of circumstances leading to a top event

Consequence Outcome of an event

Consequence evaluation Assessment of physical effects due to accidents, such as fire and explosion loads

Contingency planning Planning provision of facilities, training and drilling for the handling of emergency conditions, including the actual institution of emergency actions

Control (of hazards) Limiting the extent and/or duration of a hazardous event to prevent escalation

Cost/benefit evaluation Quantitative assessment and comparison of costs and benefits. In the present context often related to safety measures or environmental protection measures where the benefits are reduced safety or environmental hazard

Chronic release The continuous or ongoing release in the form of a discharge, emission or exposure

Defined situations of hazard and accident (DSHA) Selection of hazardous and accidental events that will be used for the dimensioning of the emergency preparedness for the activity

Design accidental event Accidental events that serve as the basis for layout, dimensioning and use of installations and the activity at large, in order to meet the defined risk tolerance criteria

Design Accidental Load (DeAL) Chosen accidental load that is to be used as the basis for design

Dimensioning Accidental Event (DAE) Accidental events that serve as the basis for layout, dimensioning and use of installations and the activity at large

Dimensioning Accidental Load (DiAL) Most severe accidental load that the function or system shall be able to withstand during a required period of time, in order to meet the defined risk tolerance criteria

Emergency preparedness Technical, operational and organisational measures, including necessary equipment that are planned to be used under the management of the emergency organisation in case hazardous or accidental situations occur, in order to protect human and environmental resources and assets

Emergency Preparedness Analysis (EPA) Analysis which includes establishment of DSHA, including major DAEs, establishment of emergency response strategies and performance requirements for emergency preparedness and identification of emergency preparedness measure, including environmental emergency and response measures

Emergency preparedness assessment Overall process of performing a emergency preparedness assessment including: establishment of the context, performance of the EPA, identification and evaluation of measures and solutions and to recommend strategies and final performance requirements, and to assure that the communication and consultations and monitoring and review activities, performed prior to, during and after the analysis has been executed, are suitable and appropriate with respect to achieving the goals for the assessment

Emergency preparedness organisation Organisation which is planned, established and trained in order to handle occurrences of hazardous or accidental situations

Emergency preparedness philosophy Overall guidelines and principles for establishment of emergency response based on the operator vision, goals, values and principles

Emergency response Action taken by personnel, on or off the installation, to control or mitigate a hazardous event or initiate and execute abandonment

Emergency response strategy Specific description of emergency response actions for each DSHA

Environment Surroundings in which an organization operates, including air, water, land, natural resources, flora, fauna, humans and their interrelation

Environmental impact Any change to the environment, whether adverse or beneficial, wholly or partially resulting from an organization's activities, products or services

Environmental resource Includes a stock or a habitat, defined as: Stock: A group of individuals of a stock present in a defined geographical area in a defined period of time. Alternatively: The sum of individuals within a species which are reproductively isolated within a defined geographical area. Habitat: A limited area where several species are present and interact. Example: a beach

Environment safety Safety relating to protection of the environment from accidental spills which may cause damage

Escalation Escalation has occurred when the area exposed by the accidental event (AEAE) covers more than one fire area or more than one main area

Escalation factor Conditions that lead to increased risk due to loss of control, mitigation or recovery capabilities

Escape Actions by personnel on board surface installations (as well as those by divers) taken to avoid the area of accident origin and accident consequences to reach an area where they may remain in shelter

Escape way Routes of specially designated gangways from the platform, leading from hazardous areas to muster areas, lifeboat stations, or shelter area

Escape route Route from an intermittently manned or permanently manned area of a facility leading to safe area(s)

Establishment of emergency preparedness Systematic process which involves selection and planning of suitable emergency preparedness measures on the basis of risk and emergency preparedness analysis

Essential safety system System which has a major role in the control and mitigation of accidents and in any subsequent EER activities

Evacuation Planned method of leaving the facility in an emergency

Event tree analysis Inductive analysis in order to determine alternative potential scenarios arising from a particular hazardous event. It may be used quantitatively to determine the probability or frequency of different consequences arising from the hazardous event

Explosion load Time dependent pressure or drag forces generated by violent combustion of a flammable atmosphere

External escalation When the area exposed by the accidental event (AEAE) covers more than one main area, external escalation has occurred

Facility Offshore or onshore petroleum installation, facility or plant for production of oil and gas

Fault tree analysis Deductive quantitative analysis technique in order to identify the causes of failures and accidents and quantify the probability of these

Fire area Area separated from other areas on the facility, either by physical barriers (fire/blast partition) or distance, which will prevent a dimensioning fire to escalate

Functional requirements to safety and emergency preparedness Verifiable requirements to the effectiveness of safety and emergency preparedness measures which shall ensure that safety objectives, risk tolerance criteria, authority minimum requirements, and established norms are satisfied during design and operation

Group Individual Risk (GIR) Average IR for a defined group

Hazard Potential source of harm

Hazardous event Incident which occurs when a hazard is realized

Immediate vicinity of the scene of accident Main area(s) where an accidental event (AE) has its origin

Individual Risk (IR) Risk an individual is exposed to during a defined period of time

Inherently safer design In inherently safer design, the following concepts are used to reduce risk:
- reduction, e.g. reducing the hazardous inventories or the frequency or duration of exposure
- substitution, e.g. substituting hazardous materials with less hazardous ones (but recognizing that there could be some trade-offs here between plant safety and the wider product and lifecycle issues)
- attenuation, e.g. using the hazardous materials or processes in a way that limits their hazard potential, such as segregating the process plant into smaller sections using ESD valves, processing at lower temperature or pressure
- simplifications, e.g. making the plant and process simpler to design, build and operate, hence less prone to equipment, control and human failure

Internal control All administrative measures which are implemented to ensure that the work is in accordance with all requirements and specifications

Internal escalation When the area exposed by the accidental event (AEAE) covers more than one fire area within the same main area, internal escalation has occurred

Main area Defined part of the facility with a specific functionality and/or level of risk

Main load bearing structures Structure, which when it loses its main load carrying capacity, may result in a collapse or loss of either the main structure of the installation or the main support frames for the deck

Main safety function Most important safety functions that need to be intact in order to ensure the safety for personnel and/or to limit pollution

Major accident Acute occurrence of an event such as a major emission, fire, or explosion, which immediately or delayed, leads to serious consequences to human health and/or fatalities and/or environmental damage and/or larger economical losses

Material damage safety Safety of the installation, its structure, and equipment relating to accidental consequences in terms of production delay and reconstruction of equipment and structures

Mitigation Limitation of any negative consequence of a particular event

MOB-boat Man Over Board Boat

Muster station A place where personnel may gather in a Safe Haven prior to evacuation or abandonment from emergency situations

Muster area Area on the platform where the personnel may be sheltered from accidental conditions until they embark into the lifeboats

Normalisation The normalisation phase starts when the development of a situation of hazard or accident has stopped

Occupational accidents Accidents relating to hazards that are associated with the work places (falls, slips, crushing etc.), thus other hazards than hydrocarbon gas or oil under pressure. These accidents are normally related to a single individual

Performance requirements for safety and emergency preparedness Requirements to the performance of safety and emergency preparedness measures which ensure that safety objectives, RAC, authority minimum requirements and established norms are satisfied during design and operation

Personnel safety Safety for all personnel involved in the operation of a field

Probability Extent to which an event is likely to occur

Recovery time Time from an accidental event causing environmental damage occurs until the biological features have recovered to a pre-spill state or to a new stable state taking into consideration natural ecological variations, and are providing ecosystem services comparable to the pre-spill services

Reliability analysis Analysis of causes and conditions of failure, inspection, maintenance and repair, and the quantitative assessment of up-times and down-times

Residual accidental event Accidental event which the installation is not designed against, therefore it will be part of the risk level for the installation

Residual risk Risk remaining after risk treatment

Risk PSA (2015): Consequences of the activities, with associated uncertainty Traditionally: Combination of the probability of occurrence of harm and the severity of that harm.

Risk acceptance Decision to accept risk

Risk analysis Structured use of available information to identify hazards and to describe risk

Risk assessment Overall process of performing hazard identification, risk analysis and risk evaluation

Risk avoidance Decision not to become involved in, or action to withdraw from, a situation that involves risk

Risk control Actions implementing risk management decisions

Risk evaluation Judgement, on the basis of risk analysis, sometimes involving RAC, of whether the risk is tolerable or not

Risk identification Process to seek for, list and characterise elements of risk

Risk management Coordinated activities to direct and control an organisation with regard to risk

Risk management system Set of elements of an organisation's management system concerned with managing risk

Risk perception Way in which a stakeholder views a risk, based on a set of values or concerns

Risk picture Synthesis of the risk assessment, with the intention to provide useful and understandable information to relevant decision makers

Risk reduction Actions taken to reduce the probability, negative consequences, or both, associated with a hazard

Risk Tolerance Criteria (RAC) Criteria that are used to express a risk level that is considered as the upper limit for the activity in question to be tolerable

Risk transfer Sharing with another party the potential burden of loss or benefit of gain (Contracts and insurance are two examples)

Risk treatment Process of selection and implementation of measures to modify risk

Rooms of significance to combating accidental events CCR and other equivalent room(s) that are essential for safe shutdown, blowdown and emergency response

Safe area(s) Area(s) which, depending on each specific defined situation of hazard and accident (DSHA), are defined as safe until the personnel are evacuated or the situation is normalized

Safety barrier A measure intended to identify conditions that may lead to failure, hazard and accident situations, prevent an actual sequence of events occurring or developing, influence a sequence of events in a deliberate way, or limit damage and/or loss

Safety function Measures which reduce the probability of a situation of hazard and accident occurring, or which limit the consequences of an accident

Safety goals Concrete targets against which the operations of installations at the field are measured with respect to safety. These targets shall contribute to avoidance of accidents or resistance against accidental consequences

Safety objective Objective for the safety of personnel, environment and assets towards which the management of the activity will be aimed

Serious accidents See major accidents

Shelter area An area on the platform where the crew will remain safe for a specific period of time in an emergency situation

Stakeholder Any individual, group or organization that can affect, be affected by, or perceive itself to be affected by, a hazard

System Common expression for installation(s), plant(s), system(s), activity/ activities, operation(s) and/or phase(s) subjected to the risk and/or emergency preparedness assessment

System basis Inputs (regarding the system subjected to assessment) used as basis for the assessment

System boundaries System boundaries defines what shall and what shall not be subjected to the assessment

Working accidents Accidents relating to other hazards than hydrocarbon gas or oil under pressure (falls, crushing etc.) normally related to a single individual

Worst case consequence The worst possible consequences to health, environment and safety resulting from a hazardous event. For this to occur, all critical defences in place must have failed

Index

Printed in the United States
By Bookmasters